TS 178.4

NEW ENGLAND INSTITUTE
OF TECHNOLOGY
LEARNING RESOURCES CENTER

AUTOMATED ASSEMBLY

NEW ENGLAND INSTITUTE
OF TECHNOLOGY
LEARNING RESOURCES CENTER

WRITTEN AND EDITED BY

KENNETH R. TREER

Copyright © 1979 by the Society of Manufacturing Engineers, Dearborn, Michigan.

All rights reserved, including those of translation. This book, or parts thereof, may not be reproduced in any form without the permission of the copyright owner. The Society does not, by publication of data in this book, ensure to anyone the use of such data against any liability of any kind, including infringement of any patent. Publication of any data in this book does not constitute a recommendation of any patent or proprietary right that may be involved.

Library of Congress Catalog Card Number: 78-75097
International Standard Book Number: 0-87263-047-1

AUTOMATED ASSEMBLY

Introduction

Automation in assembly goes back a long way, depending upon how loosely the term is used. The phrase was coined around World War II and its definition varies with the views of the user. Food, beverage, tobacco and other industries developed "automation" long before the hard goods industries and they still lead the way in degree and high speed of assembly. For the purpose of this book we will use industrial hard goods manufacture as the principle manufacturing field and any effort which replaces the human element as the definition.

Technically, assembly automation is still more of an art than a science with only patchy areas reduced to the level of formulas. Compared to other technologies such as machining and grinding, little has been published, primarily because few general rules, formulas, etc., have been developed. Additionally, automated assembly covers such a wide range of technologies that a comprehensive treatment would require many volumes. Virtually any manufacturing technique can be incorporated into an assembly system and, conversely, any assembly technique can be incorporated into a metal cutting or forming system. Defining the boundaries of automated assembly becomes a moot point.

The Society of Manufacturing Engineers formed an Assembly Division about 1970 to fill a gap in the technical coverage of this manufacturing engineering field. One of the major projects of the division has been the annual ASSEMBLEX conference and show. Each year some 30-40 papers have been presented on specific areas of automated assembly ranging from labor relations and product design through fastening methods and hopper feeding to laser applications and computor controls. With the formation of the CAD/CAM and ROBOT conferences and their coupling with the ASSEMBLEX, literally hundreds of valuable and interesting papers have been presented, printed and published in bound volumes. Unfortunately, no compilation or condensation of these papers with any continuity has been attempted until now.

The purpose of this volume is to outline the general areas of automated assembly, define their basic problems, review some of the guidelines and warn of the major pitfalls based on these papers. It is regrettable that space limitations prevent printing all of the material available.

Sincere appreciation is expressed to the many who contributed to the conferences and to this book. A list of these specialists in the many fields involved follows.

With the format established, and the publication of this volume, the Society of Manufacturing Engineers looks forward to future papers and contributions and welcomes your participation.

ASSEMBLY COUNCIL
Professor Jack Lane, CMfgE
Chairman

Richard G. Abraham
Westinghouse R & D Center

Raymond C. Achterberg
Gilman Engineering & Manufacturing Company

James S. Albus
National Bureau of Standards

R. F. Ashton
Reynolds Metal Company

Louis Baccei
Loctite Corporation

V. Badami
University of Rhode Island

Richard Barber
Photon Sources, Inc.

Anthony Barbera
National Bureau of Standards

Richard G. Barnich
Process Computer Systems Inc.

J. V. Belcher
ITT Research Institute

Paul A. Bennett
Townsend/Textron

James F. Beres
Westinghouse R & D Center

R. William Bertha
Assembly Engineering

John H. Bickford
Raymond Engineering Inc.

J. Birk
University of Rhode Island

R. A. Bishel
International Nickel Corporation

Phillip F. Bitters
TRW

Paul K. Bizilia
Ingersoll-Rand Company

Ronald D. Blunck
General Dynamics Convair Division

T. Blunt
Ford Motor Company

Stephen R. Bolin
Raytheon Company

Roger W. Bolz
Automation for Industry Inc.

Robert L. Bongo
Sonics & Materials Inc.

Geoffrey Boothroyd
University of Massachusetts

R. Brand
University of Rhode Island

Shawn Buckley
MIT

W. C. Burgess, Jr.
Burgess & Associates Inc.

Dale Byrely
Champion Spark Plug Company

C. Calengor
Ford Motor Company

William R. Childs
Essex Chemical Corporation

Rolind Chin
University of Missouri

N. Chen
University of Rhode Island

R. A. Chihoski
Martin-Marietta Aerospace

Mark Chookazian
EMA Bond, Inc.

Albert B. Church
Parker-Kalon Corporation

Janis Church
ITT Research Institute

E. Clark
General Motors Corporation

Allan H. Clauser
Battelle Labs

William L. Clippard, III
Clippard Instrument Laboratory, Inc.

H.R. Conaway
International Nickel Corporation

Pierlugi Corti
Istituo Di Elettrotecnica

Peter W. Craine
Modicon Corporation

Donald M. Crisapulli
AMF Inc.

John DiPonio
Ford Motor Company

Arwin A. Douglas
Collins Radio Co.

Robert L. Douglas
Gilman Engineering & Manufacturing Co.

Samuel H. Drake
Draper Labs, Inc.

Samuel J. Dwyer, III
University of Missouri

Jim H. Elgin
Digital Equipment Corporation

H.W. Ellison
General Motors Corporation

Joseph F. Engelberger
Unimation, Inc.

S. Eshghy
Rockwell International

John M. Evans, Jr.
National Bureau of Standards

Joel A. Fadem
University of California

Barry P. Fairland
Battelle Labs

William G. Fillmore
Richard Muther & Associates

Robert J. Finkelston
Standard Pressed Steel Company

Lee K. Fisher
Cargill Detroit Corp.

George H. Franklin
Gilman Engineering & Mfg. Co.

Otto M. Fredrich, Jr.
Collins Radio Co.

E.R. Friesth
Deere and Company

Keith M. Gardiner
IBM Corporation

William A. Gardiner
MacLean-Fogg Lock Nut Company

Carmen Giannandrea
Cavitron Ultrasonics

Eric Graham
Fisher Gauge Limited

John H. Gray
USM Corporation

Walter E. Gray
General Electric Company

Richard L. Gruber
Parker-Kalon Corporation

Uzi de Haan
University of California

Russell J. Hardiman
Standard Pressed Steel Corporation

Charles A. Harlow
University of Missouri

Robert W. Harries
Inmont Corporation

Girard S. Haviland
Loctite Corporation

Richard E. Hohn
Cincinnati Milacron

Glenn D. Hollister
Ideal Industries, Inc.

C.L. Hovey
Lockheed Missiles & Space Company

Tony A. Huber
Taumel Assembly Systems

Calvin Hulestein
Loctite Corporation

Takashi Inoue
Mitsubishi Electric Corporation

L. Phillip Jacoby
Milford Rivet & Machine Company

K.G. Johnson
IIT Research Institute

Ichiro Kawakatsu
Aoyama-Gakuin University

R.B. Kelley
University of Rhode Island

Jerry Kirsch
Auto-Place, Inc.

John R. Knoche
Chicago Pneumatic Tool Company

Donald A. Kugath
General Electric Company

Jack Lane
General Motors Institute

Gerald J. Lauer
Automated Process, Inc.

Richard S. Lesner
Chicago Pneumatic Tool Company

Harry H. Loh
AMF, Inc.

G.C. Macri
Ford Motor Company

Frank R. Marciniak
Lucas Milhaupt Inc.

J.C. Martin
Westinghouse Electric Corp.

Lawrence J. Matteson
Eastman Kodak Corporation

Gordon McAlpine
ITT Industrial & Automation
                    Systems

D.E. McCulloch, Jr.
Chicago Pneumatic Tool Company

R.W. McLay
University of Vermont

Ross M. Mehl
Plato Products, Inc.

John M. Menke
General Mills Chemicals, Inc.

Melvin Millheiser
Waldes Kohinoor, Inc.

L.E. Murch
University of Massachusetts

James F. Murphy
AMF, Inc.

Bruce Murray
Loctite Corporation

Reginald Newell
International Assoc. of Machinists

Kenneth J. O'Brien
International Harvester Company

Earl A. O'Connor
Milford Rivet & Machine Company

Edward M. Ogrin, Jr.
Thermo King Corporation

N. Olinjnek
Honeywell, Inc.

Tadashi Osawa
Noyama-Gakuin University

J.G. Palfery
Duo-Thread, Inc.

Barry M. Patchett
Cranfield Institute of Technology

Richard Paul
Purdue University

Charles F. Paxton
Weltrome Company

Vern Peck
Elco Industries, Inc.

C.R. Poll
University of Massachusetts

Theodore O. Prenting
Marist College

Kenneth Purdy
Roy W. Walters & Associates

Todd L. Rachel
Bendix Corporation

D.D. Rager
Reynolds Metal Company

L. Wesky Rearick
General Motors Corporation

E.E. Rice
Ingersoll-Rand Company

Frank J. Riley
Bodine Corporation

C.A. Roest
Reynolds Metals Company

Asbjorn Rolstadas
NTH-SINTEF

William Sahm
General Electric Company

Richard A. Searle
Jones & Lamson

Gerald L. Schneberger
General Motors Institute

Louis A. Seiberlick, Jr.
Omnisystems

Lee Seymour
Scans Associates Inc.

R. Silva
University of Rhode Island

F.A. Smith
Dow-Corning Corporation

Wesley E. Snyder
N. Carolina State University

Melvin E. Stanford
Faultfinders, Inc.

Kim Stelson
MIT

Alvin E. Stenli
Gilman Engineering & Mfg. Co.

Donald S. Stevens
Assembly Machines, Inc.

Werner R. Stutz
Taumel Assembly Systems

Dieter Sundermeyer
Leybold-Heraeus Vacuum Systems, Inc.

William R. Tanner
Ford Motor Company

Mary Tarzia
General Electric Company

Thomas T. Taylor
Gilman Engineering & Mfg. Company

Kenneth R. Treer
Gilman Engineering & Mfg. Company

B.F. von Turkovich
University of Vermont

R. Vincent
Ford Motor Company

A.R. Voss
IBM Corporation

Scott B. Wakefield
Thor Power Tool Company

Paul Watson
Draper Labs, Inc.

Charles K. Watters
Assembly Machines, Inc.

Frederic Weigl
Collins Radio Co.

David Weltman
O.K. Machine & Tool Corporation

Mark A. Wieland
IBM Corporation

Bruce R. Williams
Fusion, Inc.

Robert A. Willoughby
Ingersoll-Rand Company

L. Wilson
University of Rhode Island

Edmund J. Yaroch
3M Company

Anthony Yocum
General Motors Corporation

Fred Yodis
Universal Instruments Corp.

Dante Zarlenga
Zarlenga and Associates

R. Zimmerman
Spectrum Automation

Guenther Zittel
Maxwell Labs, Inc.

TABLE OF CONTENTS

CHAPTER		Page Number
1	PRODUCT DESIGN	1
2	HUMAN RESOURCES	29
	Labor's View of Automation	30
	Management & Motivation	38
	Job Enrichment	54
	Improving Assembly Productivity	69
	Training for Assembly	79
3	EQUIPMENT JUSTIFICATION	89
4	PROJECT MANAGEMENT	111
5	BASIC CONCEPTS	127
	Single-station	128
	Synchronous Dial	132
	Synchronous In-line	137
	Non-Synchronous	141
	Continuous	152
	Selection Factors	157
	Productivity	167
6	PARTS FEEDING	177
	Parts Selectors	183
	Parts Orienters	200
	Escapements	204
7	POSITIONING DEVICES	215
	Special	215
	Standard	218
	Robots	237
8	FASTENING METHODS	261
	Adhesive Selection	262
	Adhesives; Water Base	265
	Adhesives; Hot Melt	276
	Adhesives; Curing	286
	Adhesives; Anerobic	297
	Injected Metal Fastening	304
	Soldering and Welding	319
	Laser Applications	332
	Electron Beam Welding	347
	Ultra-Sonic Applications	362
	Threaded Fastener Tensioning	379

TABLE OF CONTENTS
(continued)

CHAPTER		Page Number
9	INSPECTION FUNCTIONS	389
	Selective Assembly	393
	Leak Test Methods	405
	Balancing	409
10	ANCILLARY OPERATIONS	413
11	CONTROLS	425
12	DE-BUGGING	449

CHAPTER 1

Product Design for Automated Assembly.

Product design is the first step in the manufacturing process. All of the opportunities for, and limitations on efficiency in manufacturing, are established at the product design level.

The Product Design Engineer has a prime goal the most functionally efficent assembly he can achieve. As he develops his design, he will probably have many points at which more than one possible design choice would achieve the same functional efficiency. Very probably only one of these choices will be most efficient in manufacturing.

The Manufacturing Engineer has as his prime target the least expensive process consistent with an acceptable quality control level. His choice of process is necessarily confined to the product design as given.

The optimum automated assembly process almost always requires some compromise between the two. In many organizations this is the function of an "advance production engineer" or his equivalent. Combining expertise in both product and manufacturing engineering, his responsibility is to analyze designs for their "producibility" and to arrive at the least costly product design and manufacturing process consistent with functional capability and quality control.

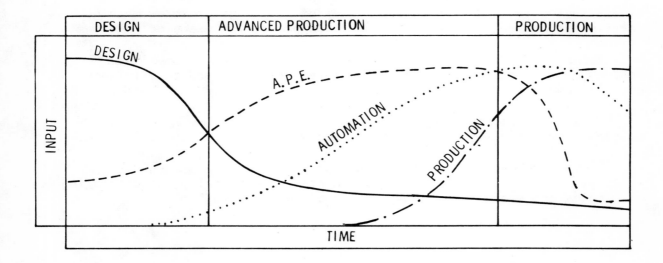

Figure 1

Figure 1 graphically portrays the input inter-relationships through time as a product is designed and phased into production. Note the overlap between the Product Design Engineer and the Advanced Production Engineer, as well as the Advanced Production Engineer and the Production Engineering group. The automation process overlaps all three.

Ideally, this design compromise procedure takes place prior to placing the item into production. Where this is not possible, or where manufacturing has been going on for some time, some form of "value analysis review" should be used. Unfortunately, this "after-the-fact" kind of manufacturing engineering is costly, both in lost manufacturing savings and in duplicated engineering time, and would be minimized if the project were properly done the first time around.

Product design for automated manufacture has three major considerations:

1. Design for simplification.
2. Design for ease of automation.
3. Design for modular construction.

While these three areas are inter-related with each other and with design for function of the part, we will consider these three areas separately with examples.

1. DESIGN FOR SIMPLIFICATION

Many times a new product design uses whatever parts are shelf items, standard, or possible to make with a minimum of experimental tooling. *Figure 2* illustrates one such possible design which, after advance production engineering, was changed to a one piece part from three pieces. Such an instance might occur when a three piece part could be made from fairly standard components, whereas the one piece part would require expensive tooling and time delay. A set of stampings or screw machine parts might be replaced by one cold headed piece, for example.

Figure 3 illustrates a timer assembly which may have been "breadboarded" for functional prove-out and later re-engineered to eliminate one gear and one pinion in the assembly.

Figure 4 illustrates a massive redesign effort which reduced an original forty-seven piece design to twenty pieces.

Figure 5 shows a printed circuit board with thirty-one components which was redesigned to thirteen components. Additionally, manufacturing work holes and space for automatic assembly were added.

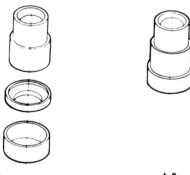

Figure 2.   Before – 3 Piece Parts   After – 1 Piece Part

Before   After

Figure 3.

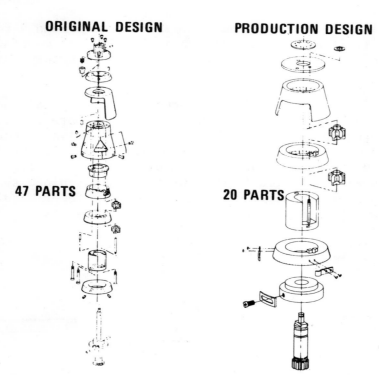

Figure 4.

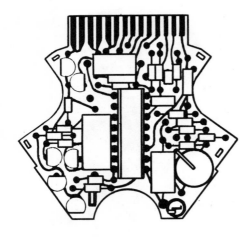

BEFORE
12 RESISTORS
6 TRANSISTOR & 3 RESISTORS
2 SGLE CAPACITORS

AFTER
1 RESISTOR
3 DUAL TRANSISTORS
1 DUAL CAPACITOR

Figure 5.

Design simplification for assembly does not always mean a reduced number of parts. It is conceivable that a more efficient manufacturing process can be achieved in instances where one part is broken into two or more pieces.

One example might be a one-piece, 2 foot long, axle shaft and flange requiring a number of forging operations, heat treat operations, grinding operations, machining operations, and assembly. An analysis should be made of the cost effect of changing to a two piece assembly involving the shaft which might be centerless ground from bar stock and spline rolled on each end, combined with a forged flange which can be machined separately and assembled with a pin or welding, resulting in far fewer handling problems such as weight, bulk, etc.

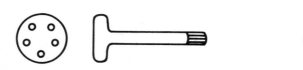
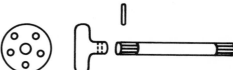

Figure 6.

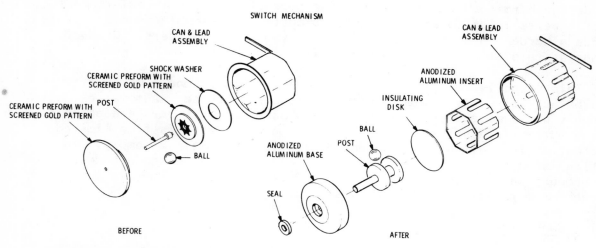

Figure 7.

    Figure 7 illustrates a reversal of the usual trend by increasing an original seven part design to an eight part redesign. It changes from a design of expensive parts which had to be precisely aligned and could not be automatically assembled economically to a design of inexpensive parts which were self-aligning and readily assembled automatically.

    Product design for simplification opens a whole Pandora's Box of opportunities for savings in manufacturing through:

1. Use of standard components or materials.
2. Use of sophisticated manufacturing processes not available to the prototype builder.
3. Minimization of inventories of components and materials.
4. Reduction in handlings.
5. Reduction in malfunctions and downtime.
6. Reduction in opportunities for errors.

2. DESIGN FOR EASE OF AUTOMATION

    This section presents general concepts and design ideas which will help simplify the automatic parts feeding, orienting, and assembly processes. Automatic assembly considerations are outlined which should be incorporated in the product design phase if efficient, reliable, automatic assembly is a production requirement.

    Automatic assembly usually includes feeding parts from bulk, orienting or hoppering these piece parts, tracking them to loading stations and placing them into a main chassis or product case. These phases each have pitfalls which can reduce overall assembly system efficiency. Typically the assembly machines came after the product was completely developed and the design "frozen." In many cases the product had been in existence for several years prior to attempting to automate the assembly process. This places a burden on both feeder and assembly machine builders. Product change at this stage of the game was virtually impossible.

This trend is reversing. Maintainability was at one time an afterthought but has evolved into a part of the design phase. Automatic assembly has also evolved and fortunately some manufacturers are anticipating automatic assembly for their new products. At this time limited information is available on product design for automatic assembly. This section is an attempt to help fill this void and provide product design and production with some examples of what can be done to improve part design for more efficient assembly systems.

Design Considerations For Discrete Parts.

Parts coming to the assembly machine are usually bulk components. They are typically placed in a hopper and then tracked to a loading station. Whether the hopper is vibratory, rotary or oscillatory, it relies on gravity and/or friction to move parts past gating or orienting features to allow only parts in the proper orientation to pass to the next orienting feature and ultimately out of the bowl into a track. Typically, non-vibratory feeders are limited in the number of orientations they can perform. Here are some of the causes of bowl inefficiency.

First is the number of orientations required. See *Figure 8*. In most cases, unoriented parts are returned to the bottom of the bowl. The odds of a part going out of the bowl on its first try decreases with the number of orientations required. The way to increase the odds is to reduce the number of orientations by part symmetry as in *Figure 9*.

Hard to sense features, such as off-center holes, or cavities, are another problem and sometimes they may require extra tooling outside the bowl to sense and orient. In many cases, external features can be added to allow the bowl to orient these parts. Make sure these features will allow the tracks to maintain these orientations. *Figure 10*.

Number of Orientations		Required parts/hr.	Required rate/hr. minimum
1	sphere, symmetrical cube, symmetrical flat washer, symmetrical tube or cylinder (2x or more)	1200	1200
2	Illustrated parts will naturally fall in one of two possible positions necessitating two orientations	1200	2400
4	Four natural selections occur in these parts	1200	4800
8	This square part can fall in any of 8 positions, only 1 of which is acceptable	1200	9600

Figure 8.

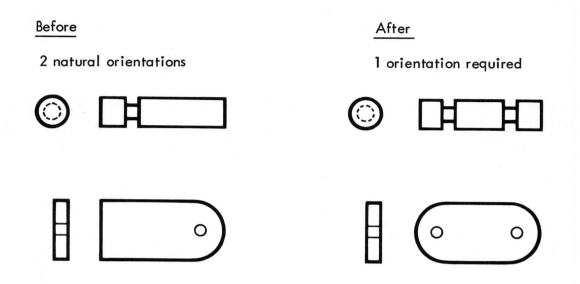

Figure 9.

Use of symmetry to simplify parts orientation.

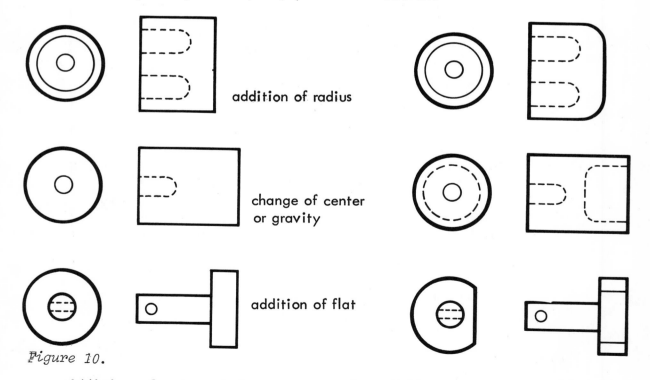

Figure 10.

Addition of external features to reduce difficulty of sensing orientation.

Closely related to the internal feature type problem is similar profiles. Profiles which are too similar are difficult to detect and orient. The obvious solutions to this dilemma is to change dimensions, or to add an external feature, or possibly change the center of gravity of the part. *Figure 11.*

Tangling is another common problem. A protrusion on the part can enter an opening in another part. Here again, the solution is obvious - eliminate the protrusion or close the opening, or both. Compression type coil springs are the most common part of this type. Close coiling the ends, increasing wire diameter and increasing or decreasing the pitch all help alleviate the condition.

Nesting such as paper cups can also be a problem. Here the solution can be either to keep them from engaging by adding ribs or changing diameters or keep them from sticking by increasing the angle of the taper. *Figure 13.*

Shingling is sometimes a problem. It may be possible to feed parts in this attitude, but it usually complicates the escaping of these types of parts from the end of the track. Increasing the thickness of the surfaces contacting adjacent parts may solve this problem, or possibly providing a different tracking surface. *Figure 14.*

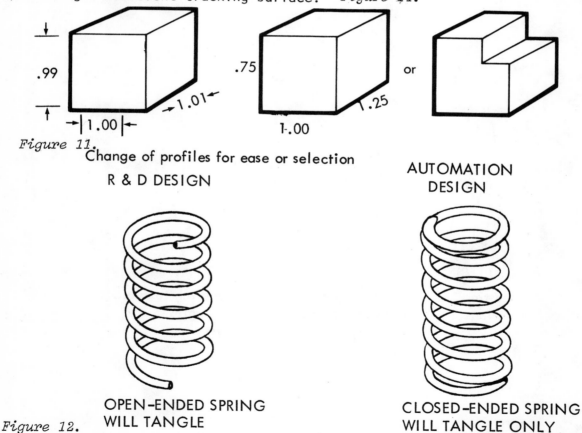

Figure 11. Change of profiles for ease or selection

R & D DESIGN

AUTOMATION DESIGN

Figure 12.

OPEN-ENDED SPRING WILL TANGLE

CLOSED-ENDED SPRING WILL TANGLE ONLY UNDER PRESSURE

Closed-End Spring Used to Facilitate Automation. Avoid heat treated or ground end springs where possible to permit winding springs on the assembly machine.

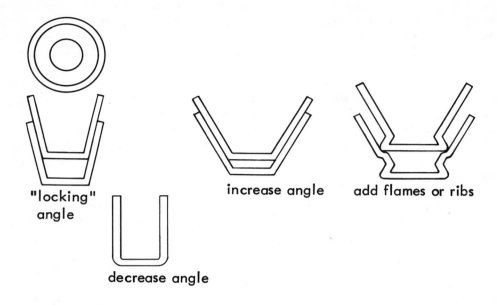

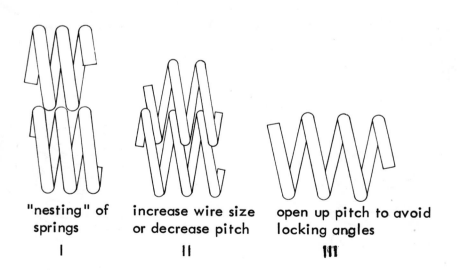

Figure 13.

Avoid "Nesting" Tendencies.

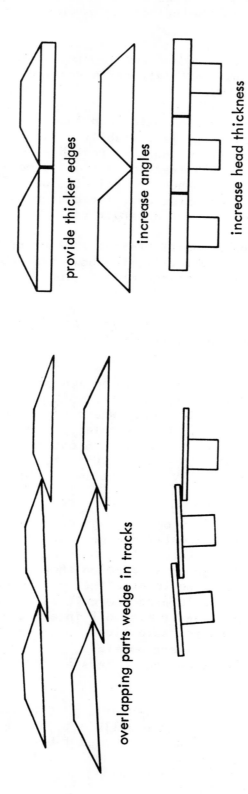

Figure 14. "Shingling"

In in-line feeders or any place where the top of the part is confined, jamming may occur. This is true of extremely thin parts and also of parts with tapered edges. Solutions are obvious once the problem is defined.

Keep the parts from climbing and jamming by making the common contacting surface large and vertical, or increasing this angle to the point where parts will not climb.

Another part configuration which can lead to poor hopper efficiency and can severely limit feed rates is instability due to the tracking surface and center of gravity. If the part is unstable in the sense that its c.g. is high relative to its tracking surface dimension, the part will try to fall over. Further, when the rejected part falls back into the bowl it tends to land wrong side up. The design change for efficiency would be either a lower c.g. or a larger tracking surface.

Soft fabric or rubber parts present another problem. They may tangle in the hopper, but, more seriously, the bowl driving forces may distort the part beyond the ability of the orientation devices to select, *Figure 15*.

Conversely, un-tempered glass, other brittle materials or fragile designs should be examined for possible redesign to parts with more rugged characteristics.

Distortion of parts due to stacking and handling can cause very serious problems. Again, plastic parts, soft parts and thin metal parts are susceptible to these conditions. Where possible, replace such designs with more stable components. For example, *Figure 16* illustrates a shim material for selective assembly in 5 categories of .001. The original part design was .010 to .015 in .001 steps. A much easier part design to handle automatically would be something over .030 to .035 thickness in .001 steps.

Flash from parting lines on castings or moldings do not always appear on drawings and frequently are not noticed on prototypes. These are analogous to burrs and cutoff tabs on stampings which grow worse as production goes on. Design consideration should be given to locating them where they will have the least effect on hopper-feeding, tracking and loading. Flash may be located where it is unimportant to the assembly machine. If not, it must be trimmed off in an accurate and consistent manner, *Figure 17*. Along the same lines, a stamping burr should be planned to assist assembly where possible, *Figure 18* suggests blanking out the part in the direction that offers a "lead-in" rather than a sharp edge when assembling.

The effects of the sensitivity of a part to moisture, static electricity and residual magnetism are usually not considered until too late. *Figure 19* illustrates a resin-impregnated fabric washer which is stamped out flat and remains so if excessive humidity is not present. When the humidity rises it tends to curl up along the lines of the "grain," making it nearly impossible to hopper feed. It is possible to stamp them into magazines and retain the grain orientation, but it would be far better to

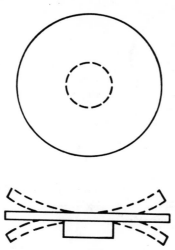

Figure 15.

Distortion of flexible part may make orientation impractical.

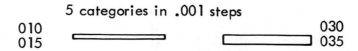

Figure 16.

    A vibratory feeder may lift and lower a part by .015. Using a thicker part minimizes "shingling" tendencies.

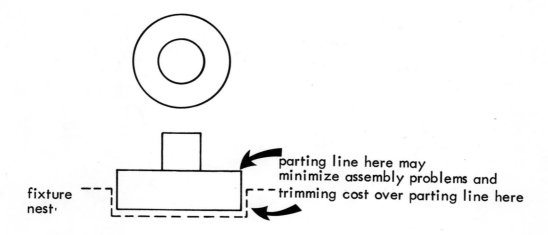

Figure 17.

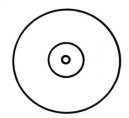

Burr up for assembly down & vice versa.

Figure 18.

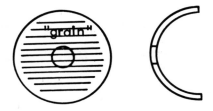

Figure 19.

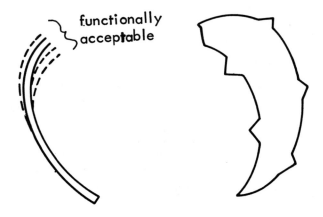

Figure 20.

use a material with cross-grain plies, or no grain at all. Static electricity and residual magnetism have similar complications, usually requiring additional equipment to counteract, which should be engineered-out in the design of the part if at all possible.

"Sacred" surfaces are those which must not be damaged in handling, such as ground surfaces. Some parts cannot be automatically fed or handled for this reason. It may be possible to protect such a surface in the product design through the use of flanges or other projections.

Multiple-cavity dies and molds create consistency problems for the assembly operation. Many times a very loose tolerance on configuration is entirely adequate for the functional requirements of the part, but is impossible to handle without excessive set-up changes in the assembly machine. *Figure 20* shows a thin sheet metal stamping which would cause serious problems in assembly if the full design tolerance occurs across several dies and through various mills runs of coil stock. Stiffer material, added ribs, etc. will help, but the problems inherent in the design will probably dictate the additional costs of far more accurate die-making.

The list of design considerations goes on and on, almost as long as there is another assembly to make. With only a little tongue-in-cheek, we offer *Figure 21*, the "UHP," (unhopperable part), as the horrible example to be avoided.

A few general rules for product design for hopperability would be:

1. Keep tracking surfaces flat and large.
2. Keep the center of gravity low.
3. Minimize the number of orientations required.
4. Keep profiles distinctly different.
5. Avoid designs which nest, tangle, shingle, jam or wedge.
6. If "hidden," or internal features are required, add external features for orientation purposes.
7. Keep parts stiff and rigid, but not brittle or frangile.
8. Avoid designs subject to variations in humidity, temperature, pressure, static electricity, magnetism, etc.
9. Design flash, burrs, "sacred surfaces," etc., into areas least likely to interfere with assembly.

Now that we have the part oriented and to the loading mechanism in good condition at the proper rate, we are ready to assemble.

The sequence in which parts can be assembled can be arrived at by logic, but more often a graphical approach is better. It should show all alternatives. One of these is a precedence diagram, *Figure 22*. The sample shows you the steps in getting dressed. Most assembly operations are more restrictive since the first loading operation is usually a single choice, a main casting or stamping. Circles may represent manual operations while squares may represent automatic operations.

During the design phase, this chart may be helpful in designing the product for maximum flexibility during the assembly process. This

U.H.P. *

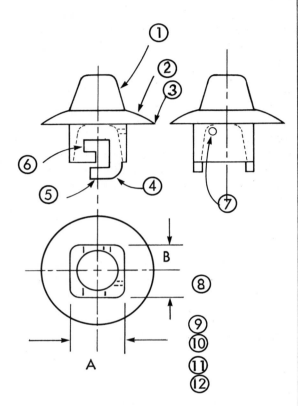

1. Top will nest in bottom hole with locking taper.
2. Parts will shingle.
3. Shingled parts will jam.
4. Ears will tangle in slots.
5. No flat tracking surface.
6. Cut-outs different on both sides.
7. Hidden feature.
8. Dimension A = B by .001.
9. Soft or Magnetized parts.
10. Flash, burrs, excessive.
11. Excessive grease on parts.
12. Foreign material mixed with parts in bowl.

* Un-Hopperable Part

Figure 21.

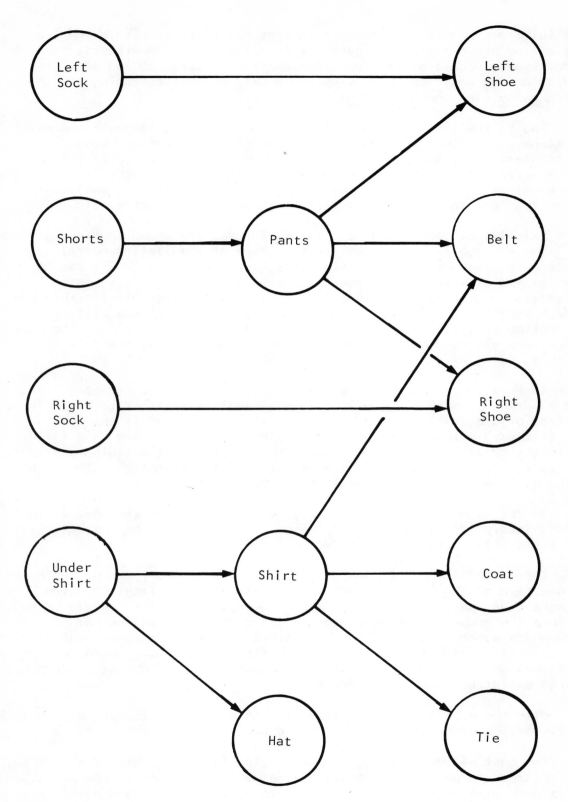

Figure 22.
Precedence Diagram "How to Get Dressed"

flexibility provides the assembly machine manufacturer with a means of optimizing his machine configuration for placement of operators, minimum floor space, accessibility for maintenance and operation and parts feeding and unloading positions. It might even help prevent designing something which cannot be assembled.

One of the first considerations in the automatic assembly process development is fixturing. Two main types of approaches are used: fixtureless and fixtured. In the first case, the main body of the part is transferrable--capable of being handled as a fixture itself. Obviously, the elimination of fixtures offers a substantial savings in equipment cost as well as a potential savings in fixture downtime, repairs, etc.

Examples of parts handled without fixtures are shown in *Figures 23 and 24*. The voltage regulator base is the shape of a pallet, symmetrical, rectangular, and with the addition of at least 2 work holes it can be shot-pinned at working stations for accuracy. Adding an off-center orientation cutout permits automatic feeding with discrimination between side A and side B. Automobile engine blocks, cylinder head castings, connecting rods, transmission converters, and disc brake rotors, to name a few, are all handled with the main part piece assuming the role of the fixture.

In *Figure 25* is sketched a die cast part, irregular in shape, which had a series of machining, plating and assembly operations to be performed. Instead of die casting and immediately trimming off the flash, the flash was thickened with a ridge cast all the way around and used as a fixture up to the final operation when the trimming turned out a finished product. Possibilities exist for doing the same thing with stampings. Shot pin holes and other orientation devices can be incorporated.

If the main body of the assembly must be fixtured, the part design should incorporate locating surfaces or holes, orientation characteristics and accessability to all assembly areas.

The ideal assembly is one in which all the assembly can be performed on one face of the part with straight, vertical or horizontal motions. If more than one side, or face, must be used, every effort should be made to keep the number of faces used to a minimum. Every additional direction from which assembly work must be performed creates additional problems and cost. It is possible to fixture main bodies for assembly operations to be performed on all six sides and from many angles, but the best product design minimizes such requirements.

With the main body of the assembly loaded and positioned, we can look at the design priorities in the loading of individual parts.

The most efficient type of load station is one in which the part may be "speared" to positively locate it, cam open tracks or escapements, continue down with a straight-line motion to place the part and positively check for the placement of the part. If the part body has a pilot hole for the spear, or if adjacent guide pin holes are available for load head guide pins, greater reliability of station efficiency is available. See

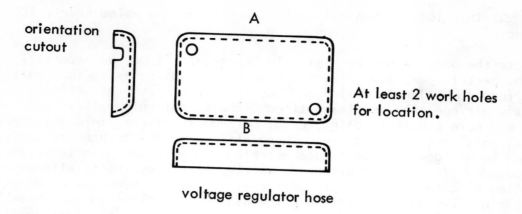

Figure 23.

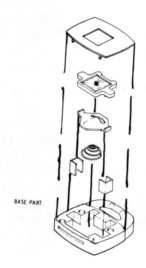

Figure 24.

Use of a Base Part for Automated Assembly.

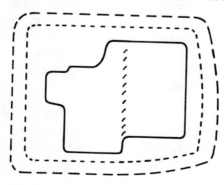

Figure 25.

Figure 26. This 100% mechanical part control should provide nearly 100% efficiency.

Where the part cannot be similarly "speared," it may be necessary to grasp the part in jaws and "pick-and-place" it into the assembly. This usually involves a jaw actuation and two or more directional motions, frequently air operated requiring timed signals, solenoid valves, cylinders, relays, and more complex tooling; ergo, more malfunctions, more maintenance, etc. All this tends to decrease machine efficiency due to product design. Parts can be designed to facilitate grasping and maintaining orientation. See *Figure 27*. Try to keep distances short. Allow for radial alignment and tooling clearances in the product design. Provide generous chamfers and "lead-ins" for easy entry into the body. Put yourself in the assembly machine's place; if you have difficulty loading the part, the machine will also.

If spearing or grasping is impractical, another approach is vacuum pickup and transfer. This is somewhat less reliable and subject to problems due to dirt or oil in the air. Provide large, flat, smooth surfaces for pickup. Add locating features if possible. Thin discs, for instance, could have raised circular ridges to locate on. Raised lips may provide an external feature for radial alignment.

The product design must be adequate for the forces involved and the stability of the parts relationships after assembly. *Figure 28* illustrates several of these problems. If the washer being riveted to the assembly has a squareness requirement, sufficient shoulder area must be allowed to insure holding the washer during riveting. If the cross-section of the shaft at the drilled hole is too little, the riveting forces may collapse this area. The same thing is true of the stem at the bottom. A taper of about 17° might provide a locking angle to jam the part in the fixture. At all times, remember that the design which was adequate during prototype assembly may prove inadequate under the forces involved during high production assembly.

Figure 29 illustrates additional considerations in product design. Can the parts be placed in the main body with simple, short, straight line motions? If not, can clearances be provided to assist the tooling motions? Avoid combined rotary and straight line motions if possible. These will usually require two (2) or more stations. Compound motions usually require a separate special motion or even a machine.

After looking at loading motions, the tooling clearances require some thought. Try to think in terms of dis-assembly. If you do not have room to grasp or push the part out of position, then you may have difficulty in loading it.

In addition to clearances, locating points for the tooling may be required, especially if precision alignments or close tolerances are required.

In all cases, chamfers, leads and any guide surface you can provide will enhance the assembly efficiency. We cannot recall a case where we had too much lead or chamfer. Rivets are a good example of this. Because

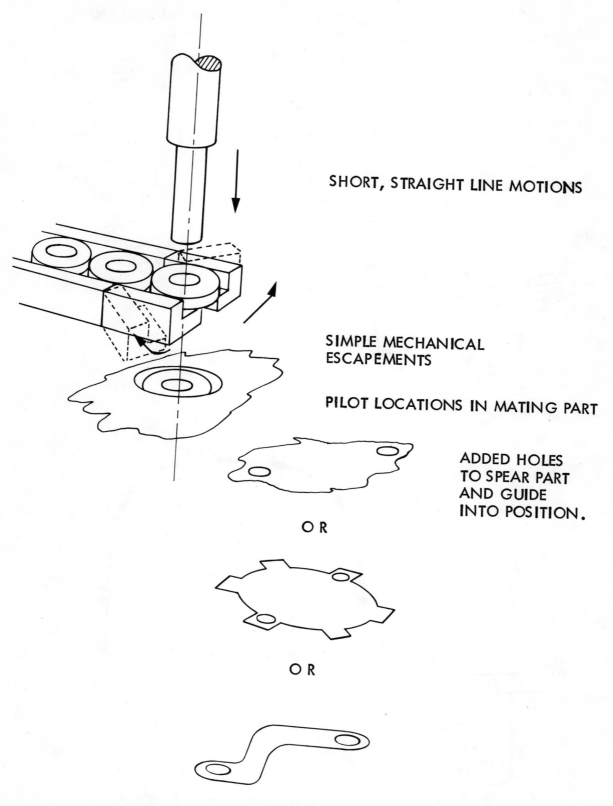

Figure 26.

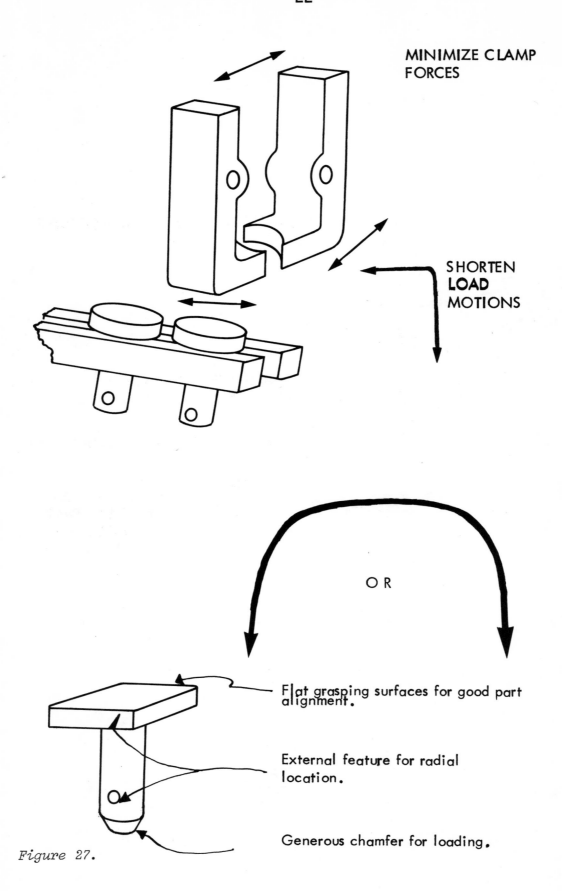

Figure 27.

PART STRENGTH AND STABILITY

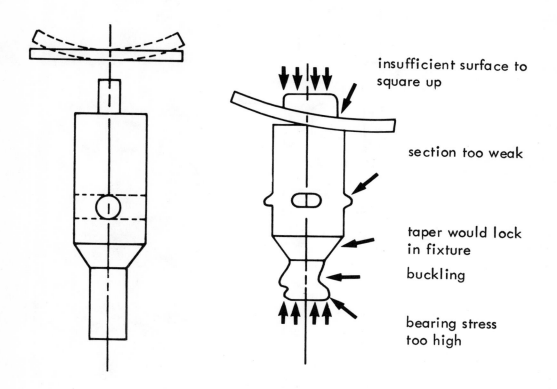

Higher forces may be required during assembly operations due to cycle time. Test process and tooling and fixturing before finalizing design. Use liberal design factors.

Figure 28.

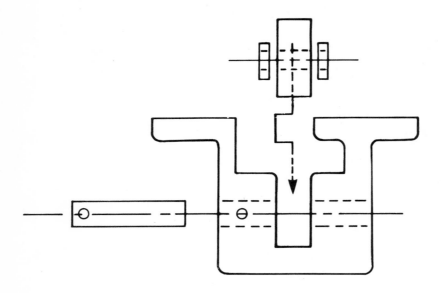

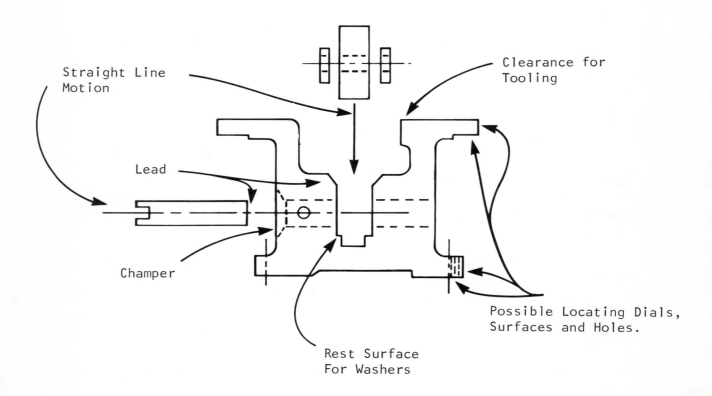

Figure 29.

of hot forming and possibly different vendors, the rivet chamfer may vary. Let's help the operation by adding a chamfer on the mating part.

On stamped parts, the direction of punching may be important. Also, the relationship of groups of holes for mating parts is critical for assembly.

In addition to hopperability design considerations, product design for ease of automation suggests that the engineer try to provide: *(Figs. 30 & 31)*

1. A base or main body which will serve as its own fixture.
2. Locating surfaces and/or holes for fixturing.
3. Tooling clearances and room to remove misloads.
4. Simple, straight-line vertical loading motions.
5. A minimum of multiple motions or compound motions.
6. Adequate bearing surfaces for maintaining assembled relationships.
7. Part strength for high speed assembly forces.
8. Chamfers, lead-ins and locating points for parts to be loaded.
9. Tooling & piloting locating features for precision operations.
10. Burrs, flashes, etc., in directions which will assist assembly.
11. Stamping, casting and machining sequences which will maintain relationships and uniform patterns of part features to assist handling, locating and assembling.

3. PRODUCT DESIGN FOR MODULAR CONSTRUCTION.

Two purposes lie behind design for modular construction;

1. Subassemblies may be standardized across a family of products resulting in substantial savings by grouping operations of those components common to all. Benefits from higher production volumes of fewer different components, line balance and standardization are possible. Inventories may be reduced. Fewer process procedures and personnel training problems are another advantage. This effort is a compromise between manufacturing economies, product cost and product function.

2. Many assemblies are far to complex and involved to make complete in one pass. There is a "state of the art" limitation on the numbers and types of operations which can be linked together in one sequence. It becomes more efficient to break down the total assembly into a series of sub-assemblies culminating in a final assembly. A thorough knowledge of the processes involved and common sense dictate the degree to which the assembly is broken down.

Figures 32 and 33 illustrate two such modular break-outs.

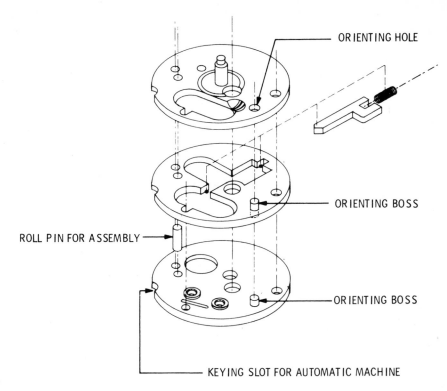

Figure 30.

Typical Built-In Features for Automation.

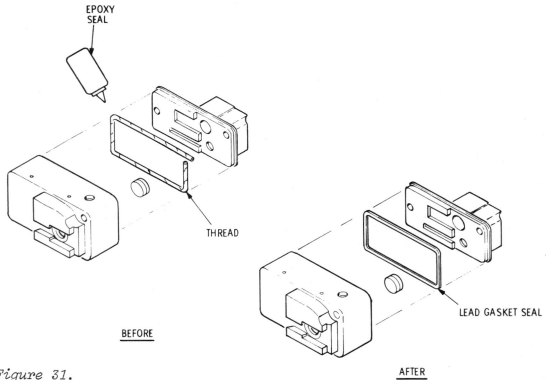

Figure 31.

Seal Redesign Simplifies Automation.

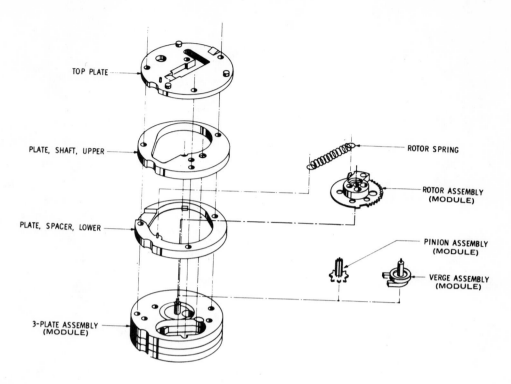

Figure 32.

Figure 32 shows how four modules and four parts are brought together to complete another module.

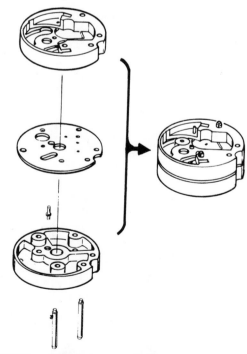

Figure 33.

Figure 33 brings seven parts together to complete a module. Note the use of the orientation notch and the roll pins.

CHAPTER 2

Human Resources and Automated Assembly.

"People potential."

Automation should not <u>ignore</u> people potential.

Automation should <u>enhance</u> people potential.

As the level of automation rises, more and more opportunities are available to everyone concerned with it. Automation also demands much more of all concerned. Some of these opportunities and demands are:

1. Removal of tedium from the production line assignment.

2. Reduction of physical labor.

3. Potential for greater use of <u>people capabilities</u>.

4. <u>People participation</u> possibilities in day-to-day decisions.

5. Restoration of pride of workmanship comes with greater <u>people responsibility</u>.

The manufacturing process pendulum is completing full cycle. Beginning with the "cottage industry" in which one man or family produced a finished product from scratch, work responsibility and authority has been steadily diminished through the breaking down of production functions until the individual placing one nut on one stud feels little or no personal interest in the product. Today, the assembly machine recombines many of the separated functions into one work center (a "cottage" industry away from home) and offers the opportunity to management to <u>permit</u> the individual or team ("family" group) to feel proprietary interest in the efficiency and quality of their work function.

Whether this resurgence of personal pride and interest takes place or not will be determined by the management and manufacturing engineering personnel's understanding of the human resource factors.

For example, almost any automation project can proceed along two divergent paths: "exclusive" or "inclusive" of human resources capabilities. An <u>exclusive</u> project further reduces the operator capability requirements to that of a material handler filling hoppers. The setup man can be broken down to a feeding specialist, a mechanical specialist, an electrical specialist, etc., ad nauseum. Maintenance and engineering personnel can also be fractionated to the point of tedium. Management will pay the price of lost time finding each specialist who will have more and more equipment to tend with less and less interest in any one operation. This trend can only result in further impersonalization of the manufacturing job.

On the other hand, the <u>inclusive</u> project planning provides for up-grading the operator and minimization of the supporting services required. Many automated assembly systems are the equivalent of a former complete factory department which required a foreman, as well as setup men and operators. If the man or team to run this "department" is up-graded through training to handle all but major maintenance, they can be "motivated" to feel a genuine proprietary interest in "their" department, its efficiency and quality of workmanship.

If we are to reverse the trend toward impersonalization of the human resources, the manufacturing engineer and his management must understand the basics of motivation and apply them to the automated assembly process development. There are probably as many theories as there are psychologists. While they may conflict in approach, they all have a common goal; efficiency through effective use of human resources.

This chapter will offer five viewpoints, verbatim, as presented at recent SME ASSEMBLEX Conferences. It is difficult to read these completely objectively; you will find yourself identifying with many of the ideas presented. In the field of utilizing human resources, this identification is essential.

The first viewpoint is "Labors View of Automation and Productivity" by Reginald Newell, Director of Research, International Association of Machinists.

Let me start out by answering a question which may be in the minds of many of you. Namely, what is a person who has spent all of his working life in the field of labor-management relations doing here, today, talking to a group of manufacturing engineers? After all, the manufacturing engineer has, traditionally, been concerned with machinery and methods by which materials are turned into finished products. His management asks him to apply machine elements and manufacturing concepts with the highest degree of efficiency possible--limited only by the state-of-the-art of technology and guided by sound business practices. Manpower needs have been the manufacturing engineer's concern only to the extent that these needs must be filled by qualified workers. The man who runs the machine is a concern of the Industrial Relations or Personnel Departments, not yours.

I am afraid that I do not share this view. I feel that the manufacturing engineers plays an important role in determining the industrial relations climate in a particular plant. Indeed, the manufacturing engineer is the key figure in the industrial relations scene. For example, the personnel manager tries to resolve labor disputes and solve personnel problems as they arise; the first-line supervisor can only work within the prescribed framework of shop organization and control; the shop steward is only concerned with getting the best for his union members out of the situation on the shop floor. The manufacturing engineer is responsible for production administration and sets the stage on which the others play out their parts. Specifically, when the manufacturing engineer is engaged in such activities as setting batch sizes, determining plant layout, establishing process specifications, ordering changes in the priority of work to be done, or when making any decision about matters involving shop operations, he influences the behavior pattern on the shop floor.

Thus, for this reasons and others that I will touch on later, I feel we share a large area of mutual concern. This is especially true when we talk about automation and technological change.

I think it vital to spend a few minutes making clear labor's view of productivity. Labor is fully aware that if productivity had remained static over the centuries, man would have never emerged from the Stone Age. Increases in the standard of living are only possible through increases in productivity. Man can only consume what he produces. Such things as wages, incomes, profits and prices are only factors in the distributive process, not direct determinants of the standard of living.

Maintaining and accelerating our historic rate of productivity increase is a vital and pervasive factor in determining the future quality of American life.

-- It is vital to a sound economy which can provide more and better jobs for everyone who wants to work.

-- It is vital to curb inflation and raise real income. Only by increasing production per unit of resources can we expect to achieve both rising incomes and stable prices.

-- It is vital to our ability to compete in world markets and preserve job opportunities. Foreign competitors have modernized faster, produced cheaper, and overcome quality differences to the point where we face the prospect of domestic employment effect.

-- It is vital to our growing concern for the environment and our ability to pay for clean air and water, without a reduction in other facets of the quality of life.

-- It is vital to freeing the resources necessary for elimination of need, hunger and deprivation, and to aid underdeveloped countries of the world.

-- It is vital to more and better community services without backbreaking taxes. Productivity increases in the public sector are a partial answer to the fiscal crisis in the cities.

Rising productivity means high levels of employment for American workers, optimum utilization of plant capacity for business and industry, and a better standard of living for all Americans.

I do not need to spend any time with this group discussing the many elements which contribute to our productivity growth. Instead, my remarks will center on the relationship between the two most significant factors in productivity, _i.e._, labor and capital.

Capital resources are the monies, machines, equipment, and technological tools which translate ideas and aspirations to products. A strong, expanding economy with attractive returns to capital insures a willingness to invest in new technology, adquate modernization, and stimulus for efficient growth. As professional engineers, it is this area which draws your greatest concern and understandably so. The fundamental question for you is how can electrical, electronic and mechanical engineering combine to produce technological advances which will lead to greater productivity?

I, on the other hand, am primarily concerned with labor, or as I prefer to call it, the human element in the productivity factor. Human resources are first and foremost the fountain of energy, drive and intelligence in advancing the fortunes of our country. A misunderstanding exists between the American labor movement--which advocates human needs--and the engineering community--which concentrates on precise technical needs of production. I hope to show you how our two groups, despite some areas of conflict can and must work together if we are to succeed in our mutual desire for increased productivity.

There can be no questions that radical and rapid changes in technology are altering the way goods and services are produced and distributed. The use of automatic and semi-automatic processes--as well as new materials, machines and the speeding up of the flow of work--has been spreading through almost all parts of the American economy. The impact can be seen in factories and offices, warehouses and large retail stores, construction and transportation.

New technology is present in computer-guided machine tools, reorganized assembly lines and other new production methods, automated warehousing, faster transportation, self-service retailing, computer-aided instruction for school children and medical diagnosis, new information and communication systems and other new products and services.

Much of the new technology has the effect of all labor-saving operations --increased production with the same number or fewer workers. As a result, the application of radical and rapid technological changes frequently poses the potential of displacing some existing jobs. Moreover, the new technology typically has additional widespread impacts. These impacts include creation of new jobs while existing jobs are eliminated, changes in job content, in skill requirements and the flow of work from one operation to another. The new technology often causes changes in industry location--shutdowns of departments and entire plants and shifts to new locations in suburban or outlying areas. Sometimes the new locations are hundreds of miles from the previous locations.

In addition, these radical changes during the past 30 years have been running at a very rapid rate. In many industries, workers are confronted by continuing changes that affect their jobs and their livelihood. No industry is immune to these changes, which are constantly shifting the structure of skills, occupations, jobs and earnings of American workers.

It should be noted parenthetically that technological changes do not necessarily increase productivity in the usual sense of output per man hour but rather bring about qualitative improvements in products, as well as the creation of new products.

Typically, however, these production systems give too little consideration to the human beings who run the machines. All too often, it appears that these designers are guided by the feeling that technological advances in manufacturing engineering relegate the worker to the status of a machine, designed to perform small tasks, precisely specified on the basis of time and motion studies. He is assumed to be motivated primarily by economic needs and classified by a known degree of strength, dexterity, and perseverance. The worker is often considered by management as incapable of dealing with variables in the production flow; any unplanned occurrences are to be handled by supervisory personnel.

Yet, by ignoring the men who operate the machinery you design, you miss a real opportunity for greater, over-all productivity. The numerically-controlled multi-purpose machine tool is an example of what could be done. The only manual skills required are to load the machine and press the button providing a good opportunity to utilize and develop other skills in the operator. However, one often sees such an operator doing nothing while waiting for the completion of a cycle. He could be planning his daily or weekly workload, modifying punched tapes, or working out other part's programs. He could, alternatively, be used to control several machines or be given the responsibility of inspecting his own work. This would not necessarily remove work from inspection or planning departments, but it would free them from day-to-day details in order to concentrate on shop-wire or long-term problems.

Yet this approach has scarcely been explored by management and, in those few cases where such an innovative approach has been tried, management steadfastly refuses to reward the worker in his new role. Instead of paying him for his added responsibility and duties, management asks him to be satisfied with "psychic income" and the knowledge that he is "part of the team."

Significant changes have taken place in worker expectations and companies and their engineers will have to adapt to and take advantage of these changes. Both those who design the machines and those who own them must recognize that superior job performance and job satisfaction are related. Further, job satisfaction is influenced by a great many factors which, together, make up the total quality of working life and most production systems and jobs give too little weight to these factors.

What are those factors? Although commentators on the quality of work life differ on precisely what to include, a typical list includes at least four factors.

<u>First</u>, is adequate pay, fairly administered. There are some who say that labor is already too well paid and that with machines which require only limited monitoring, workers cannot expect to continue getting "generous" pay hikes. Let us look at this assumption of worker affluence.

The U.S. Department of Labor recently released their Index of Urban Family Budgets with data for October, 1974. At that time the typical urban American family of four required $14,300 a year to live at a modest but decent standard of living.

This family is typical. The husband 38, works full time; his wife of 15 years has no outside job. They have two children, a son, 13, and a daughter, 8. They bought their house six years ago, drive a previously-owned car and generally have no pennies to squander.

At that time, the average American factory worker, working full-time was earning $9,089.60, including overtime. Indeed, at last report, the median income of American families was just over $12,000 a year. In other words, half of the families in the USA cannot afford what the Labor Department considers a modest but adequate standard of living. And, in many families the wife has an outside job.

The point I wish to make is that workers pay has not yet reached that utopian level where psychic income will be sufficient motivation to increase worker productivity. The worker must be paid a fair share of the worth of all production over and above what he is expected to produce for his "normal" wages. It must be an amount that he can recognize as a fair share, and it must be one that he can follow and understand. If by working smarter, workers in a department can double production, they could be paid double their wages and it would not cost the company one cent.

The second factor is a safe and healthy working environment. Each year, literally thousands of new chemicals are introduced into the work place with little if any idea as to either their short-term or long-range effects on the safety and health. We are only just now becoming aware of the health hazards of such long-time substances as asbestos, polyvinyl-chloride and many solvents used in the aerospace and other industries. In 1970, the Occupational Safety and Health Act was passed mostly because Congress felt that neither the provisions of existing law nor the voluntary efforts of industry were sufficient to provide worker safety.

Thus, the individual worker, who shows an amazing tolerance for hazardous jobs and is generally the last to know when a new hazard is suspected, through his union, now has a strong voice in insuring his safety on the job as well as long-range health.

The third factor is an opportunity to use and develop both skills and knowledge. Management and labor often jointly operate industrial training, especially apprenticeship programs. Such training is not a collateral issue but relates to the quantity and quality of the labor input. As a rule, training in or related to the work place is generally conceded to improve both performance and job satisfaction.

Traditional apprentice programs will probably have to be altered to accommodate innovations and specialties created by new technology. For example, new developments in laser technology may replace many mechanical cutting and boring operations; computers will eventually control many mechanical processes that are now done by men. The rigidity of present apprenticeship programs must also be examined. The length of the programs, the lack of alternatives to allow fast and slow learners to advance at their own paces, and the policy of training all apprentices in every aspect of a particular trade must be reviewed. The amount of classroom training and the quality and variety of on-the-job training must be updated to accommodate new and improved equipment, tools and techniques. To accommodate other workers who already possess certain specialized skills--and who may be eventually required to perform many multi-skilled operations--training must be altered and revised.

The last factor is a job that provides a sense of meaning, responsibility and a knowledge of results. This gets us into the area of alternatives to the traditional structuring of work which, in recent years, has drawn increasing attention. Last year, the Industrial Conference Board published an excellent summary of some of the human engineering systems used to improve productivity and allegedly improve job satisifaction. Among these approaches are: job rotation, job enlargement, job enrichment, plan and control, work simplification, autonomous work groups, etc. Perhaps the most inclusive label to cover the entire trend in "the-quality-of-working-life" movement.

I must, at this point, express some real concern (some might say cynicism) over management's view of increasing productivity through humanizing work. I find it hard to picture management humanizing jobs at the expense of profits. In fact, I have a sneaking suspicion that "job humanization" may be just another name for "time and motion" study. As Thomas Brooks said in a recent article in the AFL-CIO Federationist "Substituting the sociologist's questionnaire for the stop watch is likely to be no gain for the workers." While workers have a stake in productivity it is not always identical with that of management. Job enrichment programs have cut jobs just as effectively as automation and stop watches. And the rewards of productivity are not always equitably shared. I also have a feeling that what some companies call job enrichment is really little more than the introduction of gimmicks--like doing away with time clocks or developing "work teams" or designing jobs to "maximize personal involvement" whatever that means.

Still, no one can deny that labor-management relationships affect productivity in many ways. Collective bargaining agreements, as well as customs and practices at the workplace, can influence the pace of utilization and staffing practices; the sharing of the benefits of productivity gains and many other factors affecting worker morale and performance. In this broad sense, productivity considerations have historically been a feature of collective bargaining in the United States. In fact, the common interest of both labor and management in productivity improvement has been specifically recognized in two major collective bargaining agreements in the United States; the basic Steel and Auto agreements.

Further, the potential for constructive joint action by labor and management in the interest of increased productivity is far greater than is often recognized. Such joint actions can and should take many forms, industries, the complex subject of work rules needs to be addressed. These may relate to such matters as crew size, work methods, assignment procedures or other practices affecting the pace and volume of output per worker. Management contends that certain of these work rules represent unreasonable obstacles to productivity improvement. However, these rules are considered by some unions as "property rights" in jobs or as necessary worker safeguards, which have been bargained over and developed over many years.

The possible scope of joint labor-management action for productivity improvement is, moreover, much broader than the area of work rules modification. In all sectors of the economy, there is a large, mainly untapped, potential for increasing productivity in the imagination, ingenuity and know-how of workers. Cooperative arrangements by organized labor and management can help in drawing upon this potential in a wide variety of ways, such as through training and upgrading programs, job redesign, safety programs, improved work scheduling, materials conservation and improved workplace participation, communication and information procedures.

Furthermore, there are areas of labor and management practices and policies for productivity improvement which do not require any new technological developments or formal collective bargaining.

First, management in most companies does not have to wait upon technological developments to achieve marked increases in productivity. The adoption of selected existing techniques and procedures already tested and utilized by existing firms would significantly improve performance. A program pointing out to business the magnitude of the possible benefits to be derived from such modernization--e.g., the sizeable differentials in efficiency that exist between efficient and inefficient plants in the same industries--might help speed progress.

Secondly, there is considerable scope for obtaining improved performance by workers, in ways which do not involve oppressive "speed-ups" or other undesirable working conditions. Progressive personnel practices and increased emphasis on human aspects of work relations have long been recognized as having "pay-off" for management, in terms of improved worker performance, reduced absenteeism and lower turnover. Recent joint labor-management efforts in some industries, designed to reduce tardiness and absenteeism through pre-employment orientation programs and similar measures, illustrate positive approaches to achievement of progress in this direction.

One final area of worker productivity I wish to touch on deals directly with the new machinery that you design. If you want to see productivity decline, visit a plant where the workers have, through the "grapevine" become aware that the company has just purchased some new automated production equipment. Rumors are created and enlarged overnight with respect to the impact of the new equipment on jobs, wages and training opportunities. It is in this area of worker adjustment to technological change that American labor unions can play a vital role.

Collective bargaining holds a vitally important role in meeting the challenges of new technology. The flexibility of the American system of company and plant bargaining helps workers and the unions that represent them--negotiate and settle with employers on reasonable and humane protections for workers against the potentially dangerous impact of job-destroying technological innovations.

For those workers under a union contract, collective bargaining can help them gain a fair share of the wealth they produce; help democratize labor-management relations and humanize the work place and work itself. And, equally important, collective bargaining can help ease the impact of new technology on workers' jobs and earnings. Collective bargaining can provide cushions to soften the adverse impact on workers by setting up adjustment procedures and programs at the work place.

The costs to the employer of the adjustment procedures and cushions should be viewed as part of the cost of installing new machines and new work processes. The costs of business investment very reasonably and properly should include and compensate for the human costs of technological innovation.

Unions acting to protect workers against loss of jobs and loss of income make technological progress and technological change more responsive and more sensitive to human needs. The labor movement will be performing one of the great tasks posed for it by the times in which we live--the task of insuring that the current technological revolution, unlike the Industrial Revolution, moves forward with full consideration for people as well as profits.

In negotiating contracts, the IAM always seeks protective language with regard to technological change. Such language involves:

1. Advance notice and consultation with the union whenever employers plan major changes.

2. Right to transfer not only to other jobs within the plant but, to jobs in other plants as well, whenever the company is located in more than one community or state, with adequate moving allowance.

3. Training for new jobs (or old jobs which have not been eliminated) at full pay.

4. No reduction in the hourly rate of pay for workers who have been down-graded because of new technology.

5. Supplementary payments to employees who are laid off because of technological change.

6. Continuation of insurance coverage and other fringe benefits during periods of layoff and after retirement.

7. New job classifications and rates of pay wherever automation has increased skill requirements or responsibility.

8. An equitable distribution of the gains resulting from greater productivity.

While admittedly this language is designed for the workers benefit, it also benefits the company by maintaining morale and assuring that the worker is viewed as a human being with legitimate fears and needs, rather than just another expendable "part" in the manufacturing process.

In conclusion, let me once again reiterate American labor's acceptance of the principle that continued productivity growth is necessary to insure economic growth and prosperity. America's ability to produce effectively and efficiently--our high level of national productivity--has been critical to our growth and development. Our national productivity is not a mysterious phenomenon. We know it means an ever-increasing improvement in the use of our resources, and the ability to meet new targets of opportunity so that America can fulfill its promise. It is a rising level of productivity which under-lies the basic strength of our economy to defend our freedom and work for peace in the world.

In accepting the necessity for productivity growth, organized labor in

America also accepts automation and other forms of technological change. Unlike the desperate and unhappy men who roamed the English countryside more than a century ago, American labor has no desire to slow the Nation's rate of technological progress. We do not want to smash the machines. On the other hand, we do not want the machines to smash our society. While we look forward to the abundance that automation can bring--we must be prepared to face the complex manpower adjustments that are associated with it.

To say that labor can influence productivity, is one thing. To expect labor to work for increased productivity merely for the sake of producing more goods, however, is quite another. Labor is willing to work for increased productivity only if it is assured that it will receive a share of the benefits derived from such increased output. Thus, any discussion of the role of labor and management policies and practices in productivity improvement needs to be placed in a larger setting, which recognizes that advances in productivity must be equitably shared between labor, capital and the consumer.

The second viewpoint is "Management's Responsibility for Employee Motivation" by Louis A. Seiberlich, Jr., Sr. Partner, Omnisystems.

This viewpoint will define "motivation" in a practical way; provide a brief synopsis of the more well-known authorities in the field of motivation; show where motivation fits into a manager's job; list twenty factors affecting each person's motivation; and, describe how this information can be applied by any manager to his own motivation or the motivation of his subordinates.

WHAT MOTIVATION IS

When we cut through the theory and get down to practical cases, "motivation" seems to be: the internal desire to close the gap between what each of us wants and what each of perceives we presently have. In other words, it's a hole that needs filling.

While at first glance this seems overly simplified, a closer look reveals we must do a great deal of soul-searching to answer three critical questions.

1. What specifically do I want out of my job?

2. What do I presently receive from my job?

3. How important are each of these wants?

This paper is an attempt to help each manager answer these penetrating questions for himself and for each of his subordinates.

WHERE MOTIVATION FITS INTO A MANAGER'S JOB

Classically, the functions of a manager are quite similar for managing any kind of situation. Each functions is made up of specific skill activities in which we must acquire and maintain proficiency. While the list following is not all-inclusive, it nonetheless does illustrate typical types of decisions we make every day using one or more of the managerial skills shown. Note the relative position of "motivating."

Planning

- forecasting: assessing outside influences that may impact on the organization

- establishing objectives: setting specific goals

- establishing and utilizing policies: building guidelines for operating on a day-to-day basis

- programming: setting priorities

- scheduling: setting time limits in which work is to be completed

- establishing and utilizing procedures: stating uniform operating methods

- budgeting: allocating <u>all</u> resources (time, people, money, space, facilities, authority, etc.)

Organizing

- developing structure: building needed management levels

- delegating: sharing authority, responsibility, accountability

- establishing effective working relationships: coping with organizational and individual conflicts

Leading

- initiating: getting action started

- deciding: choosing among alternatives

- MOTIVATING: encouraging others through climate control

- communicating: creating understanding

Controlling

- personal observation: seeing actual results

- management reporting: allowing others to report results

Selecting

- putting the right person on the right job

Developing

--   upgrading the knowledge, skills, and attitudes of self and others

Utilizing

--   making the best possible use of staff talent and time

Evaluating

--   letting people know where they stand and how well they are doing

Analysis of the list above shows clearly that "motivating" is not the only job of managers. It is just one of many we must continuously deal with.

RE-CAP OF MOTIVATION THEORIES AND FINDINGS

Behavioral scientists' findings upset many cherished and traditional methods of management, and myths about how people behave in organizations. It has been quite easy for managers to ignore the things that behavioral scientists are saying, because American organizations generally have enjoyed tremendous success.

Now there is renewed interest in human motivation in organizations, simply because we seem to have come almost to the outer limits of offering more money, fringe benefits, security, and other things which have traditionally been considered motivational. There is a growing interest on the part of management as to whether there are principles and techniques which behavioral science has discovered which can be applied to increase productivity and profit in an organization.

If behavioral science can identify some of the techniques being used by successful managers (and if these techniques seem to be rather different than traditional management techniques), they warrant serious consideration by all managers to see whether or not these techniques should be applied in their own organizations.

What are the behavioral scientists saying? The common thread that runs through all behavioral science reports seems to be this: you are not using employees at their full capacity. You've designed jobs to take advantage of only the minimum performance human beings are capable of. You're appealing to basic needs to motivate employees, rather than higher human psychological needs.

The result is that management finds itself in a position of having to find more and more in the way of economic incentives to accomplish less and less in terms of productivity. So we hear that employees are lazy, they always want more, they're disloyal, they cannot be motivated.

Behavioral science says that people behave this way in organizations because that's the way jobs are designed, that's the way policies, procedures, and traditions have taught both managers and employees how to behave.

It certainly shouldn't come as a surprise to learn that employees at all organizational levels want challenging, rewarding, meaningful assignments. That's basically what each of us wants to varying degrees.

Management by Motivation--Saul Gellerman

An organization's employees are nearly always a significant cost factor. When they are motivated to be cooperative and creative, they can also be a significant profit factor. However, most experienced managers would agree that this happens all too rarely. Gellerman asserts that there is nothing inevitable about the lack of effective work motivation. It is often the unintended result of management practices that are either antiquated or unrealistic.

There are enough consistencies in behavioral research findings to permit some useful generalizations.

1. Motivational problems are more likely to result from the way an organization is managed than from a simple unwillingness on the part of employees to work hard.

2. Modern management has a tendency to over-manage--that is, to define the employee's job too narrowly and to make too many decisions for him.

3. If you want to understand the reasons why an employee behaves as does, you have to learn to look at his environment the way he does.

Conclusions like these emerge from many behavioral studies that have been completed for a wide variety of organizations. They are evidently not peculiar to any particular industry, geographical region, or type of job.

Today's employee is typically much better educated, more in demand, and therefore more independent, than the typical employee of only a generation ago. Management practices have not always kept pace with these changes. Consequently, motivational methods which may have worked well enough a generation ago are increasingly ineffective today. Neither the threat of losing a job nor the attraction of more money are enough, in today's environment, to assure effective motivation.

To explain why changes in the environment affect employee motivations, Gellerman cites a frequent finding of behavioral scientists: that nearly everyone regards his own behavior as sensible and justifiable. In other words, people are always behaving in ways that make sense to them, based on their understanding of the circumstances in which they find themselves. That same behavior may seem quite irrational to someone else. The difference probably lies in the fact that they aren't making the same assumptions about those circumstances.

Managers are likely to have attitudes toward their own environments (which usually seem full of challenges and opportunities) that are quite different from the attitudes employees have toward their environments (which are often more likely to seem rather dull and unstimulating). To occupy the same physical environment is not necessarily to see it the same way or to share the same attitudes toward it.

Motivation is not simply a matter of things that a manager does to influence his subordinates. It is much more complex than that. People are motivated not so much by what other people want them to do as by their own desire to get along as best they can in the kind of world they think they are living in. This is called the "principle of psychological advantage." Basically, it means that people will tend to seek whatever values they consider important to the extent that they believe it is safe and possible for them to do so.

To most employees, the manager and the organization as a whole are simply parts of the overall environment, and not necessarily the most important part. He will make concessions to his manager to the extent that he thinks he must, but he will not necessarily consider it advantageous to do more than that <u>unless</u>--and that is a big "unless"--it appears that doing so will lead to a lasting, significant gain. That gain is not necessarily monetary. It often has more to do with a change in the role the individual can play (that is, the kind of activity and the degree of self-management expected of him) than with money itself.

All people are motivated all of the time (it's more a question of degree and direction) and most of the things that management does have a motivational effect. The problem is that too often people are motivated to act in ways which are unproductive. This is usually because they see no advantage in increasing their productivity or because they are actually motivated to thwart the organization if they can.

Restriction of output can occur even in the face of incentive payments, simply because the employees regard the incentive as a bad bargain. They don't expect it to be a lasting advantage, and even if it were, it brings the risk of lasting disadvantages (the hostility of peers) which would more than offset the gains. Underlying these attitudes is a basic conviction that management's actions toward employees are likely to be against their real interests regardless of how attractively they may be packaged. Situations of this kind are very difficult to deal with effectively. The root cause is the employee's perception of the way management perceives him. Changing this requires time, patience, and above all, a genuine willingness to deal with employees as individuals.

The role of money in motivation must be seen in its true perspective to other motivations. Money is neither all-important, as some management theorists have assumed, nor unimportant as some over-simplified versions of behavioral science have implied. The effect of a possible monetary gain on an individual's behavior depends on his financial status, and more broadly, on the psychological meaning of money for him. Money is an effective motivating tool when financial needs are strong and/or when money can serve some psychological purpose for the individual. It does not tend to be effective when financial needs are not strong, or when money does not have an important psychological value. In the case of restriction of output, a potential monetary gain is often incapable of changing behavior even when there is financial need, because it conflicts with a still stronger psychological value (acceptance by one's peers, for example).

Behavioral science has not revealed a simple list of do's and don'ts for managers. On the contrary, it has revealed that the process of motivation is much too complex to be handled effectively by approaches that require no analysis or creativity on the part of the manager. It is up to the individual manager to diagnose the specific problems that he is confronted with, and to take action that is addressed to root causes and not merely to symptoms.

Interpersonal Relations--Chris Argyris

The problems of apathy and lack of effort are not, according to Argyris, simply a matter of individual laziness. Rather they are often healthy reactions by normal people to an unhealthy environment--created by common management policies. More specifically, Argyris states that most adults are motivated to be responsible, self-reliant, and independent. These motives are acquired during childhood from the educational system, the family, and communications media such as books, television, and radio. But the typical organization confines most of its employees to roles that provide little opportunity for responsibility, self-reliance, or independence. On the contrary, too many jobs are designed in ways that make minimal demands on the individual's abilities, and that place the responsibility for major decisions not in his hands but in his manager's.

In effect, such jobs create a childlike role for the employee and frustrate his normal motivations for a more adult role. The common reaction of withdrawing one's interest from the job--treating it with indifference or even with a degree of contempt--is a necessary defensive manuever that helps the individual to preserve his self-respect. Unfortunately, the cost of these reactions to the organization is heavy: minimal output; low quality; and, excessive waste.

Both labor unions and management have missed the main point with regard to employee motivation. They have concentrated largely on matters related to income: job security; pay; and, fringe benefits. These are necessary, but in themselves insufficient conditions for effective motivation. The battle for adequate employee incomes was fought and won long ago. Today the real frustration of many employees is not so much a matter of income as or utilizing their abilities in a significant way. They need a sense of pride and accomplishment from their work. Instead of this, they frequently find their work neither stimulating nor dignifying.

So, for the typical employee, work tends to become a "necessary evil." It should be a source of personal satisfaction. This is why Argyris feels that the psychological importance of income is changing for many people: they no longer regard it chiefly as a means of elevating their standards of living. Instead, it has taken on some of the characteristics of a penalty payment: a "fine" they can periodically levy against their employers to compensate them for the lack of satisfaction in their work. Since the external symptom of this resentment is a continual demand for income improvement, both management and labor tend to be blinded to its underlying causes. But both are also under pressure to match income improvements to productivity improvements. The potential contribution of motivation to increased productivity is largely neglected in negotiations over income, since these deal only with the sypmtom and do not address themselves to the underlying cause.

While "job enrichment" programs have proved successful in certain situations, such programs must be undertaken selectively, since not every employee wants to accept more responsibility, or is prepared to carry the added burden of worry that responsibility inevitably brings. Nevertheless, the number of employees that can be successfully motivated by upgrading their responsibilities is much larger than most managers would suspect.

Argyris offers these suggestions as a bare minimum in coping with self and employee motivation.

1. Each person be judged in terms of ability to make contributions to the organization, not in terms of whether he "conforms."

2. Development of general trust and confidence between organization members, instead of suspicion of competitiveness. Consequently, there would be more willingness to introduce or consider new ideas.

3. Commitment to organizational goals obtained chiefly through involvement in an effective, creative team ("internal commitment"), rather than chiefly through rewards given or withheld ("external commitment").

Basically, "organizational effectiveness" (sustained profit, efficiency, growth, etc.) is most likely to occur when the organization encourages "interpersonal competence" (a mutual ability by members to use each other's unique abilities effectively).

The Achievement Motive--David McClelland

The most convincing sign of a strong achievement motive is the tendency of a person who is not being required to think about anything in particular--that is, when he is free to relax and let his mind just "idle," as it were--to think about ways to accomplish something difficult and significant.

A second characteristic of the achievement motive reveals its importance to management. Although only about 10% of the people in the population at large have a strong achievement motive, the percentage in certain occupations is likely to be much higher. This is especially true of sales and marketing positions, managerial positions of all kinds, and independent businessmen. Further, a person with a strong achievement motive is likely to surpass the accomplishments of an equally able but less strongly motivated person--especially in one of these occupations.

McClelland's studies have identified three major characteristics of the self-motivated achiever.

1. Achievers like to set their own goals.

2. The achiever tends to avoid the extremes of difficulty in selecting goals.

3. The achiever prefers tasks which provide him with more or less immediate feedback--measurements of how well he is progressing

toward his goal.

The effect of money on an achiever is actually rather complex. On the one hand, achievers usually have a fairly high opinion of the value of their services, and prefer to place a fairly high price tag on them. They are unlikely to remain for long in an organization that does not pay them well. On the other hand, it is questionable whether an incentive payment actually increases their output, since they are normally working at peak efficiency anyway.

Provided he feels that he is being equitably paid, the main significance of additional income for the achiever is as a form of feedback: a way of measuring his success.

The achievement motive is not the only source of high achievement. Other drives can also lead to high levels of attainment in other occupations. But in the work-a-day world of business, achievers have a considerable advantage. This raises the question of whether the level of achievement motivation could be increased in people whose achievement drives are not unusually strong. This may be possible; indeed, there are considerable "reserves" of latent, untapped achievement motivation in most organizations. The key is to build more achievement characteristics into more jobs: personal responsibility; individual participation in the selection of productivity targets; moderate goals; and, fast, clear-cut feedback on the results each individual is attaining.

For achievers themselves, many standard management practices are inappropriate, and in some cases may even hinder their performance. Work goals should not be imposed on the achiever: he not only wants a voice in setting his own goals, but he is unlikely to set them lower than he thinks he can reach. Highly specific directions and controls are unnecessary-- some general guidance and occasional follow-up will do. But if the job does not provide its own internal feedback mechanism regarding the achiever's effectiveness, then it is vitally important to the achiever that he be given frank, detailed appraisals of how well he is performing his job.

Hierarchy of Needs--Abraham Maslow

Maslow's concept of human behavior includes these fundamental assumptions about human behavior:

1. human behavior is purposeful

2. human wants are insatiable

3. human wants and needs can be arranged in a hierarchy

4. different people place different priorities on wants and needs

5. people will move from the low point on the pyramid (hierarchy) to a higher point as needs are satisfied

Maslow's hierarchy of individual needs, in ascending order, is made up of:

1. physiological needs--air, rest, hunger, sex, thirst, elimination, exercise, etc.

2. safety or security needs--protection against danger, threat, deprivation

3. social needs--belonging, associations, peer acceptance, giving receiving friendship and love

4. ego needs--of greatest significance to management and man himself. Includes self-esteem, self-confidence, independence, achievement, status, recognition, respect as others see him.

5. self-fulfillment--creativity, realization of one's own potential

Maslow estimated that the percentage of people who have the above needs satisfied is:

1. physiological--85%

2. safety or security--70%

3. social--60%

4. ego--50%

5. self-fulfillment--5 to 10%

In most organizations, physiological and security needs have been satisfied to a very high degree.

The greatest opportunities for us to effectively influence employee behavior therefore, lie in providing satisfaction for the employee's social, ego, and self-fulfillment needs.

Theory X and Theory Y--Douglas McGregor

McGregor is best known for his Theory X and Y assumptions about people as outlined below.

Theory X

1. The average human being has an inherent dislike of work and will avoid it if he can.

2. Because of this human characteristic dislike of work, most people must be coerced, controlled, directed, threatened with punishment to get them to put forth adequate effort toward the achievement of organizational objectives.

3. The average human being prefers to be directed, wishes to avoid responsibility, has relatively little ambition, and wants security

above all.

Theory Y

1. The expenditure of physical and mental effort in work is as natural as play or rest. The average human being does not inherently dislike work. Depending upon controllable conditions, work may be a source of satisfaction (and will be voluntarily performed) or a source of punishment (and will be avoided if possible).

2. External control and the threat of punishment are not the only means for bringing about effort toward organization objectives. Man will exercise self-direction and self-control in the service of objectives to which he is committed and has had a hand in developing.

3. Commitment to objectives is a function of the reward associated with their achievement. The most significant of such rewards, e.g. the satisfaction of ego and self-actualization needs, can be direct products of effort directed toward organization objectives.

4. The average human being learns, under proper conditions, not only to accept but to seek responsibility. Avoidance of responsibility, lack of ambition, and emphasis on security are generally consequences of experience, not inherent human characteristics.

5. The capacity to exercise a relatively high degree of imagination, ingenuity, and creativity in the solution of organizational problems is widely, not narrowly, distributed in the population.

6. Under the conditions of modern industrial life, the intellectual potential of the average human being is only partially utilized.

Managers who belive Theory X assumptions tend to be characterized by:

1. autocratic direction
2. tight control
3. scientific management--spell out special job duties in great detail
4. employee has little latitude to think for himself
5. highly proceduralized

Managers who believe Theory Y assumptions tend to be characterized by:

1. participative management
2. management by objectives
3. employee self-control

4. integrating personal goals with organizational goals

The major lessons to be learned from this seem to be:

1. neither set of assumptions is "good" or "bad"

2. the assumptions we make about people in general tend to condition our managerial actions

3. our managerial actions will attract and retain employees who feel comfortable following those assumptions

4. to be most effective, our managerial actions must be flexible and grow out of a consideration of the forces at work in the leader, the follower, and the situation.

Motivator/Hygiene Concept--Frederick Herzberg

Man has two different sets of needs. Out of that simple observation has come both a new insight into the nature of work motivation, and a valuable new strategy for increasing employee productivity.

This classical approach to motivation has concerned itself only with the environment in which the employee works; that is, the circumstances that surround him while he works and the things he is given in exchange for his work. This concern with the environment is a never-ending necessity for management, but it is not sufficient in itself for effective motivation. Effective motivation requires consideration of another set of factors, namely, experiences that are inherent in the work itself.

The assertion that work itself can be a motivator represents an important behavioral science breakthrough. Traditionally, work has been regarded as an unpleasant necessity, rather than as a potential motivator. For this reason, it has generally been considered necessary for management to either entice people to work by means of various rewards, or to coerce them to work by means of various threats, or both. The potential motivating power of work was obscured by the fact that most jobs were not at all stimulating, and therefore some kind of external pressure, either positive or negative, had to be applied to get people to do them. Indeed, many jobs today still have this characteristic. But when a job provides an opportunity for personal satisfaction or growth, a powerful new motivating force is introduced.

Herzberg holds that the environmental approach, which he refers to as "hygiene," is inherently limited in its capacity to influence employee behavior. On the other hand, the approach through the job, which he refers to as "motivation," seems to be capable of larger and more lasting effects.

The term "hygiene" describes such things as physical working conditions, supervisory policies, the climate of labor-management relations, wages, and various fringe benefits. In other words, "hygiene" includes all the various "bread and butter" factors through which management has traditionally sought to effect motivation. Herzberg chose the term "hygiene" to describe these factors because they are essentially preventive actions taken to remove sources of dissatisfaction from the environment (just as sanitation removes potential threats to health from the physical environment). Research has shown that when any of these factors are deficient, employees are quite likely to be displeased and to express their discpleasure in ways that hamper the organization--for example, through grievances, decreased productivity, or even strikes. But when the deficiencies are corrected, productivity may return to normal, but is unlikely to rise above that level. In other owrds, an investment in "hygiene" may eliminate a deficit, but it does not create a gain. Further, it is inherent in the nature of "hygiene" needs that satisfactions are not lasting, and that with the passage of time a feeling of deficiency recurs. Just as eating a meal does not prevent a person from becoming hungry again in the future, a wage increase will not prevent him from becoming dissatisfied eventually with his new wage level. "Hygiene" is a necessary but thankless task for management. There can be no end to it, since inadequate "hygiene" will surely lead to inefficiency. But even a fully effective "hygiene" program will not motivate employees to sustain a higher than usual level of efficiency.

The term "motivation" is used to describe feelings of accomplishment, of professional growth and recognition, that are experienced in a job that offers sufficient challenge and scope to the employee. He chose this term because these factors seem capable of producing a lasting increase in satisfaction, and with it an increase in productivity above "normal" levels. This has been found to be true in a wide variety of jobs and organization settings.

Herzberg's analysis focuses attention on job design. In most cases, jobs were either not consciously "designed" at all, or were designed primarily from the standpoint of efficiency and economy. To the extent that these steps have taken the challenge and opportunity for creativity out of a job, they have probably contributed to a demotivating effect. Therefore, apathy and minimum effort are the natural resolts of jobs that offer the employee no more satisfaction than a paycheck and a decent place to work. The "hygiene" factors may keep him from complaining, but they will not make him want to work harder or more efficienctly, but they will not make him want to work harder or more efficiently. Offering still more "hygiene," in the form of prizes or incentive payments, produces only a temporary effect. Investments in "hygiene" reach the point of diminishing returns rapidly and do not, therefore, represent a sound motivational strategy.

Much has been done with job enrichment as a means of introducing more effective motivation into jobs. This term draws a clear distinction between job enrichment--the deliberate enlargement of responsibility, scope and challenge--and job rotation--the movement of an individual from job to job without necessarily increasing responsibility at all. Herzberg has found job rotation unsatisfactory as a motivating tool, but has achieved impressive results with job enrichment. After an intial adjustment period during which productivity temporarily declines, efficiency tends to rise well above previous levels--and more importantly, it stays high. At the same time, employee satisfaction with the job reaches very high levels.

Herzberg regards money (and various substitutes, such as fringe benefits) to be a "hygiene" factor. Studies have shown that most people are not strongly affected by their income except when they consider it inadequate--in which case they will be dissatisfied and will take whatever steps they can to correct the situation. This might include working harder, if they thought that would have a reasonably prompt effect on their incomes, or working less effectively, if they felt that management was unresponsive to any other kind of pressure. The important point is that the positive effect on effort is likely to be brief; the individual will either get his increase, thereby eliminating the incentive for the extra effort, or he will decide that the extra effort was futile and will revert to a normal working pace. Further, satisfaction with monetary increase does not last long, and in time a new feeling of inequity will set in if new wage increases are not forthcoming.

These observations led Herzberg to conclude that the principal effect of money is to create dissatisfaction (when pay is perceived as inequitable), and that the principal effect of pay increases is to remove dissatisfaction and not to create satisfaction. Therefore, careful attention to wage and salary administration is an absolute "must" for management, but we should recognize that we are only performing a "hygiene" function and cannot expect this to be sufficient in itself for the attainment of effective motivation.

Conclusion

The major conclusion we can draw from the research of these men might be stated thus: effective long-range motivation is likely to occur when the work itself provides opportunity for achievement, recognition, responsibility, and professional growth.

FACTORS AFFECTING EACH PERSON'S MOTIVATION

The writings of these and other distinguished behavioral scientists clearly show at least 20 motivational factors of varying importance to each person:

1. authority

2. self-esteem

3. personal growth and development

4. pay

5. prestige inside the organization

6. opportunity for independent thought and action

7. security

8. self-fulfillment

9. prestige outside the organization

10. worthwhile accomplishment

11. chance to help others

12. participation in goal-setting

13. being informed

14. participation in setting methods and procedures

15. recognition

16. pressure

17. responsibility

18. organization policies and the way they are administered

19. working conditions

20. competence of our management people

Our major task is to determine for ourselves and our employees:

1. how much of each factor we want

2. how much of each factor we presently have

3. the relative importance of each factor

Only after we have done this can we begin to build meaningful solutions to "motivation problems."

ON-THE-JOB APPLICATION OF MOTIVATION FACTORS

The motivational factors can be used best in a simple problem-solving format as follows:

Self

1. list the 20 factors

2. using a scale of 1 (low) to 7 (high), rate how much of each factor you want

3. using the same scale, rate how much of each factor you have

4. using the same scale, rate the relative importance of each factor

5. asterisk those factors with the greatest discrepancy and importance

6. note in writing the primary specific cause of the gap

a. self-actions or attitudes
   b. actions or attitudes of the manner
   c. specific factors in the environment (policy, rule, procedure, regulation, etc.)

7. note in writing <u>what</u> could be done, <u>when</u>, and by <u>whom</u>--to close the gap

8. discuss with yor immediate superior to determine practicality and finalize your plan

9. set a date for repeating the process

Your employees

1. have each employee follow steps 1 through 7 above

2. complete a list as you believe the employee will complete it

3. get together to compare notes

4. decide what will be done when and by whom

5. set a date for repeating the process

Added thoughts for application

In working with the factors shown above, one or more of the following techniques may be helpful.

Employee Development

-- Provide on-the-job coaching for all employees

-- Conduct employee appraisals regularly, formally or informally

-- Use job rotation where possible, making sure to increase responsibility with each move

-- Schedule people for appropriate education and development programs

-- Provide understudy training

-- Encourage individual development efforts inside and outside the organization

-- Plan for guided experience to implement specific individual plans

-- Encourage participation in job-related correspondence courses

-- Have others represent you at internal/external meetings

-- Appoint employees as members of special committees, task forces, etc.

-- Permit and encourage discussions with other inside experts
-- Use employees as vacation replacements for managers and supervisors
-- Include employees in discussion of job-related problems
-- Schedule employees to serve as instructors, conference leaders, etc.
-- Involve employees in setting goals, objectives, procedures, etc.

The Job

-- Enrich the job to provide a larger, more identifiable scope of responsibility
-- Redesign jobs to better suit capabilities and experience of present people
-- Clarify responsibilities so that each employee knows what is expected of him
-- Assign jobs that are challenging yet attainable
-- Reward results rather than activities
-- Make special assignments to provide experience in new areas
-- Develop reliable, realistic, and meaningful measures of job performance

Communications

-- Spend time with your employees--be accessible
-- Allow people to express their opinions and feelings
-- Show people how they work they do fits into the overall picture
-- Consistently let people know how well they are doing
-- Provide information on developments which may affect your employees
-- Give as much advance notice as possible of overtime or special work
-- Regularly give people information related to the overall performance of the organization in light of stated objectives

Employee Utilization

-- Place people on jobs that fully use their skills, experience, and interests
-- Identify paths of promotion and see that they are understood and used

-- Use lateral transfers as a way to prepare people for increasing responsibilities in the future

-- Schedule people for testing if questions exist about how well their skills and aptitudes are being used; find out what really motivates your people

Salary Administration

-- Grant merit increases to reward people for exceptional <u>performance</u>

-- Use some kind of bonus system--tangible or intangible--to reward outstanding <u>performance</u>

-- Call attention to the values of the organization's benefit programs

-- Consider and use the wide range of non-monetary rewards available

Recognition

-- Give individual commendations for outstanding performance

-- Give oral commendation (where appropriate) in front of a group

-- Provide written commendation for the employees' information and for his personnel folder or record

-- Publicize in your organization's internal publications

-- Encourage individuals to use your suggestion program--formal or informal

-- Evaluate suggestions immediately and take appropriate action

CONCLUSION

In spite of our best efforts, we may fail. Our only sensible recourse is to KEEP TRYING. The solution to the motivation issue requires an uncommon amount of faith, patience, and willingness to dig in even more deeply to uncover the key/s in each person's makeup that will release the untapped potential we <u>know</u> is available.

The third viewpoint is "Job Enrichment and Human Resource Management for Peak Productivity in Manufacturing" by Kenneth Purdy, Sr. Associate, Roy W. Walters & Associates.

Reaching and sustaining peak productivity, an issue as old as industry itself, is an even more pressing concern in these times. Great strides have been made toward improved technology in new processes and equipment most industries. Far less attention, however, has been devoted to the human motivational factors that influence productivity. This paper will describe progress made in recent years toward improved understanding of work motivation. The importance of the content of work itself as a key motivation determinant will be explained along with details of motivational job design in industrial settings--often called job enrichment. Emphasis will be given to the crucial importance of integrating motivational job design with planning of physical layout, equipment and materials handling when designing new plants and assembly lines in order to stimulate optimum human response for high productivity and product quality.

INTRODUCTION

Ask any manufacturing manager what constitutes his major challenge, and the answer is almost certain to be some variation on a single theme: "To reach and sustain peak productivity." This is the classic statement of the mission of manufacturing management, and it has never been a more pressing challenge than it is in today's threatening economic environment. For many companies today, productivity is a survival issue. Maintaining costs for payroll and materials, which threaten to bleed away all profit.

Just what constitutes "peak productivity" differs from one plant to another. But all the definitions come back to a central idea--minimizing unit cost so that the product can be sold at a price that is both competitive and profitable.

A simple calculation of units produced per hour is not an adequate index of productivity. To get even a relatively complete idea of direct cost, we must add to this basic output rate such factors as the quality of the product, the reject/rework rates, and scrap rates and costs.

Traditional methods of pursuing "peak productivity" are very often limited to three basic approaches, alone or in combination:

-- Technological improvement in equipment and processes

-- Control systems based on work measurement and work standards

-- Incentive systems, in which the sole or main incentive is monetary reward based on performance judged by the work standards.

All these approaches have been successful in improving productivity, but the degree of success has been variable. While technological advance is always a key factor, many companies have found that the state of the art--or the cost of buying it--have imposed limits on productivity improvement through technology alone.

At the same time, it has become clear that in recent years the manufacturing workforce has changed significantly--most notably in the educational level and certain aspirations of workers. In response, both management theorists and working managers have become aware that an alternative to traditional productivity approaches is needed.

In summary, the pressure on manufacturing managers to find innovative approaches to productivity has become tremendous.

This paper describes such an approach, through various aspects of human resources management. Its central assumption is that many manufacturing organizations, in the name of productivity, pursue policies that are literally counterproductive because they undervalue the role of human factors. By concentrating on traditional, machine-oriented methods to increase productivity, they ignore the emerging awareness that a manufacturing organization is a <u>sociotechnical system</u> that performs at optimum only when its human and technological aspects are in balance.

Specifically this paper focuses on:

-- The design of production jobs for maximum motivational effect

-- The structure of subunits in the production process, and the relationship among them.

-- The sources of motivation in work

-- The principles of motivational job design

-- Motivational job design as a key element in work layout planning

Many of the specific principles to be discussed here will be familiar to this audience as principles of job enrichment, and thus not strictly "new" in themselves. The general concepts are fairly well-known by now. But while they are finding increasing application in manufacturing, the concepts have been much more widely used in clerical and service operations.

The paper describes some specific applications of the sociotechnical approach in manufacturing. These are still rare enough to be available.

Besides the case material, there are several elements that are new in this paper. One is a more precise definition of the principles and techniques of job design for motivation.

Another, which should be significant in assembly manufacturing, is an attempt to integrate the principles of motivational job design with more "traditional" industrial engineering concepts. Some experts have found an inherent conflict between industrial engineering and motivational job design. I believe not only that it doesn't have to be so, but that a balanced sociotechnical system can be designed, and maximum productivity achieved, only when both disciplines make their distinctive contributions.

Job enrichment and the behavioral approach to organizational challenges have had both successes and failures. So, an additional purpose in this paper is to raise some issues that every manufacturing manager should consider at the assembly process design stage if he wishes to find that optimum way to balance people with technology and avoid the mistakes that have led to so many jobs that discourage high productivity or drive people away.

DIMENSIONS OF THE PRODUCTIVITY PROBLEM

It has been fashionable for some time to say that industry in the United States has reached a crisis of productivity. The phrase has much more truth in it than most fashionable catch-words. The signs are everywhere. The official statistics on output per man-hour have shown no signficiant increase for several years. Deprived of their major means of offsetting soaring payroll and materials costs, company after company has seen its cash flow and profits eroded to the danger point. The costs have been passed on to the consumer in most cases. The lack of productivity increase is the start of a vicious cycle. As the cost of living rises, workers ask for more pay, and the gap widens.

Sheer quantity is not the only problem in the productivity crunch. A vigorous consumer protection movement has seized on the declining quality of many American goods as ammunition for demanding increased regulation. And the cost of that regulation, in terms of lawsuits, product recalls, and bad publicity, is tremendous.

In the international perspective these effects are magnified. Years of faster productivity increase in other industrial nations have made it harder for United States goods to compete in price in world markets.

For about a decade this trend seemed a problem somewhere off in the future. But recent history has made clear that the U.S. depends on the rest of the world for much of its raw materials and energy sources--and must pay for them with the income of its manufacturers.

For all these reasons, the manufacturing manager is under vastly increased pressure to control unit costs and get a better return on the direct labor dollar.

Another aspect of the crisis is that the traditional means of increasing productivity don't seem to apply in today's economy. With demand in many industries at depression levels, even the almost automatic improvement that comes of full-capacity operation is not available. And there is little incentive to invest in more modern, more productive equipment--especially with profits down and interest rates high.

The principal remaining way to improve productivity is in the use of human labor resources. The traditional approach to this area has been to pattern workers' activities on those of machines, which can perform a single, highly specialized task with speed and accuracy. Each worker's fragment of the total process can be standardized, and engineers can clock it to determine how many times the task can be performed in a given unit of time. This approach, of course, is what is popularly called "Taylorism." It is the classic, traditional approach to productivity in manufacturing.

It has long been clear that the fragmented, repetitive tasks resulting from this approach offer no possibility of psychological reward. But the underlying concept of Taylorism is that workers seek no such reward. They work for economic reward, and management's best strategy is to tie that reward to their output, and remove any decisions or other responsibilities that might slow their output.

F. W. Taylor formulated his ideas at a time when the workforce was far different than it is now. Since then, both the educational level and the general economic security of the workforce have increased considerably. Doubts about the power of economic rewards along to motivate, and about the willingness of workers to settle for mind-numbing jobs, have become very serious.

The problem facing manufacturing managers today was well summarized by C. Jackson Grayson, former chairman of the Price Commission: In an article calling for a new, broader definition of productivity, Dean Grayson asked:

"How can this nation hope to increase machine productivity if there are no workers willing to operate the machines? Men and machines are equal constituents of productivity. A complete understanding of the concept demands the recognition of this relationship."

SOME PROBLEMS OF TRADITIONAL WORK SYSTEMS

Recent research in the psychological laboratory and in the work place leaves little doubt that human motivation is a crucial factor in productivity. Yet because psychological responses are harder to understand and to measure than technical systems, the specifically human factor in productivity has often been ignored or distorted. Management has often assumed that the human component can be viewed as an extension of the mechanical component. It has assumed that most of the work force cannot master tasks of any complexity, and that for maximum productivity the more complex tasks must be fragmented to suit a lowest common denominator.

Similarly, management has underestimated the psychological needs workers bring to their jobs. On the assumption that work can never possibly be satisfying and motivating, management has sought to maintain high productivity through the combination of fragmented tasks with line pacing, production quotas, and pressure as devices of stimulation and control.

Consequences of Traditional Organization

The best-known example of traditional manufacturing organization is the assembly line. By breaking the complete job down into many small tasks, and having them performed repeatedly in sequence, management is able to put individual workers on the line with less training. The other side of the ledger is that the desire to perform the task well decreases very fast once the task is mastered. Once the challenge of learning the task is past, the worker quickly succumbs to mental fatigue, shorter attention spans, and all the other symptoms of boredom.

It is only a short step to a feeling of being shackled to the machinery and involved in a meaningless activity. Many workers revert to restlessness and almost juvenile behavior. Getting no psychological rewards from the work itself, they spend time devising ways to beat the system, and get their satisfactions that way.

The indications of this state of meaninglessness in work are many: absenteeism, high turnover, interpersonal conflicts on the job, excessive grievances, quality problems--even, in some extreme cases sabotage.

Supervisors, as well as the actual production workers, are affected. Extreme specialization of tasks often means underutilization of higher-salaried technicians. In addition, more supervision if often needed to "police" work behavior and work flow and management often resorts to negative, repressive management.

All these problems add up to a loss of productive resources and higher costs for direct labor.

A BALANCED STRATEGY OF PRODUCTIVITY

The manager who seeks to reach and sustain maximum productivity has two basic alternatives. On the one hand, he can follow the traditional methods of conforming workers to the technological system, suffering the consequences in unproductive work behavior, and trying to overcome these problems with increasingly restrictive and repressive control systems.

The other alternative is to find ways of designing work to fit the workers and allow them to greater personal satisfaction. This approach need not abandon all that has been learned of industrial engineering, work measurment, and their related disciplines, but it should employ these tools in the light of what behavioral science has learned of workers' psychological needs.

This section outlines such a strategy. In many aspects, it is not brand new. Many European manufacturers have used some of the techniques mentioned here to combat severe labor turnover and attract higher-caliber workers. Although less widely applied in the United States, the techniques are finding greater application among some highly innovative manufacturers. In fact, similar techniques were used, out of necessity in times of labor and supervisory shortages as far back as World War II.

A simple way to describe the concept is to say that its goal is to reintroduce into mass-production industry some features of work associated with the individual craftsman of the past. This need not involve total rejection of work specialization, but it does seek a middle ground where workers either alone, in pairs, or in small teams, can experience some of the feeling of entrepreneurship of a small job shop, operating as a supplier to other parts of the assembly process. The idea is aptly suggested by the title of Scott Myers' book, <u>Every Employee a Manager</u>, based on job design experience at Texas Instruments in the 1960's.

If we wish to regain some of this much-sought "pride of workmanship," we must rescue workers from fragmented, repetitive tasks and give them some meaningful product they can proudly identify as their own.

Work and Psychological Needs

Before we can discuss motivational job design, we need some basic understanding of psychological needs as they apply to the job. A psychologist, the late Abraham Maslox, posited a <u>hierarchy of needs</u>, and his concept underlies almost all work in job enrichment. Needs in the hierarchy range from basic physiological needs, through ghose giving security and belonging to the "higher" needs for psychological growth and selffulfillment. The lower needs must be met first. But Maslow theorized that once they are met, they no longer have power to motivate.

Many of the later theorists of job enrichment have said that today's increasingly young work force has generally had its security needs met. Therefore, management cannot rely on financial incentives and fringe benefits alone to motivate today's workers to superior productivity. It is necessary to give workers jobs that are meaningful.

A leading psychologist of work, Prof. J. R. Hackman of Yale, has identified three psychological states that a worker must experience in order to find his work meaningful:

- —A feeling that through his work he is involved in something meaningful and worthwhile.

- —A feeling that he is personally responsible and accountable, through his job, for results.

- —Knowledge of the results.

What characteristics of a job cause the worker to experience these psychological states? Hackman has identified the following five, the characteristics of highly motivational jobs:

<u>Task variety</u>. The job should give the worker the chance to exercise several different kinds of knowledge and ability.

<u>Task identity</u>. The worker should have a chance to complete a whole piece of work, from beginning to end, with an identifiable outcome or product.

<u>Task significance</u>. The worker should be able to grasp the effects of his job on other individuals and their jobs and lives, either in the organization or beyond it.

<u>Autonomy</u>. The worker should exercise the greatest feasible independence and discretion over important aspects of work--particularly scheduling time, choosing methods, and other aspects of planning.

<u>Feedback</u>. The worker should receive regular and timely information about performance. It is most effective when it comes from activities of the job itself, but it can also come from supervisors, fellow workers, or customers.

Applying the Principles to Manufacturing

While every work organization must find its own particular ways to apply these principles, there are certain general guidelines for major job categories. In assembly manufacturing, a useful strategy is to analyze the end product in terms of its subassemblies. Then each subassembly or subproduct becomes the focus of a small team or shop, turning out an identifiable whole part of the final product. Let us look more closely at the possible steps in this organization process.

1. <u>Whole products or subproduct segments</u>. If a product is relatively simple, it may be possible for an individual to assemble it completely. If it is more complex, look for those points in the assembly process where there is an identifiable subproduct. These are often the points where a subproduct or subassembly undergoes a test--even a simple go-no go check.

Depending on complexity, some of the processes may be accomplished by individual workers, while others might logically require collaboration of a self-managing pair or team.

The idea is to find in the assembly process the various points that mark closure or completion. A working subproduct is the best benchmark. However, if there are not enough of these in the process, it may be logical to group certain related fabrication processes into "shops"--for example, machining, punching, drilling, etc. These can then be made largely self managing.

2. <u>Distribution of functions</u>. To the greatest possible extent, determine the possibilities of one worker/one unit organization, with individuals completing a whole process or subproduct. Where this is not feasible, the next logical approach is to have a small team handling the process and acting as a supplier shop to the next step in the line.

It is very important to keep the shops or teams as small as feasible, for the closest possible identification with the finished subproduct.

3. <u>Discretion in methods</u>. Timing, sequence, and other aspects of assembly methods are often determined in advance, in considerable detail, by manufacturing engineering. For motivational strength, however, operators should have the greatest feasible latitude to set up procedures, layouts, schedules, etc.

In a shop containing several one-operator machines, for example, letting operators rotate assignments on the machines may be the most practical way to provide skill variety.

An extension of this idea, in shops where several machines are used in sequence, may be to have one operator or pair move with an item or batch from station to station all the way through a process to completion.

Methods, once set, need not remain unchangeable. Operators should have latitude to suggest method changes as product changes are introduced, as materials vary, or other factors develop. Naturally, these ideal concepts must always be tested against the reality of product complexity and time factors.

The ideal is always to organize tasks so that people can complete something. Resist the tendency to expand team or shop size and specialize functions, except where there is no other solution.

4. <u>Integration of testing and inspection</u>. Wherever the technology of testing permits, the testing function should be incorporated into the assembly job, rather than left entirely in the hands of an external diagnose and test function. It may be necessary to maintain a separate test function, but experience shows that fixing at least some responsibility for quality within the assembly process itself improves motivation, performance, and quality.

It is also important, wherever possible, to have people rework their own errors--provided the repair process is not prohibitive in required time or skills.

When it is not feasible to integrate testing and/or repair function with basic assembly, at least try to establish close communication and contact between the two functions. Locate them physically close together. Set up channels for feedback of data on errors and rework and for consultation. If practical, give certain inspectors or test technicians continuing responsibility for the work of certain assembly groups.

5. <u>Integration of materials handling and maintenance</u>. It may be possible and desirable to integrate into the basic assembly process certain responsibilities often left to materials handling or materials management functions. These responsibilities may not include bulk handling of materials or operation of materials moving equipment but could include parts storage areas. This kind of arrangement not only increases task variety; it has also been shown to increase operators' sense of being in control of a complete operation.

Similarly, consider integrating at least some maintenance responsibilities into the operator's job. Besides increasing variety and the sense of responsibility for equipment, this step often eliminates delay due to waiting for a maintenance man to make a routine repair. Feelings of concern and "ownership" of equipment are likely to be increased as well.

6. <u>Establish effective feedback channels</u>. The importance of feedback cannot be overemphasized. By setting clear standards of performance and letting employees know how well they meet those standards, management lets employees know what is important in the organization. In the process, management opens the primary channel of employee learning. It is only through effective feedback--knowing the results of efforts at short intervals--that employees gain both the competence and the motivation needed for superior performance.

There are numerous ways to establish good feedback. Two of the most effective ways have already been cited: having workers do their own testing, and having them do rework as well to the extent that is feasible.

It is also possible to set up feedback loops among teams or shops that serve one another. These might take the form of simple reports of errors or recurring problems that one shop finds in the work of another. Such informal QC reports should become positive tools for the various groups to use for exchange of information. They will not be valid or effective if they are used in a negative, punitive manner by supervision.

Whatever their form, these feedback channels need careful planning, especially as regards the choice of pertinent data among many possibilities. In many cases it is advisable to have a single individual responsible for keeping the feedback data flowing among sub units.

All these feedback channels have day to day value among production workers themselves. But it is also important for management to compile periodic summaries of trend data for each team or shop for posting and frequent review with the work force. These data can be reviewed at periodic meetings along with information on the broader business picture--product performance, demand, quotas, profits, and the like.

A good feedback system depends on standards that are clearly defined, set in advance, and discussed frequently. Once such a system exists, it becomes a basis for frequent goal-setting and performance-review meetings for both teams and individuals.

Whenever possible, the standards should have a dollar value attached. Other possible criteria include: meeting the production needs of other teams; reject or rework rates; and scrap rates, including dollar value of scrappage.

THE PRINCIPLES AT WORK

The foregoing job design principles were recently employed by a manufacturer of computer peripheral equipment in the design of a system for assembly of a new product, an alph-numeric cathode ray terminal. Some brief but specific descriptions of the job designs used may clarify some of the principles which have been presented in the abstract:

1. <u>Printed circuit board fabrication</u>. Two men operate a shop where five different machines are used in sequence to make PC board bases for component insertion.

2. <u>Silk Screen shop</u>. Eight operators work in pairs. Each pair does the complete silk screen circuit printing and etching on batches of boards, moving with them through five different operations. Each pair finishes and inspects a complete subproduct.

3. <u>Machine insertion shop</u>. A four-man team operates automated equipment for inserting certain components on the circuit boards. These four can rotate assignments at will or on a schedule as they prefer. This shop has less of a feeling of completing some subproduct than the others have.

4. <u>Hand assembly area</u>. Assemblers work in teams of four or five to install the remaining small components. Besides three or four assemblers, a team includes a final assembly/test function. On the simpler boards, one assembler completes the entire insertion. On the more complex boards, three or four workers assemble the board in zones, passing it along to the next assembler, and ultimately to the final assembly/test man. The test is a simple "go-no go" test but it is done close to the assemblers and feedback is immediate to the team.

5. <u>Basic assembly</u>. Pairs of assemblers put together the main components of the terminal, including the boards that have come through the other shops. The pair then does a "go-no go" test with power turn-on, before the terminal gets its outer shell.

6. **Basic unit test.** A pair of assemblers is responsible for monitoring a 12-hour test of each unit in an environmental chamber.

7. **Final assembly.** Four-man teams complete final assembly and test the units under normal operating conditions. Team members exchange functions at will and determine their own assembly methods.

8. **Copier assembly.** An optional hard-copy printer is assembled on a one man/one unit basis. Each individual turns out complete, tested copiers.

LIMITS OF MOTIVATIONAL JOB DESIGN: THE NEED FOR TRADEOFFS

As I have already indicated, the kinds of job designs described here result from attempts to approach a motivational ideal. However, other realities of the work system and of management objectives often impose limits and demand compromises. The manufacturing manager and everyone involved in design of the production system should acknowledge these limits, and be prepared for tradeoffs that lead to the best functioning of the system as a whole. Let us consider some of the factors of the productive system which impose these limits.

1. **Materials-handling requirements.** In conventional assembly, materials or components are delivered to the points on the line where they are actually needed. If the line is replaced by several individuals or team modules, each requiring all the materials and components, the materials-handling task is more complicated--and often, more costly. This fact can require compromises with the ideal organization of work. But the problems are rarely insurmountable if alternatives and their costs are studied carefully--especially where motivationally sound job design is considered before plans for the materials-handling system are made final.

2. **In-plant fabrication of subassemblies.** Where management has decided to make, rather than buy, certain subassemblies, the logic of good job design might suggest that the fabrication be made part of the mission of the team using the subassemblies. However, if such a setup would overtax the skills of the team, a compromise may be advisable; for example, a separate fabrication setup off line.

3. **Line-balancing requirements.** Proposed job designs for various parts of the production process may be optimal internally, but create imbalances of work flow with the rest of the line. To avoid these, it will be necessary to analyze volumes and cycle times for each phase carefully, and to compromise with the motivationally ideal work design where necessary. The line-balancing problem exists with multi function job or team build concept but is usually less of a rigid limitation that with a heavily specialized pass down line.

Sometimes the embalance between work rates of the various phases can be accommodated by means of buffer stocks, which allow a team or a shop to change its pace or schedule without creating a shortage in the next area.

It is important to remember, though, that the improvements in error rates and other sources of cost may more than offset time lost through balancing adjustments.

4. <u>Duplication of equipment</u>. Where several team or individual work stations need the complete set of tools or equipment, there is an added cost for duplicate equipment. In the case of very expensive equipment-- for example testing and diagnostic equipment--it may be necessary to compromise on job structure. However, the problem can sometimes be sidestepped by arranging for several groups to share time on a single, centrally located piece of equipment.

OTHER ADVANTAGES

A full discussion of advantages and limitations of job design should include beneficial side effects, as well as the improvement in motivation and productivity which is the main objective. The kinds of job designs discussed here have been shown to produce such benefits as the following.

1. <u>Greater flexibility in adjusting manpower to fluctuations in production needs</u>. It is easier to add or drop entire self-contained teams than to adjust a whole line when volume changes necessitate staffing changes. And it is less disruptive than shifting individuals on or off a line, with the consequent upset to the line balance and the need to move work stations.

2. <u>Better use of higher-paid technicians</u>. Since many of the simpler test and repair functions are handled by the assemblers themselves, higher-grade technicians can spend more time on nonroutine jobs. Often, fewer such technicians are needed.

3. <u>Cross-training</u>. Team workers can get training in, or familiarity with, a number of skills. This is valuable in terms of covering for absences, and it can provide a path for advancement as a worker acquires each new skill.

4. <u>A more palpable sense of teamwork</u> and concern for effective joint effort is more likely to develop when people are linked together by job structure to collaborate in completing something. Physical arrangement of work stations for ease of interaction and eye contact are important factors in this regard.

EFFECT ON SUPERVISORY JOBS

The characteristics that make jobs satisfying and motivating apply at the supervisory level as well. The organizational approaches outlined here to increase task variety, task identity, autonomy, and feedback in the basic assembly jobs provide a logical frame of reference for changes in supervisory structure as well.

In the traditional approach to manufacturing assembly, where tasks are highly specialized, it is common to have a foreman or supervisor in charge of a rather narrow functional specialty. This arrangement may be necessary when the foreman's most vital qualification is mastery of a very complex technical function. Where technical demands are not so stringent, however, management should try to give supervisors responsibility for groups including several functions. The organizing principle should not be one function, but one product or subproducts produced by several functions in coordination. The cross-functional foreman, supervisor, or manager is accountable for all aspects of his product or subproduct. In effect he is a "mini general manager," and might logically be responsible for an entire cost center.

A SYSTEMATIC PLANNING APROACH

The discussion so far has at least implied the various steps leading to manufacturing systems that are productive and morivationally sound. At this point it may be useful to recapitulate the steps, and show how the proposed sequence differs from what is usual.

Motivational job design principles can be applied to change existing systems. Many of the most successful cases of job enrichment have followed this procedure. Especially in manufacturing, however, the best opportunity to set up motivationally sound jobs occurs when a new line is in the design stages.

Unfortunately, however, all emphasis at that point is likely to be placed on the purely technical aspects of product design, manufacturing process design, materials handling, and the like. Fitting people into the system is usually almost an afterthought. The result, in many cases, is an assembly system almost guaranteed to generate problems with motivation and productivity in the long run.

The typical sequence is as follows:

(1) Product design.

(2) Flow planning of the manufacturing process.

(3) Equipment selection.

(4) Assembly procedures planning.

(5) Human factors engineering (not always included).

(6) Planning of physical layout.

(7) Planning of materials-handling system.

(8) Fitting people into the resulting manufacturing system.

In this regard it is suggested that a manufacturing system can make better use of its human components and optimize its total performance better, if the planning sequence is expanded by the insertion of two additional steps before assembly procedures are determined:

--<u>Identify the logical subproducts</u> or processes that can be the basis of whole job modules for individuals or teams. If possible, find the point where there is something that works and can be tested.

--<u>Within each area, whether based on subproduct or process, apply the principles of motivational job design</u> to create jobs that recapture some of the spirit of the independent craftsman or job shop operator.

These two steps assure that additional acknowledgment is given to psychological needs and their motivating power in the job design process. Then the details of the assembly procedure can be planned. Some tradeoffs with motivational factors may be necessary. But it should be possible to create jobs that can create and sustain the motivation--and the productivity--of the people who must live with the system daily.

When a system is planned in this sequence, it is possible to make motivational factors part of the "specifications" of the manufacturing system. The addition of human factors engineering completes the planning of the integrated work system, and offers the greatest possibilities for use of technological and human factors to their full productive potential.

SOME RESULTS OF MOTIVATIONAL JOB DESIGN

The ultimate test of the value of motivational job design--as of any management method--is the results it gets. Therefore, in conclusion with brief accounts of several cases of motivational job design in assembly processes, and the results gained.

IBM Corporation: A Typewriter Assembly Plant in Amsterdam

In response to growing production demands at this plant, IBM had followed a standard procedure of lengthening assembly lines and reducing the scope of each worker's task. Eventually each line was more than 230 ft long and employed more than 70 workers. Each worker put an average 3 minutes of work into each typewriter.

The highly fragmented approach required a degree of standardization that was poorly adapted to the plant's production needs. The typical typewriter had 2,500 parts. Besides 18 standard models there were 25 specials, with more than a hundred type heads and keyboards for different languages.

The production line was fast, but its price was high. Because the workers felt like robots, they had high error rates, and the plant spent 12% of all man-hours in overtime to repair defective typewriters. The turnover rate was 30% a year.

Management surveyed the workers and found that they wanted more responsibility and better relationship to the product. Management's response was to replace the long lines with 9 "mini'lines" each with 20 workers, each producing a complete and recognizable unit.

The results were dramatic. Within three months after the changeover, production was up 18%. After two years, it was up 46%. Quality improved sharply, overtime dropped, and absenteeism and turnover declined.

Because the workers themselves were now more deeply involved, management was able to eliminate some controls that had been sources of annoyance. The workers were given the discretion to decide how product design changes should be incorporated into the assembly process. Whereas the old system had forced workers to adapt themselves to a set task, the new system tempers the assembly system to the needs of the workers who man it.

Corning Glass Works: Plant in Medfield, Massachusetts

A hot plate for laboratory use had been assembled along conventional assembly line standards. It was decided to replace this arrangement with individual workers, each responsible for assembly of entire hot plates.

Within a few months of the change, absenteeism in the operation had dropped from 8% to 1%. Productivity, measured in units per man-hour, rose quickly by nearly 50%--then rose later to 84% above its original level. The number of units rejected for poor quality fell from 23% to 1% of total output.

Mepco Div., Electra/Midland Corporation

This manufacturer of electronic components had been organized in the traditional way. Different kinds of products were produced in a single work sequence, moving from one functional department to the next for assembly, welding, etc. Each department had a foreman, and there was a production manager over the whole operation. Delays were widespread, and overdue orders accounted for fully one-third of the workload.

During a summer shutdown, the plant was rearranged so that each product would have its own production line. Each line has a product operations manager--a truly cross-functional manager. He sets all goals for his line, confirms a delivery date before accepting an order, schedules, supervises, expedites materials from the purchasing department, suggests improvements in product or methods, requests capital equipment, and prepares budgets. He frequently has contact with customers. If his line has some slack, he can try to drum up business.

The product operations manager keeps the workers on his line posted regularly on the line's results versus its goals. He is evaluated--and compensated--on his line's gross profit. If his line is overloaded, he can "buy" the services of workers on lines with slack. The manager gets plenty of feedback on the line's performance. A daily computer printout tells the status of every order on the line. He is credited for all sales of the line's products, and charged for costs of labor, materials, and overhead.

Since the product line and product operations manager concepts were introduced, Mepco's backlog of overdue orders dropped from about one-third to about 3% of all orders. Net sales increased 40%, and there was no addition to the workforce.

This highly successful case illustrates the main principle of job design for supervisors referred to previously: use an identified product or market as the main orientation for the supervisory job. Attempt to gather all the main functions required to produce that product or subproduct under one individual, giving each at least an approximation of his own product line to manage rather than a single function to oversee.

The fourth viewpoint is "Improving Assembly Productivity" by Theodore O. Prenting, Associate Professor of Business, Marist College.

The last major area in which significant opportunity for productivity improvements exists in the fabricating industries is assembly. Using the 4 M's of manufacturing management; men, methods, machines, materials, as a framework, techniques for improving assembly productivity are discussed. Manpower management is improved through attention to the individual and recognition of the job environment. Assembly methods are more productive as they move toward more continuous production and upgrading of line assembly techniques. Mechanizing assembly by identification of economically feasible applications is seen as the single greatest avenue for productivity improvement, while materials changes are likely to be rewarding by-products of analysis of the assembly process.

The economist studies the production function and the factors of productions, while the manufacturing manager is typically most concerned with _improving_ that production and making better use of those factors. Never has this been more true than today. But what are these factors, and how may they be better utilized? Taking a line from books on management, one can consider the 4 M's of manufacturing management--men, methods, machines, and materials and examine each for productivity improvement possibilities. In doing so for assembly, it becomes quickly evident that men and methods are of most immediate interest, because they are the most "controllable" resources of management. Machines and materials are more directly given by the technology and take longer to change, though the resultant productivity improvements can be dramatic. Following are productivity improvements in these factor areas that are new in potential and use for some industries and plants; overlooked in other cases because of lack of understanding or oversight. In common, they all work well in a variety of industries and firms and are the result of much that has been written and observed by consultants in countless plants.

IMPROVING HUMAN RESOURCE MANAGEMENT

Aside from wholesale mechanization of the assembly process, no single resource has as much potential for productivity improvement as the human one. Yet, despite, or perhaps because of, the abundance of literature on the subject, no other area is likely to be quite as difficult for the practitioner to improve upon, for it is often unclear just which of the latest theories to believe. Hopefully what follows will not add to the confusion, but will instead place some things into better perspective.

Leadership Function of Management

Management is classically defined as having certain functions--planning, organizing, staffing, directing, controlling. Of these, directing may indeed be the most important especially for the lower level manager, who must lead his workers on a day to day basis, and who can _only_ accomplish work through them. One writer sees as a part of this leadership function, that the manager must also be the agent of the workers.[1] He says that the really effective manager/leader will not only serve the interests of his management superiors, but to the extent possible the best interests of his workers as well.

This line of thought can easily lead to a concept of leadership style; one that leans toward a participative, consultative approach. The value of this approach lies in the better worker-management communications, greater willingness on the part of workers to accept change, and generally better working climate that results. In fact, the results are sufficiently good that what is frequently overlooked in attempting to employ the approach is the recognition that no one leadership style is good for every situation. The style employed must take into consideration the organizational acceptability of it, workers' expectations, nature of the work, and, very importantly, the manager's capabilities. The more experienced manager/leader may be able to use different styles in different situations, but the most important thing is that it be genuine, that the manager be himself. The participative style is fine if it fits; if it does not, it can be a disaster. Workers, as managers, hate deception. A manager only capable of an authoritarian, "bull of the woods", style, but who respect and looks out for his workers, may be far more effective than the manager who uses the participative approach poorly because its the "in" thing, or because workers can be manipulated through it. It is the manager's attitude toward the workers as they perceive it, that spells the difference in leadership effectiveness.[2]

Motivation and Productivity

As recently as ten years ago, most management books, reflecting the naive thinking of the early human relations movement, continued to suggest that high levels of job satisfaction inevitably lead to high productivity. Today, we know, as many industrial managers knew all along, that this is simply not true. Trying to make people ever happier does not necessarily make for a more productive work force and lower labor cost. In fact, we have learned that in many cases, a high level of productivity is itself often the cause of greater job satisfaction.[3]

It seems a similar error is being made today where countless headlines tell us of bored, poorly motivated, unproductive assemblers, and the only antidote, job enrichment. A reassessment of the cause and extent of worker discontent, and the validity of job enrichment as a response, is hereby suggested.

Studies of workers in such activities as assembly do not bear our the sweeping views of the many popular writers, nor those of the researchers, who are often personally repelled by the minute nature of assembly tasks. As an example, in a recent study covering 3800 factory workers involved in assembly in five different industries in five states, only about 20 percent expressed dissatisfaction with their jobs. Depending upon the industry, from 70 percent to 88 percent of the assembly line workers, whose comments were obtained first hand, liked assembly work. They did not want added responsibilities or quality requirements, nor any significant changes in job content.[4]

Perhaps of more basic importance, and despite considerable controversy surrounding it, business management publications have generally not given the shortcomings of the Herzberg two-factor theory of motivation on which job enrichment is based much publicity.[5] So many studies appearing primarily in the psychological journals have raised questions about the casual relationship in Herzberg's theory that they cannot be lightly dismissed.[6]

Most criticism of Herzberg's theory is that it is an oversimplification of the true complexity of human motivation. While Herzberg does not de-emphasize the importance of his hygiene or maintenance factors (dissatisfiers) e.g. salary, supervision, working conditions, the danger is that in his emphasis on the motivator factors (satisfiers) e.g. recognition, responsibility, work itself, and his very use of the words motivator, satisfiers, and dissatisfiers, management may be misled in its personnel policies and practices.

There is a question about whether the dichotomy between hygiene (dissatisfier) and motivator (satisfier) factors can be supported,[7] and if this cannot be, then it is impossible to test which set of factors contributes most to worker motivation.[8]

It should also be noted that job specialization is not a whim of management but a more economic production method, which leads to even more economy through mechanization of the simplified tasks. Thus, for society to rush headlong into job enrichment is to retrogress technologically. What we should be doing is to mechanize out of existence those jobs that are truly distasteful to the worker. Until then, however, the obvious economic answer is that we will simply have to pay workers what they believe the job is worth, based on what society is willing to pay for the end product. Further, we should pay much more attention to the selection of workers for particular types of work.[9] As Herzberg has put it, "If you can't use him (employee) on the job (meaning full utilized) get rid of him, either via automation or by selecting someone with lesser ability"[10]

The question arises, if workers are so unhappy with the quality of their work, why aren't they seeking redress through the collective bargaining process. An American labor union leader says, "If you want to enrich the job, enrich the paycheck...begin to decrease the number of hours a worker has to labor in order to earn a decent standard of living...give working people a greater sense of control over their working conditions."[11] Note nothing is said about the job content, in fact, it is by inference in the article denied as a factor.

Management Opportunities

The implication for management from the job enrichment controversy is that management cannot expect all people to react the same to changes in job design and enrichment. Herzberg's contribution is to call attention to the work itself as an important job factor, but individuals react as differently to expansion of their work as they do to other factors surrounding their jobs.

Not all workers react favorably to enlarged or enriched jobs, nor are all jobs amenable to such enlargement. Apparently overlooked by many, Herzberg himself says, "Not all jobs can be enriched,"[12] and this holds true for much assembly and other repetitive work. Given the technology of the product being produced, the degree of operator control over the work pace, and responsibility for methods used and quality can only be relative, never absolute. One assembly task cannot be made much different from the next, and no amount of juggling them around is going to change them.

There are less expensive, more direct responses than job enrichment to solve such worker discontent as may exist. Studies have shown that job rotation can be successful on existing assembly lines,[13] and that operators can participate effectively in determining tooling to be used, the work methods, the speed of the line, and the quality of their output. Incentive wage payment systems seem to reduce the repetitiveness of a job,[14] and the encouragement, rather than discouragement, of socializing on the line is now more widely recognized. As Friedmann commented, "a social life emerges and enlivens the hours of work on the conveyor which appear so 'monotonous' to the external observer."[15] Companies can also do much to personalize the work place, and very importantly, they can work out avenues for employee growth and advancement, if they choose to do so.

Many researchers familiar with the industrial scene today are becoming convinced that the answer to such worker unhappiness as exists does not lie in more theories about its causes, nor in the expenditure of millions or billions on new plants and equipment. It rests on grounds of genuinely involving the worker in more of the day to day management decisions affecting his working life. Consultative management has been preached for too long; it is high time that it is practiced.

Such investment and operating cost risks as at least one Swedish automobile firm is taking, are not necessary either for the company of the consumer, who may have to pay more for the company's product.[16] Nor is there a need to subordinate productivity, so desperately needed in our economy today, to achieve human needs. The two goals can, and should be, reasonably compatible given a management desire to truly recognize the potential of the worker as an individual. This recognition of the individual is really at the heart of much of what Herzberg is saying.

METHODS IMPROVEMENTS

Because of the continuing manual operations in most assembly work today, much of the improvement in productivity must come from better use of labor time; an area in which the assemblers themselves may profitably contribute as part of a participative management approach. It will be seen, however, that many of the methods improvements suggested here will be steps taken to provide a more standardized, continuous flow of the work. This is natural, because while continuous flow production is not the most appropriate for all plants, it is the most efficient where it can be practiced. To the extent that it can be utilized, it should be!

Toward Continuous Production

Meaningful effort toward greater standardization and continuous flow production requires more advanced design and manufacturing planning, and the use of such concepts as modular design. Through modular design all, or many, of the models in a product family use a number of common modules, resulting in the standardization of extensive numbers of subassemblies. Standardization which can, in turn, result in higher combined production volumes, and the economic feasibility of using continuous and perhaps even mechanized assembly where it could not be justified before.

Where modular design is not feasible, the emerging manufacturing concept known as group technology should be considered. Here, parts are classified into families, and the parts families are then sent through the various operations as essentially a continuous process. Instead of equipment grouped by type of process (process layout), it is grouped for the most direct processing of the family of parts and assemblies that it is intended to serve (similar to product layout). In addition to a shorter total processing time, greater productivity, reduced material handling time, and lower in-process inventories, where the volume is great enough, group technology permits ready utilization of automatic equipment, since the parts can remain oriented throughout the process.

A relatively new product of interest for low volume, mixed model assembly is a special conveyor system first introduced in Europe, which offers improved flexibility for such production. With this system, assemblies in process are moved on carriers equipped with station selectors permitting any operator to sequence work, as required, to any other operator on the line. A console operator can see at a glance the queue before each work station, and, thus, can introduce appropriate models to the system to keep all operators supplied with work.

A major problem found in many low to medium volume assembly operations is simple negligence or the failure to see that volume has grown or could lend itself to more continuous production. This occurs especially in companies where low volume assembly is a relatively minor part of their total operation, or where their entire operation is low volume, and they continue to think only in terms of low volume, job shop assembly methods. What they fail to see if that as their business grows so does their volume, and that as this occurs assembly techniques should also change. Further, they frequently overlook multiple use sub-assemblies, which lend themselves so well to higher volume assembly techniques, even mechanization, if they are combined in production runs.

Consultants commonly achieve at least ten percent increases in productivity in low volume plants by grouping families of like assemblies and using short bench type progressive lines. One company with fifty different product models improved productivity and decreased learning time from about three months to three weeks by this simple technique.

Line Assembly

Much attention in the literature has been given to this topic, and expecially to an important part of it, line balancing. What has been frequently overlooked however, is the choice of the line cycle time itself; the time the operator has available to work on each unit. Given the required number of units per day, and the total work content per unit, this choice of cycle time determines the number of operators, the length of line and the number of lines needed. Research suggests that many companies are paying penalties for failure to recognize these important relationships, because of the higher than necessary imbalance in operator workload and idle time resulting on the line.[17] A corollary benefit foregone is the possibility of greater output using the same number of operators. In the referenced research study, an actual TV assembly line used as a model could produce 164 units in eight hours with 15 operators at a 2.92 minute cycle time and line imbalance of about six percent. If the cycle time were only slightly altered to 2.75 minutes, the same line with zero balance delay could produce 174 units with the same number of operators---174 units instead of 164 with no increase in direct labor cost!

When designing new, or improving existing, facilities even a broader view of the total assembly system that takes into account not only the line imbalance issue discussed above, but such things as non-productive work in assembly, the wage cost of skill and learning cost should be considered. All these factors have been examined in a model that has been developed to determine the optimum cycle time.[18] This is defined as that time yielding the lowest unit cost of assembly.

The optimum cycle time is found by minimizing the total of two classes of cost factors: intrinsic and extrinsic. Since the extrinsic factors, such as the cost of space and tooling, the available personnel, and the assembler wage rate structure, are relatively fixed over a range of production requirements, the model developed considers only the intrinsic cost factors inherent in the man-work relationship itself. Two of these costs, learning and the wage cost associated with skill level, argue for a shorter cycle time, while the other two, imbalance of work and non-productive work cost (product/tool handling and operator movement time), argue for a longer cycle time. The minimization of the total cost of these four factors, the least cost point on a total cost curve will yield the optimum cycle time with respect to these components of direct labor assembly cost.

Application of the above model provides a surprise. The optimum division of labor does not continue to increase indefinitely with increases in the volume of the production run. Imbalance of work, non-productive work costs and wage cost of skill do not vary with size of production run, while learning cost does. Thus, as production run size increases and unit learning cost decreases, it makes sense to maintain or even <u>increase</u> the cycle time to avoid short cycle imbalance of work and non-productive work costs. In practice, as assembly volume increases it may be more economical to set up more lines with fewer operators each, than to simply add more operators and reduce the cycle time on one line. Through this alternative, management also gains production scheduling flexibility by having more lines available, while possibly making the work more appealing to operators. Two firms applying the optimum division of labor concept and the line balancing reported over a fifty percent reduction in direct labor cost. A significant part of the savings in both applications was due to the proper choice of cycle time.

Line Balancing

As suggested by a recent study,[19] a principal method of improving assembly productivity for many companies should be to begin making use of some effective line balancing technique. So much has been developed and written about this subject, that it is indeed surprising the number of firms who are still using either no, or very poor, line balancing methods.

Good methods need not be complicated, and a simple precedence diagram depicting the actual assembly line situation can be used almost as is to balance a line well.[20] Such diagrams have several advantages over other simple paper and pencil or "pigeon hole" techniques of line balancing:

1. The logical, systematic analysis of the precedence relationships of the work elements leads to greater combinatorial possibilities and better line balances.

2. Preparation of the diagrams leads to clearer understanding of the product and line being studied, and improvements in product and line designs to reduce assembly costs.

3. The diagrams are useful in training new industrial engineering personnel, and can be used to balance a line by people familiar with the method, but not the product or line itself.

The precedence diagram is the basis for many, if not most, manual and computer line balancing methods. These methods and much more on assembly systems are now conveniently documented in a new book, which discusses single and mixed model techniques, line balancing where different product models are batched through a line, the measurement of similarity among models for assembly purposes, and the calculation of learning time for single and mixed model lines.[21]

While the advantages of mixed model assembly are numerous; continuous flow of each model, reduced finished goods inventories; elimination of line change-over, and greater flexibility in production; they do present some serious problems. These include balancing the line for a given production schedule, sequencing the build of particular models, and scheduling the various sub-assemblies. In general, mixed model assembly is more complex and costly than single model, but where a large variety of models must be assembled, it may be the most logical and least expensive way to build and should be considered.

Operator Assists

It is estimated that two-thirds of the manual work stations on otherwise mechanized lines and much more on manual assembly lines employ only very simple containers to store parts and other components used by the operator. Mechanical feeding devices, such as bowl feeders or component supply belts moving parallel to and at the same speed as the assembly conveyor, are not common. Parts should be oriented for the operator and delivered as close as possible to the assembly point.

The use of work platforms, such as benches, or "outriggers" as they are called, alongside conveyor lines, should be questioned. Similarly, carrousel type supply conveyors used to store partially completed unites between operators working at benches is questionable. For whatever human relations value they may have in removing the operator from the appearance of a straight assembly line, the operator must work harder to move units into, and out of, the work position, and obviously, management must pay for the non-productive time spent in this product handling. In one study where operators simply pushed completed units to the next operator, rather than using a powered conveyor, eight percent of the total work time at a half minute cycle was spent for product handling; at a one minute cycle, product handling was close to four percent of the total.[22]

Experience also shows that when conveyors of any type are used to provide a buffer, or intermediate storage between operators, balance delay is often high. It persists because the buffer does not allow for easy observation of idle time at underloaded stations. Management also foregoes the "traction" effect of a moving line whereby the natural pace of operators is frequently increased. If has further been shown that assemblers frequently prefer the constant rhythm of a mechanical conveyor to the sporadic rhythm of an operator paced line.[23] In the study cited, 84 percent of the television assemblers involved preferred powered conveyors.

IMPROVING PRODUCTIVITY THROUGH MECHANIZATION

What is popularly referred to as automatic assembly, but is generally mechanized assembly, lacks rigorous definition. Thus, many plants using mechanized techniques do not consider themselves as assembling anything automatically, and what is bad from a productivity point of view, is that the term seems to keep many plants from exploiting its potential. Unquestionably, the mechanization of assembly, as for most other processes, is the ultimate answer for improving productivity. Labor cost reductions of two-thirds, one machine doing the work of twenty previous operators, and machinery pay off periods of five months have been reported.[24]

Defining automatic assembly as an operation in which two or more parts are oriented in relationship to one another, so that they can be mated or assembled, and often physically fastened together, without human intervention, may help to broaden perspective on the possibilities of its use. Clearly, this definition encompasses much of what is really mechanization of components of the assembly process and includes such potentially rewarding and very simple mechanization as using a bowl feeder(s) to make up bolt and washer assemblies of various types for subsequent use in manual assembly.

Unlike the process industries, in which entire operations must often be automated at one time, either for control purposes or to reap the economic advantages, assembly can be automated in logical steps. Thus, the combined use of automatic assembly machines and manual assembly is customary. Manual operations on the line remain those too difficult or costly to automate. The assemblers then perform the additional roles of machine tenders and visual inspectors. As the assembly process is periodically reviewed with improved knowledge of automatic assembly, the remaining manual operations can be analyzed for automation.

Identification of Operations for Mechanization

Borrowing a concept applied early in the automotive industry's steps toward automatic assembly, companies should pay special attention to subassemblies having multiple use in a product or in various product models. Examples from the automotive industry include the early mechanization of the assembly of piston connecting rods and spark plugs, used 6 or 8 per car, and the "black boxes", such as alternators and voltage regulators, used in a variety of models.

It appears that operations performed on progressive assembly lines are easier and less costly to automate than those performed by the single station bench or batch method. The relative ease and lower cost derive from the easier analysis by the machine designer of the more simplified tasks performed on a line, the better methods and better designed parts usually existing on lines, and the better orientation of product and components on assembly lines.

Specific applications for automatic assembly are considered by one company wherever several operators perform the same operation, the job requires difficult or continuous product handling, a production bottleneck exists, a high number of reject occur due to the human factor, or large in-process inventories are involved.[25] The presence of any of these conditions shows an area for study, but does not alone justify automatic assembly.

One machine tool representative says that prospective assemblies should generally contain at least four components. Otehrwise, the small number of stations relative to basic machine (transfer mechanism) costs makes automatic assembly uneconomical.[26]

Factors influencing the ultimate feasibility of assembly automation include present and anticipated future volume of production, labor cost and time of assembly, model design variations and frequency of build of each model, the stability of product design, product complexity, and a receptive management/labor climate for machanization. Clearly, companies seriously seeking the productivity rewards of mechanization must often take active steps to satisfy some of the above factor requirements. These can include such things as product design modifications to achieve greater component standardization and to make orientation easier, or to improve component reliability.

The savings that can be expected from mechanization are not limited to labor, and, thus, make efforts to automate doubly worthwhile.[27] Included are savings through better quality, better working conditions and safety, and better inventory--production control. Through better quality there are frequently savings in materials costs and less scrap and rework. Some space savings may also result.

Simulation

A very useful analytical tool for automatic assembly is simulation. While it has not yet received extensive use in the field, its value is clear. One major assembly machine builder uses a general purpose simulation program developed specifically for automatic assembly.[28] This program enables designers to determine what machine types will be most suitable for a specific application, what production rates will be, and the most critical stations on the machine. A fifteen minute computer run is able to simulate an actual machine run of four hours. Included in the simulation model are such things as the probablilty of component difficulties, machine breakdowns, service times, size of queue, and so forth. Other companies in developing their own machines have used simulation, but much more can and will be done with it.

Through simulation it may be discovered, for example, that what is called for is differing degrees of flexibility in the automatic assembly equipment.[29] This may suggest an integration of the common rotary or in-line indexing machines with some type of pick-and-place or programmable manipulator. Standard indexing type equipment would be used for those parts of the assembly that are relatively fixed in design, while the latter equipment would be used to provide the flexibility needed for assembly components subject to design change.

An outgrowth of the use of simulation for automatic assembly machine design is a Generalized Assembly Line Simulator (GALS) available as a computer software package.[30] As with automatic assembly, a manual line is also a complex system composed of many components, which can be arranged and controlled in a variety of ways. With so many factors to consider, it can be difficult to predict the performance of a proposed line and envision what problems could arise. Mistakes in design of a system can be very costly in terms of idle direct labor, lost production, and idle equipment.

With simulation, the engineer can find out what the expected performance of the total line will be during the initial design stages. The resulta can also indicate where in a proposed line a bottleneck might occur or where there might be idle time, before the line is put into operation.

MATERIALS IMPROVEMENTS

As the assembly process is analyzed, hopefully by management and workers, productivity improvements through changes in materials will become evident. These may result from, or accompany, efforts toward standardization, modular design or the use of group technology. The analysis required in determining appropriate cycle time and/or line balancing will also lead to materials changes to remove "bottleneck" operations or to otherwise decrease assembly time.

Perhaps the most significant materials improvements suggestions will come, however, from steps taken to mechanize existing or new assembly processes. Here, materials changes may be recommended to reduce the number of components, to make orientation, insertion or fastening easier, or to improve reliability. This is exemplified by changing from individual piece parts to integrated components through plastic molding or die casting, or by shifting from threaded fasteners to adhesives or fasteners requiring only simple, straight motions for insertion, such as rivets. Using simulation, a variety of materials improvements will most assuredly appear, though the nature of the changes is highly dependent of the product itself.

CONCLUSION

As can be seen, techniques for improving assembly productivity are available to serve a variety of industries and needs, and the methods discussed here are not an exhaustive list. It is up to management, however, to decide how serious it really is about improving productivity and then devoting the effort and money needed to do it. Without question, most companies and the national economy itself depend upon such steps being taken, as in the fabricating industries there is simply not much left that can be improved in the other large production area, machining. Assembly remains the labor-intensive area most amenable to substantial productivity improvement.

The fifth and last viewpoint is "Manufacturing Training for Large Volume Integrated Circuit Production" by Keith M. Gardiner, Staff Engineer, IBM Corporation.

Good quality, high productivity, cost-effective manufacturing is dependent upon the commitment, cooperation and skill of the manufacturing operator. This is true, more than ever, in today's highly automated, but technically sensitive environment. The fact that economically competitive, high-performance semi-conductor computer memories can be produced is convincing evidence of success in both mastering the subtleties of the many involved manufacturing technologies, and also in training manufacturing personnel. This paper describes some of the philosophies and methods used in a plant-wide manufacturing training program which accompanied the introduction of a totally new (high technology) mission.

The introduction of high performance System/360 Series computers in the late sixties required semiconductor memory components. Market projections indicated that a large volume manufacturing capability would be required on-line on an urgent basis. The technologies to be exploited for constructing the fast integrated circuit memory modules were already well established as the basis of System/360 logic.[31,32] However, the designs for memory called for substantial extrapolation of "state-of-the-art" in terms of both circuit/device densities and manufacturing performance. The IBM plant in Essex Junction, Vermont, had the responsibility for supplying semiconductor memory products. This was an entirely new high technology activity, which had to be developed using a relatively innocent workforce. A comprehensive and well-developed education program was an essential concomitant to the successful fulfillment of the manufacturing task.

HISTORY

The Essex Junction plant was opened in 1957, and most of the 450 people employed were involved in hand assembly of the wire contact relay, an electromechanical switch which is still widely used today. People and product were added at such a rate that by 1967, the plant had a workforce of over 3000 people turning out a multitude of products ranging from the Reed Switch, Wire Contact Relay, Standard Modular Systems, Transformer Read Only Storage, Card Capacitor Read Only Storage, Flat Film Memories, Solid Logic Technology and the 64 bit bipolar monolithic memory chip -- the plant's first introduction to large scale integration. Then, in 1970, with the introduction of the 128 bit bipolar memory chip, main memory for the early versions of the System/370 Model 145, and the 1024 bit MOSFET product, the plant population increased to just under 4000 people where it has remained to this day.

Before the introduction of semiconductor memory manufacturing in 1967, the manufacturing concepts were basically hierarchical. Parts were processed in batches, placed in racks or trays, and passed on, sometimes through a hatch, to the next department in the manufacturing sequence. There was no necessity for extensive mutual understanding of the integrated, or interactive, role of each department in the total manufacturing output. In general, each department conducted a relatively high yield process which was independent of the prior or future history of the component under fabrication. Manufacturing operators had responsibility for performing their unique functions and passing the part along to some unappreciated destination. The workforce was dedicated but required little technological sophistication, while all processes were manually controllable with few varying parameters.

The demands of silicon integrated circuit manufacturing, however, called for a new approach. It was now essential for all operators to be aware of the consequences of their contribution to the success of the process. This new awareness required an overall and total comprehension on the part of all manufacturing personnel.

The following discussion outlines the education program that was instituted in order to achieve our more rigorous manufacturing goals. And, although this paper describes the approach to our particular manufacturing situation, it is felt that similar educational philosophies should apply to any advanced technology manufacturing mission.

OBJECTIVES

A well defined objective is a mandatory prerequisite of any educational program. This objective must encompass an appreciation of student capabilities, attitudes, and needs, both real and imagined. This latter concept is worth emphasizing because there are possibly as many, if not more, individuals who underrate their intellectual and skills capabilities rather than the converse. Such people can become willfully ineducable if the education regime is aimed too far above their imagined threshold. Thus everything was planned initially as very low key, with capability being built in to upgrade in accord with student responses. The objective definition accepted for the technology portions of the training program was: "To generate an overall appreciation for the nature of silicon, the properties of semiconductors, and an understanding of the manufacturing processes and the close interdependency of the processing."

METHODS

The method was to assign employees full time for three or four day classroom sessions. The population of each class represented a spectrum across existing manufacturing departments. This selection method was not too disruptive of existing production, and also assisted in improving interdepartmental attitudes and communications. The personnel were, in the main, unaccustomed to a classroom environment and formal teaching. Relatively brief concentration and effective teacher communication spans were assumed. Classroom sessions would have to have variety. Consequently, different instructors were employed for short sessions on individual topics, and the schedule was broken up by use of related movies.

The total change from the work environment tended to confer a positive student attitude. Sessions covered not only technological understanding and process but administrative concerns, production control, safety aspects, clean room discipline and associated topics. For the key technology sessions where prolonged exposure was unavoidable a range of techniques was exploited and humor was deliberately injected. Student involvement was enhanced by means of simple questions to develop themes, and continual interruption and interchange with the instructors was encouraged. Examples:

"What is a metal?"

"What are the differences between a metal and a nonmetal?"

Such questions were thought-provoking and provided good leads into elementary exposition of the Periodic Table ("A chart that individuals who pretend to be scientists hang on their walls"). Further stimulation was gained by taking humorous advantage of the system. Students were encouraged to pay attention because a final test *(See Figure 1)* would be administered. Failure, it was threatened, could result in elevation to management! Throughout the presentation it was emphasized that success, or yield, depended totally upon what had gone before. Simple analogies were used to explain the complexities of the manufacturing process. For example: crystal pulling was likened to a cooled rotating broomstick drawing icicles from a pail of water on a Vermont winter day. Similarly, diffusion was compared to placing a waterproof mask around a patch of blotting paper and allowing some ink to penetrate it, with the drive-in phase easily being represented by a simulated redampening of the absorbent paper which permits the ink stain to lessen in intensity (concentration) but spread to a greater depth and uniformity.

The regions of differently diffused silicon which made up the assemblage of transistors, diodes and resistors were at the most elementary level referred to as different flavors of Neapolitan ice cream with crucial differences both in flavoring properties and flavor strength which gave the exact electrical switching characteristics desired.

DATE__________ NAME__________

NEW PRODUCTS MANUFACTURING TRAINING PROGRAM

PROCESS TECHNOLOGY

Please enter name at the top of each sheet; indicate each answer as appropriate.

Place these processes in the correct sequence:

1. a. Lap
   b. Polish
   c. Grind

   1. Grow
   2.
   3. Slice
   4.
   5.

2. a. Spin Photoresist
   b. Diffusion
   c. Etch

   1. Oxidation
   2.
   3. Expose
   4. Develop
   5.
   6. Strip Photoresist
   7.

3. a. Evaporate aluminum
   b. Photoresist removal
   c. Develop

   1.
   2. Spin Photoresist
   3. Expose
   4.
   5. Etch aluminum
   6.
   7. Sinter

4. a. Evaporate Lead/Tin
   b. Evaporate Chrome:Copper:Gold
   c. Pad reflow

   1. Metal Mask H align
   2.
   3. Metal Mask I align
   4.
   5.
   6. Mounting
   7. Dicing

Place these layers in correct order (they are not necessarily consecutive):

5. a. Sub-collector diffusion
   b. Aluminum
   c. Lead/Tin

   1.
   2. Epitaxy
   3.
   4.

6. a. Sputtered oxide
   b. Aluminum

   1. Emitter reoxidation
   2.
   3.
   4. Chrome:Copper:Gold

Indicate the correct statement:

7. Epitaxy concerns: (a) continuous crystalline growth; (b) adhesion; (c) stacking faults; (d) deliberate structural change.

   ☐

8. When gases pass into solids, this is: (a) fission; (b) transmutation; (c) diffusion; (d) dissolution; (e) emission.

   ☐

9. Add these materials to the appropriate group: (a) Silicon; (b) Quartz; (c) Aluminum.

GROUP I	GROUP II	GROUP III
Beryllium Copper	Germanium	Sulphur Wood
☐	☐	☐

10. Rank in order of size, smallest first:

   a. Aluminum conductor thickness
   b. Micron
   c. Mil

   1. Atom
   2. 1000 Angstrom
   3.
   4.
   5. Aluminum conductor width (average)
   6.
   7. Human hair

Figure 1. Sample of final test.

The format of a typical technology session would be:

a) Historical background, earlier products
b) The new products
c) Elementary materials science; atoms, molecules, elements, compounds
d) Nature of elements; metals, non-metals
e) Descriptive properties of semiconductors
f) Preparation of silicon, growth of single crystals
g) Manufacture of wafers

h) Process from wafer to final mounted encapsulated ship
   Oxidation -- Photoresist -- Etch -- Diffusion -- Expitaxial --
   Growth -- Reoxidation -- ..... -- Evaporation -- ..... --
   Passivation -- ..... -- etc.
i) Possible slide review of key process characteristics
j) Slides of finished product
k) Student questions not dealt with during presentations

It is obvious from this process listing that the manufacturing operation has to be considered as an integrated whole. It was emphasized in the lectures that poor performance during a photo or etching step could have dramatic impact upon subsequent final test yield. Additionally, a marginally different process setting in evaporation, or some other coating operation, could grossly affect photoresist adhesion and etch definition. Cleanliness was stressed as being of supreme importance through the entire process flow. It was emphasized time and again that the application of advanced technology to high volume manufacturing requires a very different kind of skill, care, application and dedication on the part of the operators.

To conclude this training regimen, a final test was administered separately by the education specialist running the course. This was used as a measure of the quality and success of the instruction. In addition, all students rated the performance of all instructors with respect both to material, level, communication ability and session applicability. In this way deficient instructors could be counselled or changed and irrelevant course material replaced.

Training Techniques

Visual aid techniques developed extensively during the period under discussion. For emmediacy, understanding, and good communications between the class and the instructor a dual overhead projector presentation using colored transparency foils was most effective. This was preceeded by a brief slide show (2x2) to give realism and perspective with past and present product pictures. The overhead projectors were arranged to project side by side pictures. The foils *(Figures 2 - 4)* were made to represent each process operation being performed on a specially designed product model shown in plan, cross-section and isometric view. Thus it was possible to show before and after pairs, and proceed alternately from one projector to the other going sequentially through the process. Plan and view foils suitably color coded to represent the stage reached in the manufacturing process were interposed regularly to explain the correlation of the steps.

The set of ninety sequential process foils were also reduced to 2x2 slides and a tape/slide package was assembled. This did not prove to be particularly popular or effective. The sterile, boring, and academic nature of a lengthy tape/slide sequence was an undoubted handicap. Even when used without the tape the requirement for darkness,(back projection was not available),along with a loss of flexibility impeded communication. The manual nature of the foil display permitted total instructor control, and the high light level was suitable for use of chalkboard diagrams to explore questions on details. An additional advantage was that the discussion accompanying the foils could be precisely tailored to match the audience needs.

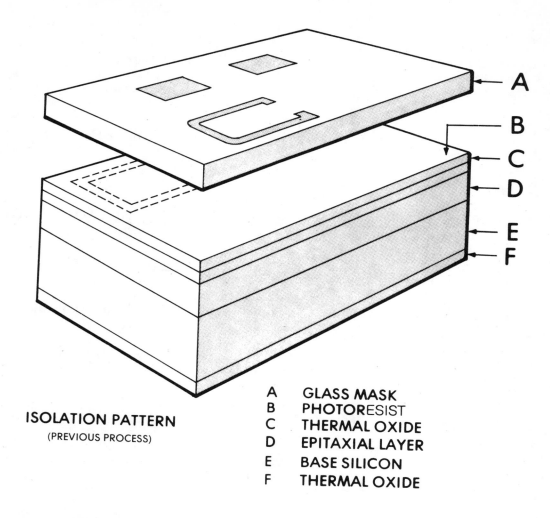

ALIGN BASE AND UNDERPASS RESISTOR MASK

Figure 4. A typical isometric teaching model (color-coded in the original foils).

CONCLUSION

The early recognition by the education professionals at IBM Burlington that a new, integrated semiconductor process demanded a totally new manufacturing training program not only satisfied the immediate objectives of getting manufacturing personnel "on board" with the least disruption to plant schedules, it also led to far-ranging and, in hindsight, more important benefits. The employees, by being exposed to a serious and comprehensive education program, developed an almost proprietary attitude toward the success of the whole manufacturing project. It led to an across-the-board increase in manufacturing operator understanding which, in turn, developed a strong esprit de corps. The team spirit thus engendered became a watchword throughout the rest of the corporation and contributed significantly to excellent manufacturing performance as measured by yield at final test, defect levels and reliability results.

SUMMARY

"People Potential" is a potentially powerful tool for the manufacturing engineer if he will study the basics of human motivation and apply those principles in the assembly automation process. He can be far more effective in his function in management. It probably will be necessary for him to "sell" his own management group on many of the uses of people potential. This is an opportunity, challenge, and a legitimate tool for the assembly automation manufacturing engineer.

REFERENCES

1. D. Yoder, Personnel Management and Industrial Relations, Fifth Edition Englewood Cliffs, New Jersey: Prentice Hall, Inc., 1962, p.6.

2. D. Beach, Personnel: The Management of People at Work, Second Edition, New York: The Macmillan Company, 1970, p. 515.

3. E. E. Lawler, III, L. W. Porter, "The Effect of Performance on Job Satisfaction," Industrial Relations Vol. 7, No. 1, October 1967, p. 21.

4. A. A. Imberman, "The Blue Collar Blues-Today's Academic Hit Tune," Assembly Engineering. July 1973, p. 26.

5. F. Herzberg, "One More Time: How Do You Motivate Employees?" Harvard Business Review, January-February 1968, pp. 53-62.

6. M. D. Dunnette, J. P. Campbell, M. D. Hakel, "Factors Contributing to Job Satisfaction and Job Dissatisfaction in Six Occupational Groups," Organizational Behavior and Human Performance, May, 1967 p. 173. (An original study, plus reports on ten others.)

7. M. D. Dunnette, Et al, op. cit. p. 174.

8. A. C. Filley, R. J. House, Managerial Process and Organizational Behavior, Glenview, Illinois: Scott, Foresman and Company, 1969, pp.385-386.

9. T. O. Prenting, "Better Selection for Repetitive Work," Personnel, September-October 1964, pp. 26-31.

10. F. Herzberg, op. cit. p. 62)parenthesis supplied by author).

11. W. W. Winpisinger, "Job Enrichment: A Union View," Monthly Labor Review, April 1973, pp. 54055.

12. F. Herzberg, op. cit. p. 62.

13. N. R. F. Maier, Psychology in Industry, Second Edition, Boston: Houghton Mifflin Company, 1955, p. 476.

14. G. Friedmann, Industrial Society, Glencoe, Illinois: The Free Press, 1955, pp. 145-146.

15. G. Friedmann, "What is Happening to Man's Work?" (Excerpts), Human Organization, Winter 1955, p. 31.

16. T. Wicker, "A Plant Built for Workers," The New York Times, May 21, 1974.

17. M. D. Kilbridge, L. Wester, "The Balance Delay Problem," Management Science, Vol. 2 No. 1, October 196., p. 84.

18. M. D. Kilbridge, L. Wester, "An Economic Model for the Division of Labor" Management Science, Vol. 12 No. 6, February 1966, pp. B255-B269.

19. R. B. Chase, "Survey of Paced Assembly Lines," Industrial Engineering, Vol. 6 No. 2, February 1974, pp. 14-18.

20. T. O. Prenting, R. M. Battaglin, "The Precedence Diagram: A Tool for Analysis in Assembly Line Balancing," The Journal of Industrial Engineering, Vol. 15, No. 5, July-August 1964, pp. 208-213.

21. T. O. Prenting, N. T. Thornopoulos, Humanism and Technology in Assembly Line Systems, Rochelle Park, New Jersey; Spartan Books (Hayden Book Company), 1974.

22. M. D. Kilbridge, "Non-Productive Work as a Factor in the Economic Division of Labor," The Journal of Industrial Engineering, Vol. XII No.3, May-June 1961, p. 155.

23. M. D. Kilbridge, "Do Workers Prefer Larger Jobs?" Personnel, Vol. 37 No. 5, September-October 1960.

24. T. O. Prenting, M. D. Kilbridge, "Assembly: The Last Frontier of Automation," Management Review, Vol. 55 No. 2, February 1965, pp. 7,11.

25. R. H. Eshelman, "Automatic Assembly-Victory for Production," The Tool Engineer, September 1956, p. 116.

26. A. R. Wiese, "Why You Should Consider Automated Assembly," Machinery, March 1964, p. 93.

27. T. O. Prenting, "Automatic Assembly: The Economic Considerations," Mechanical Engineering, Vol. 88 No. 8, August 1966, p. 30.

28. C. C. Holloway, "Resurgence of Automated Assembly: Computer Simulation Has Helped to Improve Predictability of Performance," Steel, August 29, 1966, p. 48.

29. T. O. Prenting, "Some Missing Links in Improving Assembly Productivity," Assembly Engineering, November 1973, p. 44

30. J. Church, "Simulation-Aid to Improved Assembly Operations," Assembly Engineering, Vol. 14, No. 8, August 1971, pp. 18-20.

31. E. M. Davis, W. E. Harding, R. S. Schwartz, and J. J. Corning, "Solid Logic Technology, High-Performance Micro-Electronics," IBM Journal of Research and Development, 8, 102 (1964).

32. P. A. Totta and R. P. Sopher, "SLT Device Metallurgy and Its Monolithic Extension," IBM Journal of Research and Development, 13, 226 (1969).

33. E. g., K. M. Gardiner, "Semiconductor for Computer Memories" talk submitted to SME for publication (1974).

CHAPTER 3

Automated Assembly Equipment Justification.

Industries elect to acquire automation for a variety of reasons. Some of these reasons are:

1. Manual Skills Inadequate.

   Precision machining or grinding to .0001 inches is simply impractical for a man with a file. Chemicals may not be safely handled by manpower. The manufacture of fuels and chemicals demands control beyond the human capability.

2. Quality Requirements.

   Required reliability of product and process is becoming greater than human capability. Warranty costs and supplier liabilities encourage automation and the elimination of the human element.

3. Rate of Production Required.

   Food, beverage and cigarette manufacturers require manufacturing rates well beyond human potential. Our communication systems, for example have expanded to the point that there would not be enough operators in the United States to handle all the telephone calls made daily if the equipment was manually operated. Many manufacturing operations take place at a speed beyond human ability to control.

4. Labor Productivity.

   Competition demands the lower costs of mechanized operations and the consequent reduction in labor costs. Inflationary wage spirals demand the maximum productivity increases to absorb even part of the unit cost increases. See *Figures 1* and *2*.

5. Labor Utilization.

   Mechanizing tedious portions of the manufacturing process can release the human operators for higher level, judgmental functions and greatly increase the utilization of the human capability.

6. Space Utilization.

   Consideration should be given to cost and availability of manufacturing space when comparing automation to manual operations. A factor should be given to value-added dollars per square foot density. See *Figure 3*.

7. Dollar Utilization.

   Consideration should be given to possibilities of reduction in inventory through the use of automated systems not requiring as large in-process inventory.

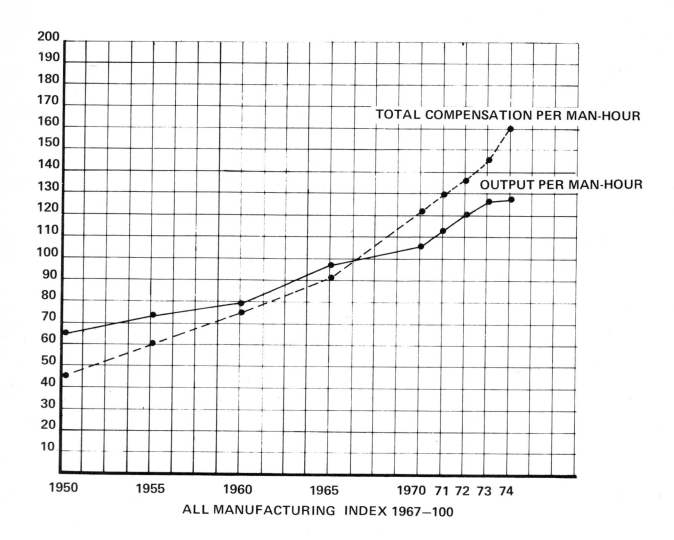

Figure 1. Compensation per man-hour has consistently out-stripped out-put per man-hour. While out-put has doubled, compensation has nearly quadrupled in a 24 year period. Automation offers one of the most effective means of bringing these curves closer in line.

Note that a few industries have even had a <u>decrease</u> in out-put per man-hour since 1973: a trend which could have disastrous effects if it spreads or continues. (Data source: U.S. Fact Book, Bureau of Census, Dept. of Commerce.)

Note also, replacing old equipment with new is not enough. The new equipment must not only be more productive to cover the price increase, it must also be more productive to cover the labor increases as well in order to merely break-even. This is more practical in assembly than in any other manufacturing area.

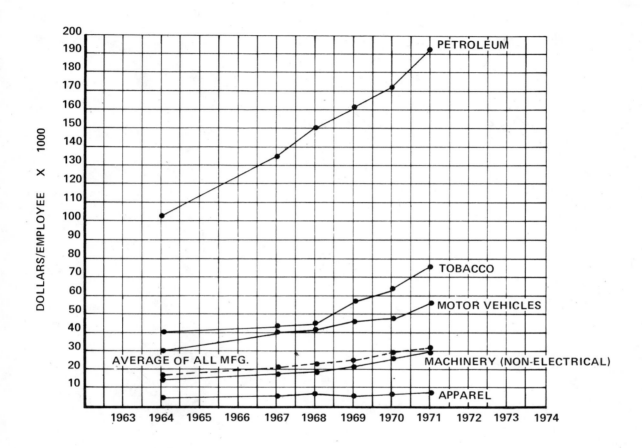

Figure 2. Capital investment in dollars per employee is charted for five major manufacturing industries along with the average for all manufacturing industries for a seven year period. (Source U.S. Fact Book, Bureau of Census, Dept. of Commerce.)

If space and data permitted chsrting fifty years back, the trends would be consistent. A substantial drop in the investment rate in the 1973-1975 period would be indicated if more recent data were available.

The statistically-minded will find many corollaries between the above curves and corporate growth, profits, corporate stability, product price changes, product value per dollar, etc.

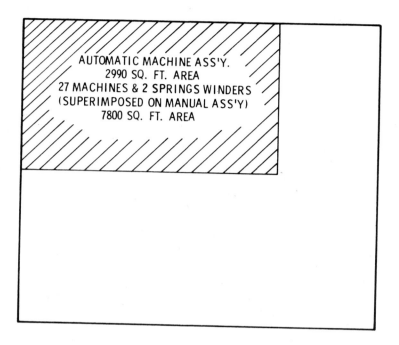

Figure 3. Comparison of Manual Versus Automated Assembly Floor Space Requirements. Assuming new construction @ $30/sq. ft., an initial savings of $144,300 would be justified in "avoidance" of additional plant area.

Capital investments, such as automated assembly, are always limited by available funds. It is a rare case where there are more funds available than there are projects and proposals requiring funds. Some form of justification formula or procedure is used to determine the priorities of the desired projects. In the case of a small, individually owned business, this may be as simple as the desires of the owner. In the case of major corporations, highly complex calculations have been evolved.

We will not labor the point of justification formulas. Much has been published on the subject, ranging from the MAPI formulas of the fifties through current calculations taking into effect tax ruling changes of the seventies. One typical form used by a major corporation is shown in *Figure 4*. Some form of "return on investment" is usually computed and the best return on investment is purchased first, with the least return being last or disregarded. Depending upon economic conditions, most major corporations look for an amortization of two to four years after taxes. The method of computation will vary with each Comptroller and the desires of the corporate management.

Figure 5 suggestions some of the variations possible in a typical hypothetical assembly machine justification. One automotive manager, with

APPROPRIATION REQUEST

Program _____

1. Division _____
2. Location _____

Date __June 23, 1976__

	Appropriation No. __1__	Project No. __A-637__
3. Fixed Assets Special Tools	$ 16,000	$ 16,000
Total Capital Expenditures	16,000	16,000
Operations - Other	164,300	164,300
Total	$ 180,300	$ 180,300

4. Description - Title: Install New Computerized Time and Attendance System

5. Necessity of Proposed Expenditures, Explanation and Economic Justification:
(If space is insufficient attach explanation on 8½" x 11" sheets)

This Appropriation Request provides funds for Data Processing equipment to install a new computerized hourly time and attendance system at _____. The proposed system will utilize electronic badge readers to reduce timekeeping costs $154,500 annually (as shown in Exhibit A reduce production supervision paperwork, and provide timely information on absentees to divisional management.

Expenditures Based on Net Increase in Rent

Recurring Annual Rental of Data Processing Equipment	
Included in Project Amount	$118,300
Released Equipment	0
Net Increase	$118,300
Capital Expenditures	16,000
Non-Recurring Expense	46,000
Expenditures Based on Net Increase in Rent	$180,300

6. Estimated Costs	A. This Request	B. Appropriated To Date	C. Further Requests	D. Total Project
Land-Land Imp. _____ Acres	$	$	$	$
Buildings _____ Sq. Ft.				
Equip. Incl. Furn. & Office Equip.	16,000			16,000
Sub-Total Fixed Assets	16,000			16,000
Special Tools				
Total Capital Expenditures	16,000			16,000
Operations - Other	164,300			164,300
Total	180,300			180,300

7. Allied Transfers (Memo)
Special Tools $ _____ Net Fixed Assets $ _____ Total $ _____

8. Total Project (Incl. Transfers) $ __180,300__

9. Forecast Expenditures	Facilities	Tools	Operations
Spent to Date	$	$	$
1 Quarter 19 77	16,000		75,500
2 Quarter 19 77			29,600
3 Quarter 19 77			29,600
4 Quarter 19 77			29,600
_ Quarter 19 __			
_ Quarter 19 __			
_ Quarter 19 __			
_ Quarter 19 __			
Subsequent			
Total	16,000		164,300

10. Extent to which proposed project is a replacement of existing property:
Replacement Cost $ _____
Replacing Present Capital Values:
Gross Book Value $ _____
Depreciation Accrued $ _____
Net Book Value $ _____
Estimated Salvage $ _____
Proposed disposition of replaced property:

11. Annual Return on Investment (See Exhibit for detail)	Fixed Capital	Working Capital	Total Capital	Estimated Savings or Return	Return On Investment
	$ 14,800	$ --	$ 14,800	$ 33,800	228.4

Payback Data	Total Expenditures	Net Annual Cash Savings	Payback Period	Economic Life
	$ 62,000	$ 36,200	1.7 Yrs.	10 Yr

Figure 4.

	Present	Proposed
Investment (net realizable value)	18,000	
Investment (installed and "launched")		150,000
Yearly interest on investment at 9%	1,620	13,500
Yearly tax, insurance allowance at 3%	540	4,500
Yearly depreciation at 14%	2,520	21,000
Yearly maintenance cost	7,500	15,000
Yearly power and supplies cost	4,500	7,500
Yearly space cost	1,200	400
Yearly materials cost	300	600
Yearly direct labor cost	108,000	18,000
Yearly indirect labor cost	6,000	4,500
Yearly fixed charges	1,000	6,000
Total costs	133,180	91,000
Annual savings		42,180
Before taxes return on investment (R.I.V.)		32%
Before taxes amortization (turn-over time)	3.13 yrs.	

Notes:

A. If "after taxes" savings are used, (50%) turnover time becomes more than 6 years and R.I.V. less than 16%.

B. If the operation is two shift, the savings become 122,880, turnover time is 1.07 years, and R.I.V. is 93%.

C. If labor costs are computed for an average over a 10 year period with a 7% increase yearly, the savings become 85,680/year, turnover time 1.54 years, and the R.I.V. becomes 65%.

D. If the assembly is an automotive engine part causing 1% warrantly claims at $100/claim and the automated process adds an inspection station at $25,000 which reduces warranty claims to .1% of an annual production of 2,400,000 assemblies, an additional $216,000 justification could be legitimately claimed.

E. Sometimes a "minute cost" is determined per piece. At 1200/hr. with 6 operators, the minute cost would be .3 vs. .05 for the automated operation. If these numbers are extended by the projected production schedule and a $/min. factor assigned such as $0.2, labor savings only can be calculated in terms of expected production rates.

F. A major fault of accounting systems which use labor hours as a basis of budgeting maintenance dollars is that an automation project which eliminates labor hours also reduces maintenance allowances when, in fact, maintenance dollars must go up sharply. A "machine-hour" basis is much more accurate, but care must be taken in either case to insure proper allowance for equipment maintenace.

Figure 5. Hypothetical justification computation comparing an existing 6 man, 6 machine operation with a 1 man, 1 machine automated operation.

tongue-in-cheek, commented on the "games that can be played" with the "numbers racket" saying, "It's much easier to justify the project if you start with the conclusions and work backward!" In any case, the manufacturing engineer must be familiar with all of the combinations and uses of the "numbers" if he would be effective in convincing management of the value to the company of an assembly machine project.

Some recent statistics suggest the following for your consideration:

1. The rate of productivity in the United States is now declining. See *Figure 6*.

2. The largest percentage of direct labor contact in manufacturing is in assembly. It exceeds all direct labor costs for fabrication, forming, molding, plating, finishing and metallurgical treatment combined. See *Figure 7*.

3. The area of assembly and related testing is the only area of manufacturing that is capable of broad-scale significant reductions in labor cost.

4. The expenditure by United States industry on automatic assembly equipment recently has been declining both in constant and in current dollars. See *Figure 8*.

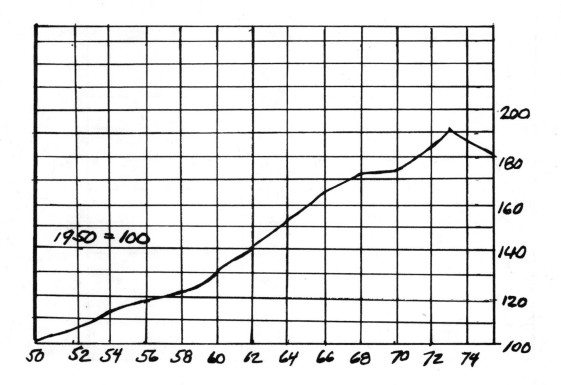

Figure 6. Output Per Man Hour

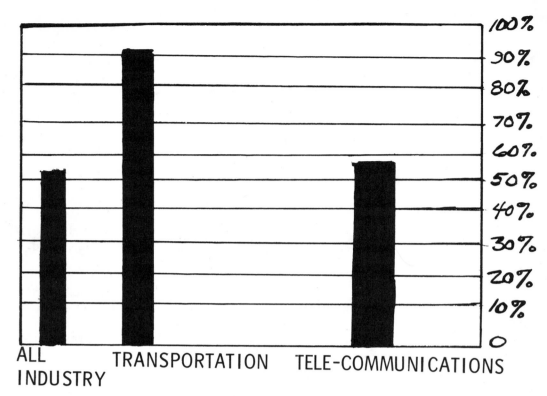

Figure 7. Assembly Percentage of Direct Labor Hours.

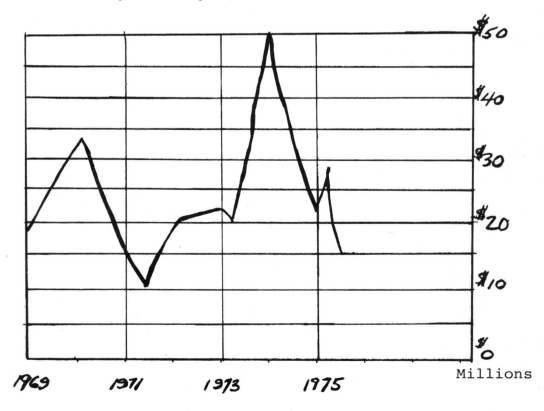

Figure 8. NMTBA Assembly Group Backlog.

The problems of American industry which are shared by our free-world counterparts are created by social changes unparalleled in history since the industrial revolution of the mid-eighteen hundreds. These problems are compounded by severe changes and increases in governmental controls and obstructions to capital formation. Automatic assembly can often meet these changes. On the whole, however, manufacturing engineers have been less than completely successful in convincing management of the value of automatic assembly.

No manufacturing engineer can expect to operate successfully in selling top management an appropriation request for mechanized assembly systems in today's economy without directly addressing himself to two basic topics and their corellaries.

First: <u>Purchase of Automatic Assembly equipment is completely optional on the part of management.</u> No one expects to mold, form, fabricate, heat treat, plate or finish high production components manually. A decision to make the product implicitly means a decision to buy or use the necessary equipment and tooling to manufacture the component or buy it from a source with the necessary capital equipment. Assembly, however, in almost every case can be done by hand with little or no capital expenditure.

Secondly: <u>Management is faced with so many mandatory capital expenditures, optional capital expenditures must be actively sold as worthwhile, profitable and essential.</u> It may seem contradictory to say that an optional expenditure can be essential. Obviously to make components on a production basis equipment is necessary. When the equipment is combined into a plant, expenditures for fuel, taxes, environmental and waste controls are also mandated. These expenditures are necessary to making the product. The ultimate goal is to make a product profitably. In varying degrees material and labor are optional. Selection of plant sites, labor pools, and material choices are management judgments concerning costs. A greater degree of management options concern advertising and marketing, financial controls, data processing, inventory levels, etc. The management judgment in each of these areas is optional, but may prove essential to ultimate or continuing profitability.

In any corporate entity, the Chief Executive Officer of that entity whether he be titled president, vice president, divisional manager or whatever, ultimately must make the decisions and accept the responsibility primarily for those options he approves and to the degree he supports them. To his desk those responsible for marketing, production control, purchasing, accounting, data processing and manufacturing, come with their options and plans. Each strives to not only do his share, but to excel in contributing to the strength and profitability of that facility or division.

In the nature of things, many Chief Executive Officers are marketing or financially rather than manufacturing oriented. If they are not, the executive or financial committee of the board of directors is so composed. It is also in the nature of things that those with the responsibility for administrative matters tend toward greater communicative skills. It is no wonder then that manufacturing engineers often feel that their appropriation requests receive scant consideration, while other requests seem to sail through.

Cost justification for assembly equipment must include selling skills usually not required for other types of manufacturing equipment.

The day is past when a short presentation on direct labor reduction is sufficient to guarantee approval for capital expenditure.

Essentially, the justification document should be realistic in terms of procurement and operating costs, conservative in stating probable returns on investment. It should face up the possible objections to such procurement, factual and fallacious, and respond to these objections in a way that is clear, concise and objective. But most of all, it should emphasize those elements familiar to and of importance to those with authority to approve such projects. These include not only contribution to profit, but specific answers to the social and governmental challenges to business continuation and prosperity.

Is Automatic Assembly Practical

Approval cannot be expected for unrealistic programs. Automatic assembly is a <u>systems</u> approach. Additionally, the proposed system must operate within the limits imposed by the manufacturing environment for that system.

Before even going out for quotations, there are areas which must be considered in depth:

a) Do the components of the assembly lend themselves to <u>mechanized handling</u>? If not, can design changes be affected? What are the cost factors involved? Parts configuration, surface considerations and fragility are among those reasons that may preclude automatic handling of individual parts from bulk condition. If there are such parts involved, is it practical to manufacture, mark, or configure these parts during the assembly process?

b) Is there <u>sufficient volume</u> on a continuing basis to justify the large capital outlay of automatic assembly. A machine which assembles only 10 pieces per minute, produces 1 million assemblies per year. (See *Figure 9*.) Is required volume affected by seasonal production demands which mandate high production rates for a limited portion of the year.

c) Are model changes and parts <u>designs stabilized</u>. Is projected model life long enough for sufficient production after machine installation. If the model life is short, there may be insufficient time to recover invested capital. Our experience indicates that people are unrealistic about the life of a model run. Most products tend to run much longer than expected. With increasing

Assemblies per minute gross	Annual Production		
	Single Shift	Double Shift	Triple Shift
10	1,000,000	2,000,000	3,000,000
20	2,000,000	4,000,000	6,000,000
30	3,000,000	6,000,000	9,000,000
40	4,000,000	8,000,000	12,000,000
50	5,000,000	10,000,000	15,000,000
60	6,000,000	12,000,000	18,000,000

Above based on 40 hour week
50 weeks per year at 83% net production

Figure 9.

government regulations, it can usually be assumed that approved or certified products will continue in production for rather lengthy periods.

d) Can manufacturing operations and <u>material handling</u> before and after the assembly equipment be coordinated to provide continuous, controlled quality input and efficiently and safely store the production. An assembly machine tends to be inflexible in its demand for sufficient component parts available at all times. This means disciplined production and inventory control if parts coming to the machine, and efficient material handling at the output side of the equipment.

e) How much <u>direct labor</u> can be saved by such mechanization? How much direct or <u>indirect labor</u> of the higher grade will have to be added for supervision and maintenance? Most assembly builders evaluate requests for quotations on the basis of present or projected labor content. Direct labor reduction will continue to remain a substantial portion of any potential return on investment. Do not neglect possible reduction in various types of indirect labor and support personnel. Reduced numbers of inspectors, material handlers, payroll clerks, can significantly contribute to the overall return on investment. It must be remembered however that such complicated or sophisticated machines will require supervision and maintenance skills superior to that usually required on manual assembly lines.

f) To what extent can in-plant engineering and tool room capabilities be used in designing, installing, tooling and debugging automation units? It requires sound judgment to evaluate honestly the existing in-plant capabilities that can be coordinated or utilized in system development, thus reducing the overall cost of the project. Our experience indicates a common failure to fully utilize in-house test capabilities. At the same time we often see a tragic over-estimation of capability of designing large assembly systems. Frank discussions with prospective builders can often bring about a happy balance of the talents in the builder and user organizations.

g) If component parts to be assembled are expensive, will the relatively high cost of parts used in tooling, debugging and pilot runs be recoverable? An insufficient supply of parts means sketchy and incomplete debugging. This puts the burden of debugging on the production facility at enormous burdens of time, frustration and missed schedules.

Answers to the above questions will determine overall feasibility. A brief outline of this examination suggested above is a valid prelude to outlining the actual cost bids by proposed vendors. It also is an excellent aid to obtaining the lowest possible bids for good production systems, since concise answers to these questions will indicate to the builder that all aspects were reviewed before issuing the request for quotation.

While it may seem excessive to outline the feasibility study, it should not be forgotten for a moment that automatic assembly is an alternative method of manufacturing. Management will want a full statement of the options.

Determining Project Costs

Once it has been determined that there is sufficient volume, product stability, technical feasibility and potential cost reduction, the next step is to determine the overall project cost.

There are many cost factors that must be considered above and beyond the purchase price of the assembly machine. Consider the following:

a) <u>Project the overall salary and travel expenses for engineering liaison.</u> Automatic assembly machines usually require substantial liaison between the builder facility and the user. Station designs must be reviewed, component part vendors sources may have to be visited, and trips taken to the building site during preliminary and final acceptance runs.

b) <u>Review end item quality control requirements, present and achievable.</u> Increasing concern for product reliability should result in a determination to include in the automatic assembly machine, every practical quality control inspection. Inspection stations should provide not only for the presence and position of each of the component parts, but, as far as possible, to determine the function and reliability of the completed assembly prior to ejection. Inspection station design should be coordinated with those responsible for quality assurance.

c) <u>Consider the cost of redesigning parts to facilitate automated handling</u>. It may be necessary to redesign individual component parts or total assembly configuration in order to make automatic assembly practical. Take into account necessary die, mold and tooling changes in order to incorporate the required changes. These may require substantial additional costs, but at the same time, provide excellent opportunities for further cost reduction. For example, new technologies such as ultrasonic welding may remove dermatitis problems, reduce or cure bonding times, and work in process storage requirements for parts previously bonded with solvents and adhesives.

d) <u>Explore the cost of efficient integration with present plant equipment</u>. Work flow requirements may mandate conveyors, storage elevators, and new pallet, or work carrier designs not required for hand assembly. This cost will be balanced against reductions in material handling found in mechanized assembly.

e) <u>Determine the cost of converting mechanized assembly for possible future redesign of component parts or product</u>. Be sure that quoted machines have salvage value and can be reused in the event the product is redesigned, changed or eliminated.

f) <u>Determine personnel requirements</u>. Supervisory emphasis will shift to technical skills rather than people handling skills. It may even be necessary to locate automatic assembly equipment in departments other than the present or potential manual operations.

g) <u>Check the various methods of depreciation and available tax credits</u>. Tax credits are very important since they reflect a direct credit on taxes paid and would have a fairly immediate impact on cash flow. Various depreciation methods can be used to show maximum results depending upon corporate investment goals. Much of the cost of assembly equipment lies in the engineering and tooling areas and hence can be quickly written off, should this prove useful in a given fiscal year.

h) <u>Determine the degree of salvageability</u> in the equipment should the particular product be phased out. Can the quoted machines be used for other products or usefully transferred to other divisions. This particular aspect is becoming increasingly important as major technological changes are imposed by environmental or consumers concerns.

The cost of these additional expenses must be considered in determining the validity of using the mechanized assembly approach. The total costs may preclude the use of mechanized assembly as a viable option, but they cannot be swept under the rug and must be part of the consideration.

A prime concern is the capital cost of the assembly system itself. The options in assembly machine design available to a process engineer soliciting proposals are so varied as to require great analysis. Obviously the less the capital expenditure and the greater the savings, the easier the justification.

In order to insure the best possible cost quotations, there are several useful guidelines.

Outline the problem rather than the answer. So often special machine builders receive requests for quotations so specific in detail of machine construction and configuration as to preclude any creative input from the builder.

The process engineer faced with choices in automatic assembly equipment might well take the advice of Thomas Edison, "Genius is 1% inspiration and 99% perspiration." Uttered by one of the most creative minds in our technological history, this comment is certainly worth considering. Automatic assembly still remains as much art as science and experience is an essential factor of successful mechanization. Recently we had cause to distinguish between a builder's reputation and a builder's record. Reputation consists of impressions and subjective judgments, whereas records are based on factual evidence of success of failure. A potential buyer has the right to insist from any builder a list of prior customers and installations. Guarantees as to performance are worth as little or as much as the builder's past performance record.

It should not ever be forgotten that once a process engineer commits himself to a builder, he has placed his own judgment on the line with his company. It is only human nature to protect ones position, should the builder fail to live up to the explicit or implicit promises of his proposal.

This may mean taking incompleted or untried machines into the plant, providing advanced cash payments, and involve unplanned expenses for tool room, maintenance and engineering time at the user facility.

In examining proposals, a buyer should expect to find a detailed statement of the work to be done by the machine and the equipment to be supplied to perform each stated function. The cost section of the proposal should include an itemized breakdown of the cost of the basic machine chassis, the electrical package, the cost of the various inspection units and inspection funds, and itemized costs for work holding fixtures, each feeding, joining, transfer or fabricating function, and any specialized ejector or conveyance device.

Such a cost breakdown rather than a lump sum price is some indication that those involved in preparing the quote have been realistic in determining some method of action for each of the described functions. It also indicates the builder can justify on a detailed basis, the proposed station costs. More important however, should it become necessary to modify any of the functions or stations during the course of the machine building, there is a point of departure cost increases.

The builder should also be able to describe approximately floor space, power requirements, payment terms and gross production rates.

Delivery times are an essential element of the quotations. A buyer has a right to know the builders current backlog and monthly delivery capability. He should consider lead times on purchased items, and availability of the builders own standard components necessary for meeting the projected delivery schedule. Usually the builders backlog and annual shipment levels are good indicators of the validity of the projected delivery promise. On the builders side however, there is also an assumption that prints and/or samples are available to commence engineering at the earliest possible time.

The builder should be prepared to discuss in the proposal or in related discussions, what elements determine the cyclic rate of the equipment and whether or not any stated manual operator functions can be done in a mutually agreeable time span.

The builder should be prepared to discuss a valid, rational concept for each of the stated functions.

In evaluating quotations, the buyer must consider the basic machine chassis proposed. Generally, mechanical machines have a greater initial cost, lower maintenance and operating costs and less downtime. Fluid power machines generally cost less initially, but require higher maintenance and have less uptime.

Control systems and control options are chosen for a variety of reasons. Mechanical machines generally need extremely simple controls, but memory or logic circuits may be suggested for various reasons. If the machine requires sequencing, logic systems are indeed useful tools. If there are a large number of quality control aspects to be considered with alternate modes of operation, control circuits using logic functions again may prove invaluable. On the other hand, if memory systems are quoted as a means of overcoming parts quality problems, their cost is usually indefensible in practice, even if attractive on paper.

The buyer's problem, of course, is to insure that the chosen assembly system has sufficient funding to be built properly, to be thoroughly debugged prior to shipment and to provide all necessary installation services that may be required to put the machine properly on line. On the other hand, the buyer must beware of unnecessary expenditures and also proposals which appear inexpensive at inception, but continue to grow in size as the proposal develops. Over specifications is main reason for excessive costs.

Customer specifications for machine design are the end result of specific problems in the operating plant. Specifications attempt to make a general answer to specific problems. They may unnecessarily complicate good machines or exaggerate costs. The check list type of specification in which the process engineer selects those specifications important to his specific project are most useful.

The justification request should detail briefly the consideration given to various quoted machine systems and the basis for the choice of the selected system. One of the basic problems in selling assembly systems to management lies in the fear, conscious or subconscious of possible failure. A concise statement as to the actual record of the selected vendor would be a valuable addition to the appropriation requests. Details of some similar successful installations of that vendor can go a long way to easing the fears of non-technical management.

A statement regarding planned provision in the selected machine to handle future model changes, product changes, will aid in relieving fear that the machine may be obsolete before significant return on investment.

Justification for the Investment

Up to this point we have discussed technical feasibility and total project costs. There is another important consideration, probable net production.

Probable Net Production

Few builders are willing to guarantee specific net production levels on large assembly systems unless such machines are essentially duplicates of proven and tried equipment. There are four controlling elements in determining that portion of gross capability, realized as net production.

The controlling factors are:

a) The machine as a system.

b) The reliability of the individual station functions.

c) The motivation of the operating and maintenance personnel.

d) The uniformity of component parts supplied to the machine.

For those with a mathematical bent of mind, *Figure 10* expresses a relatively simple formula for explaining the basis of net production on a specific machine application. Notice the large variable in this formula is the length of the average downtime. This factor is controlled primarily by operator motivation and component part quality, rather than machine design, assuming that the machine at time of acceptance had reached satisfactory levels of production, and has a strong design compatible with sustained production operations.

Most builders are willing to indicate probable levels of net production as indicative of attainable goals without being specific guarantees. An experienced builder will vary his estimates based upon his assessment of the customers technical capabilities, quality assurance capabilities, and personnel motivation factors. He will also factor into his assessment yield

Expressed mathematically net production can be expressed in two steps. It is first necessary to determine machine efficiency in terms of the percentage of machine cycles that will produce acceptable assemblies. When C is the percentage of acceptable parts in each lot of component parts coming to the machine, S the efficiency level of each work station in performing its own task of selection, transfer or joining and M the efficiency of the basic machine control system in coordinating all of the individual station operations, the probable percentage of machine cycles that will produce acceptable assemblies can be expressed:

$$\text{Machine efficiency} = (C_1 C_2 C_3 \ldots C_n)(S_1 S_2 S_3 \ldots S_n) M$$

It is possible that on a short run basis, or if the quality or efficiency is quite low, a particular machine cycle can be deficient for more than one reason, bringing up the efficiency level. However, on a long run basis the probable efficiency level indicated by the above formula is quite realistic.

Net production across a reasonable length of time is determined in the following way. The number of inefficient machine cycles (C_d) in a given period (obtained from determining machine efficiency levels as described above) is multiplied by the average downtime (T_d) that each malfunction causes. The resultant is subtracted from the total time (T_t) in the period and the result divided by gross cyclic rate (R_g).

$$\text{Net production} = \frac{T_t - (C_d T_d)}{R_g}$$

Figure 10.

levels that will be expected in various industries. Yields generally are lower in electronic assemblies than in mechanical assemblies. Yields will be better in assemblies specifically designed for mechanical assembly than would be true in cost reduction projects on existing assemblies.

Again, the machine purchaser must tread a careful path between exaggeration of possible yields from the machines and understated net production because of a concern over machine productivity. If such concern does exist, the project probably should be held in abeyance until there is a greater certitude as to probable net production.

Again, a concise statement of the rationale for the projected net production levels will do much to preclude unnecessary resubmissions for appropriation approval.

Determining the Justification

There are many areas of cost reduction that can be utilized in justification for assembly machinery. Direct labor reduction is a most important consideration, but not always the main reason for justification. Reduction of indirect labor, such as lead men, area supervisors, payroll personnel clerks, inspectors, are all potential areas of indirect labor savings. Additionally, good mechanized assembly should substantially reduce work-in-process inventories. By its substantial productive capacity, manufacturers can keep completed parts inventory to the lowest reasonable levels. Additionally, there are savings to be realized through the uniformity of assembly quality, a reduction in possible warranty or product liability claims, and the unique tendency of automatic assembly to increase overall plant efficiency. Examine each of these considerations in detail. All of them can be substantial factors in justifying mechanized assembly. Some of the indirect cost reduction aspects may be of more current interest to corporate management than the direct labor savings.

Direct Labor Cost Reduction

Most engineers consider present direct labor costs as the prime area for justifying mechanized systems. If the engineer goes only this far, and does not include such factors as probable increases in payroll costs and continuing <u>declines in productivity</u>, the justification becomes a study in past history, instead of projected actual labor costs.

With highly cyclical employment, there are other considerations today in the costs incurred when hiring and laying off personnel. Most states attribute direct costs of unemployment compensation to specific corporations. The cost of guaranteed annual wages is another indirect cost. Each person hired in the assembly area creates a potential liability for future operating costs, should it become impossible to maintain their employment on a permanent basis. On the other hand, automatic assembly equipment can be turned off and incur no additional costs other than the depreciation expenses incurred during the period of idleness. Again, cyclical employment involves training of new workers and retraining of reassigned workers, with substantial cost penalties in learning curves and in assembly quality. Mechanical assembly systems can avoid a great deal of these expenses, by substantially reducing the amount of personnel turnover.

When production requirements are cyclical in nature, such as requirements of items sold on a seasonal basis, e.g. outboard motors, lawn mowers, etc., mechanized assembly allows maximum seasonal production capability with a minimum of restaffing. Some plants are faced with exceptional sales opportunities if fast delivery can be accomplished. Again, placing assembly machines on additional shift operation can easily meet production surges with very little additional hiring.

Indirect Labor Reduction

Reduction in required floor supervisors, inspectors, and support personnel, such as payroll, personnel, food handling and health personnel, can be substantial. In one instance, a battery of ordnance assembly machines effected a reduction of 56 people in direct labor, but also reduced the required number of supervisory, quality control and maintenance people by 62 people. Since many of these costs are buried in factory burden, they may be more difficult to isolate, but these are truly savings. Overhead personnel incur the same type of fringe hiring and training costs as direct laborers.

One of the most frustrating problems is to determine the new overhead costs when there is substantial reduction in direct labor forces in a given department. The ratios will change drastically, and it may be necessary to add new technical skills in a department that was essentially people oriented. Failure to realize and state this may vitiate an excellent cost reduction program.

Product Liability and Warranty Expenses

Automatic assembly equipment has a unique tendency to force upgrading of overall parts quality and requiring the maintenance of uniform levels of quality by the very fact that wide variations in components quality will inhibit the productivity of the machinery. However, there are broad capabilities for superior assembly quality in such machines beyond this fact. It is possible and most inexpensive to inspect for the presence, position, attitude and orientation of each of the component part pieces during the assembly process itself. It is also possible to functionally inspect for such functional characteristics as capacitance, impedance, resistance, vacuum, leak rate, dimensional characteristics and torque. Without any additional labor content, 100% inspection can be realized, data recorded, and product so coded as to be able to document that care in production required of prudent men by current liability legislation and judication. Such inspection stations are usually moderate in cost. Parts are already fixtures. Fault counters can give objective statements regarding quality trends in component parts. This data is extremely useful in adjusting prior fabricating and forming operations to required quality levels.

One of the most dramatic justifications for such inspection expenditure occurred in our plant last year. The general manager of a large Japanese automotive firm responded to his engineers concerns with costs by asking them what the postage alone might be on a registered letter mail campaign for recall of their automobiles. The cost of the necessary inspection functions was minor in comparison with the possible risk. Emphasis on the uniformity of quality, the ability to document quality and the ability to inspection 100% of all production, should be of substantial interest to those concerned with the marketing and legal aspects of the corporates' operation. Such intangibles tend to be excellent aspects of an overall justification request.

Production and Inventory Control Consideration

Automatic assembly systems are highly productive where volume requirements are high. Multi-shift operation is indicated by the high capital expenditures. Their ability to respond immediately without learning curves to marketing opportunities, give salesmen in the field the advantage of being able to accept orders on the basis of quick delivery or high volume, knowing that the capabilities exist within the plant to meet these goals. This can be done without the delays incurred in hiring, staffing and training assembly departments. Governmental directives concerning such hiring and training have become so regulated that in many instances, operating managers are reluctant to do hiring on a short term basis. Possession of automatic assembly equipment means that the new production goals can be achieved with a minimum of hiring or rehiring.

Additionally, work in process can be dramatically reduced. In one instance, conversion from solvent bonding to ultrasonic welding of a small plastic pump assembly permitted work-in-process times to be reduced from approximately 6 days to 45 seconds. It was possible to eliminate the in-process storage of several hundred thousand incomplete assemblies during their curing and material handling cycles. Again, since machines can quickly respond to production goals, completed parts assembly can be kept to absolute practical minimum.

The limitations of assembly equipment present counter-balancing requirements. Sufficient quantities of all necessary components must be available to the machine at any time production is required. This involves strong inventory controls and smooth material handling operations to insure uniform work flow to the machine itself.

The Mythology of Capital Expenditure

After presenting the technical and financial justification document strongly indicating the value of purchasing a mechanical assembly system, the manufacturing engineer often runs into frustrating road blocks. Much of this has to deal with statements concerning the availability of capital for investment.

While each corporate entity has its own long range goals for development, very few deviate from the investment principle of obtaining maximum return on minimum capital investment. Each side of this statement bears on the other. Too small an investment may produce no return whatsoever. Over-investment without subsequent gains in productivity diminishes the potential return on investment. There are practical upper limits on available capital and return on investment certainly is a vital criteria of investment priority. Too often, however, manufacturing engineers accept without a struggle rejection of their appropriation requests on such statements as the following, "This machine fails to meet our return on investment goal." "Our corporate policy will not permit the purchase of any equipment having a return on investment over "x" months." "We can only utilize half of the indicated savings since we will have to pay income tax on any profits realized from this investment."

These statements assume that the corporation has a wide variety of choices of investments, each offering major returns on investment. Under close investigation, however, many of these so-called major returns turn out to be particularly intangible. Many well run conglomerates are now saddled with profitless investments purchased on false hopes. Had this money been spent on improving productivity, their current position might be substantially better. It is much easier for the manufacturing engineer to document the reality of potential return on his investment. These are real returns and in an inflationary period tend to grow in value. As <u>a part of</u>

<u>management, manufacturing engineers have both the right and the duty to question the validity of some of these negative comments.</u> At the present time, capital is broadly available and at the lowest interest rates we probably will see short of severe financial collapse.

Management can expect very short periods of return on investment only when the alternate manual method is inefficient and overly expensive. Competitively operated plans must expect longer return on investment at each increasing stage of mechanization.

Plants who make investment decisions based on present market conditions, tend to insure that such market conditions will tend to be poor, since they limit their ability to recover quickly and purchase equipment only in boom periods when deliveries are extended and prices high.

The manufacturing engineer must sell the tools of productivity aggressively and with intelligence and skill. Filling out the figures on simple forms will not suffice under todays present conditions. Appropriation requests must be presented intelligently, attractively, concisely and clearly.

CHAPTER 4

Project Management

The ultimate success of any assembly machine is judged by how effective it is in creating profits from its use. Success is also measured by the ease with which the machine becomes a part of the total manufacturing operation. To help insure this success, two distinct groups within the company must become involved: management and technical people.

First, company management should be totally involved at all levels. Second, the technical people within the organization must be fully aware of the nature of the product to be assembled. Are they certain that it is in a condition that lends itself to automatic assembly? Both of the above groups have responsibility for initiating assembly projects. Some of these projects are approved after careful and deliberate study, while others come about through a wish to automate, even though the product may be marginal or the company itself may not be ready for automation.

The success of an automatic assembly machine can be assured to a large degree by meeting the following requirements:

1. Complete involvement by management on all levels.

2. Careful planning by the engineering people.

3. A product that lends itself to automation, with consideration given to expected production volumes and the product life necessary for automatic assembly.

4. A detailed plan for subsequent production, operation and maintenance.

5. A desire by all concerned to make the project succeed.

The involvement of management and technical people is essential whether building in-house or going outside to a reputable assembly system manufacturer. Remember--and perhaps you have read this before--an automatic assembly machine is one piece of capital equipment that has to be justified on its own. It is not a mandatory machine that has to be purchased, like a lathe, drill, welder or any other standard piece of equipment. Assembly can always be by hand. The idea of automatic assembly, therefore, is critically scrutinized by all of the people involved. The machine has to stand on its own feet, not only from an economic point of view, but from an operational standpoint as well.

Management should understand that even though the builder may quote a standard machine and components, the ultimate machine will be tailored to an individual need and product. In addition, there may be technical problems with the assembly of certain parts, and the parts may not completely lend themselves to automation, or there may be a difficult part in the assembly that has to be redesigned or changed.

This has to be understood by management very early in the game. If the company goes to an outside builder there must be an understanding between them. There has to be integrity. If the company has a particular problem with the assembly under present production methods, the builder should be made aware of it. If there is a quality problem or quality specification it should be noted at once. Do not wait until later.

As an example, a prominent builder recently installed a system that involved assembly, machining, testing and welding operations. The builder, along with the customer's technical people, realized that the machining sequence could become a definite problem, and took proper steps to insure its success. It was only later that the customer's technical people and the builder discovered that there were quality specifications on the part that had not been mentioned prior to the design and construction of the machine. This caused internal problems with the customer and, of course, added costs for all involved before these problems were solved. If there had been complete involvement by the quality control people early in the game, the difficulty could have been eliminated with changes in the basic planning stage.

Management is also involved in the make-or-buy decision. Should the company build the machine itself or go to an outside supplier? Much has been written on this question, and it seems that the trend in either direction depends largely on companies and business cycles. There is no clear cut answer on what the decision should be. From experience we have found, as with anything else, it is a matter of people. Successful in-house operations normally have one or two outstanding technical people with the ability to design, build and debug an assembly machine. But this type of dedication and competence is becoming more and more difficult to find. For the large, complex machines handling different processes and many parts, the building of a successful machine in-house can tax to the utmost -- and even destroy -- a good technical group. A recent trend indicates that many in-house builders have gone to reputable outside suppliers as their assembly needs have grown more complex. Slow business conditions also force management to look closer at internal building costs.

PROGRAM TIMETABLE

When it's clear that management is committed to the development of the project, a timetable should be put together showing some of the various steps to be taken in procuring a successful automatic assembly machine. For example, the chart produced here (Figure 1) indicates approximate time percentages required for average new product program implementation. Product information and groundwork comes first. As shown here, this initial planning stage calls for 8 percent of the total time allowed. This is not a chart of personal time that one might spend on the project. Percentages will vary from one type of manufacturer to another, particularly in cost reduction programs.

		Accumulative
Product Information/Ground Work	8%	
Request to Quote/Supplier Selection	2%	10%
Receive Quote and Review	7%	17%
*Obtain Approvals	3%	20%
Process Purchase Order	2%	22%
Purchase Order Placement	3%	25%
Tool and Equipment Design Phase	25%	50%
Fabrication/Production Proving Phase	34%	84%
Acceptance at Vendor	5%	89%
Delivery	1%	90%
Installation Completion	3%	93%
Debug In Plant	3%	96%
Tryout In Plant	2%	98%
Production Sign Off and Completion	2%	100%
Total Program Time	100%	

*In cost savings programs, approval time percent varies greatly due to justification problems. New product programs more closely follow chart.

Figure 1. Time Percent for New Product Assembly Programs Implementation

Figure 2 is an actual program timing chart for a new product to be put into high production. Although this chart does not include product design time, it does include some pre-project functions which extend the lead time from an average 18 months to more than three years.

PRODUCT INFORMATION/GROUNDWORK

Anyone who proposes automatic assembly for a specific product must get back to some real basics and have this information ready for management's project approval. This information is also necessary when briefing the assembly machine builders prior to their quoting. There are five (5) basic areas of investigation and each one poses a number of questions. Accurate answers will help provide the necessary data.

THE PRODUCT

1. Success of a new product for automation in the design stage. It is important that there is communication and cooperation between company process and product design people.

2. In what stage of development is the product? Consultations with builders early in the game can assure the best design for automatic assembly by providing features for holding, tracking, orientation, fitting, etc. Also, if the product is in its infancy it could mean additional changes prior to order placement, and these could follow into the actual design and build stage of an assembly machine. Starting to build too early in the product development stage can cause problems.

-114-

PROGRAM ACTION	1976	1977	1978	1979
	4 5 6 7 8 9 10 11 12	1 2 3 4 5 6 7 8 9 10 11 12	1 2 3 4 5 6 7 8 9 10 11 12	1 2 3 4 5 6 7 8 9 10 11
PREL. DRAWING	04-15-76 – 08-2-76			
COST & FEASIBILITY	08-2-76 – 04-2-76			
DEVELOPMENT PROJECT TO STAFF	04-2-76 – 10-2-76			
PROCESS DEV. PROJECT APPROVED	10-2-76			
PREL. PROCESS SHEETS	08-2-76 – 10-24-76			
ADVANCE P.N. COST & TIMING	10-24-76			
P.P.R. COST AND TIMING	08-28-76 – 11-29-76			
PLANT LAYOUT	11-29-76 – 03-19-77			
COMPLETION OF PROCESS DEV. PROJECT		03-19-77		
12/12/50,000 MILE WARRANTY TEST		02-10-77 – 04-04-77		
OK TO TOOL PRINTS RELEASED		04-04-77		
FINAL PROJECT TO STAFF		02-14-77 – 04-18-77		
PROJECT APPROVAL		04-18-77		
PLACE P.M. FOR TOOLING & EQUIP.		05-16-77 – 08-08-77		
CONSTRUCTION OF CLEAN ROOM		06-24-77 – 10-3-77		
ENGINEERING RELEASE		10-3-77		
MAKE PARTS I.S.R.		11-17-77		
PURCHASE PARTS I.S.R. & S.Q.A.		11-10-77		
START 3,000 PCS/MONTH PRODUCTION			01-23-78	
PROCESS CAPABILITY & P.V. TEST			01-16-78	
BUILD & INSTALL TOOLING & EQUIP.		USE PROCESS DEV. EQUIP.		
POST LAUNCH REVIEW			03-20-78	
MACHINE CAPABILITY & P.V. TEST			04-10-78	
180,000 PC/YEAR START UP			08-5-78	
COST, REDUCTION LABOR & MAT'L			08-19-78	
2.1 MILLION PRODUCTION START UP			08-5-78	
PROJECT FOR 100% PROD. VOLUME				
PROCUREMENT OF ADDITIONAL EQUIP.				03-3-79
FULL PRODUCTION 4.5 MILLION				05-6-79
I.S.R. & P.V. TESTING				04-16-79

LEAD TIME IN MONTHS

Figure 2. *Typical Program Timing Chart for a New Product.*

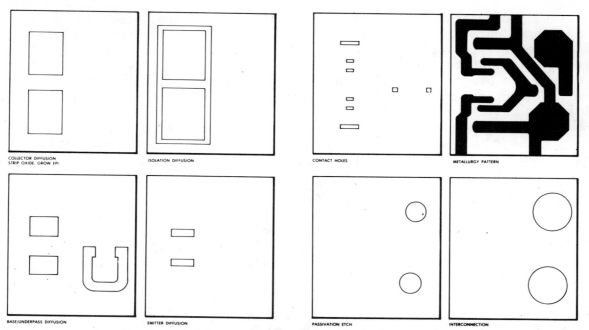

Figure 2. Examples of foil overlays: sequential process layers for the bipolar teaching model.

After the first wave of the training period had passed there were repeated requests for supplementary presentations. Thus, the foils were used by many levels of instructors for presentations to familiarize non-technical, secretarial and clerical employees with the nature and complexity of the new manufacturing regimes. Then, in 1970, technological and market developments called for the creation of a manufacturing line for field effect, or MOS, memory products. Training requirements for the FET product were similar for portions of the workforce, and a development of the existing foil package was created. This was used very effectively employing another specially designed teaching model displaying the essential device features. In addition, several color video tapes were put together using combinations of foils, slides and direct on-camera personal display or semiconductor components. Tapes were slanted towards specific audiences. For example, one such tape compared both bipolar and FET processes intended as an introductory presentation for less technical people, while another tape defined and communicated the nature of the FET process for manufacturing operators.

It is interesting to note that, although these tapes were made to describe the processes and products existing in 1970, they are still being used today with a 20 minute live lead-in which describes the latest processes and modifications required by current products. Suprisingly enough, in directly comparable situations, the personally introduced technically unsupervised video presentation received higher student ratings than the same instructor giving the "same" performance live. This suggests that the advantages of taping a well-prepared presentation may lie in the generation of instructor adrenalin, and, on the receiving end, indicates the ability of students to better absorb a color TV message.

A further, and perhaps inevitable, by-product of this change in manufacturing direction to silicon technology was the greater employee interest in sciences and technology in general. This led to large student populations for voluntary programs on "Manufacturing Processes of the Seventies" and for Computer Assisted Instruction courses based upon the instruction material cited above. The CAI course was structured to provide great flexibility and a program routing in direct relation to the students' information level. The programming was structured to provide options to candidates of all educational levels from Ph.D.'s to those who didn't graduate from high school -- the lower the level the longer and more detailed the "teach" routine. Passage to the next section was accomplished by passing a multiple choice computer-operated test sequence.

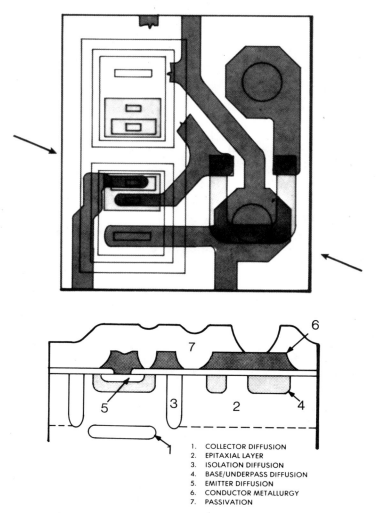

1. COLLECTOR DIFFUSION
2. EPITAXIAL LAYER
3. ISOLATION DIFFUSION
4. BASE/UNDERPASS DIFFUSION
5. EMITTER DIFFUSION
6. CONDUCTOR METALLURGY
7. PASSIVATION

Figure 3. Plan and cross-section of the bipolar teaching model.

Further refinement of the educational work product has resulted in a library of slides, and experienced speakers capable of addressing diverse audiences.[33] Additionally similar planning applied to higher level education has also produced the LSI Institute, a cooperative Masters program in Electrical Engineering developed for mature employees with the collaboration of the nearby University of Vermont.

3. What is the estimated life of the product? Even with the use of standardized basic machines and components, minimum product life should be 4 to 5 years to get payback on an assembly machine. Short lived products are not readily automated.

4. Specifications -- A close look should be taken at product specifications to see if they have to be modified for operation of the proposed assembly machine.

5. Quality requirements of the end product -- The quality people should be aware that no machine builder can do more than put together the parts provided or integrate processes or tests. The end product is the company's responsibility.

THE ASSEMBLY

1. How many parts are in the assembly and what is the material composition of each? This information is important to the machinery builder. It enables him to check drawing lists vs. parts. The material composition can make a difference in hopper feeding and handling. *(Figure 3)*.

2. What technical problem areas can be recognized in the project? If there have been problems assembling the parts manually in the past, then anticipate these problems in a new assembly system. This information should be passed along to the builder.

3. Will an adequate number of parts be available for production proving and hopperability studies? It is very important that parts are available in quantities. Lateness in this area can cause problems in delivery and impair the success of a machine. A sufficient quantity of debug parts helps insure smooth installation and start-up. Management should also be aware of the large number of parts that will be needed. In cases of high individual parts cost, this can be very expensive.

4. How do production proving parts differ from theoretical drawings? The builder will need samples of the extremes that can be expected as well as the average run of parts. Now is the time to be critical, so that anticipated problems with parts quality can be revealed.

5. Have plans been made for handling parts to and from the assembly machine? Material flow should be worked out to make sure there will be no stoppages either to or from the assembly machine.

6. Assembly/drawings -- It is essential that the drawings clearly identify any sacred surfaces, tolerances, burr conditions, fixture and gage points, etc.

PRODUCTION NEEDS

1. How is the product currently processed? Sequence, production rates, manpower, methods, shifts, etc., should be included.

NUMBER AND COMPOSITION OF PARTS

The builder of an automatic assembly machine needs a great deal of preliminary information, including a detailed description of the assembly. Full knowledge of the different sizes, shapes and materials helps the designer solve many difficult problems early in the game, when problem-solving is most economical. A good example is this refrigeration valve-in-receiver. It contains 27 parts, some of which are delicate and hard to handle. In addition to the major components, there are large-diameter thin-bodied O-rings, a delicate filter screen, a spring-loaded valve subassembly a slight glass with washer and retainer, plus an assortment of seals, plugs and fasteners.

Figure 3. Powerfloat, Valve-In-Receiver Assembly

2. What production rate will be required for the new product? Net, gross, shifts, etc. Continuing volume is a must for automatic assembly success.

3. What are the apparent justifications of the program? Do cost savings include manpower savings? The various direct, indirect and support people? Will indirect labor have to be added as additional maintenance help may become necessary? Dollar per man, shift, year, current rates and anticipated rates should all be noted.

4. An estimate should be made of the amount of dollars the project can reasonably expect in funding.

5. What is the timetable? When will the quote be due? When will the order be placed? Delivery? Start-up? All of this should be in line with the original implementation schedule.

PROCESSES

1. What processes are involved in the project beyong the normal parts handling, loading and checking? Example: gaging, testing, sorting, force or torque control monitoring, welding, inspecting, machining, staking, etc.

2. Have these individual processes been investigated? Have you worked with any vendors on these processes? Frequently, assembly builders are asked to become experts in areas in which they have no expertise. Asking assembly builders to become experts in specialized process areas may impair the project. The customer can help a great deal in the processing area by making his own expertise available to the assembly builder.

3. Occasionally, a "proof of principle" on a process or station becomes necessary. One particular operation may determine the success of the entire machine. In cases like this, a "proof of principle" contract should be issued to prove the operation of process. Two examples are illustrated and described here (*Figure 4 & Figure 5*).

CONTROLS

Depending on requirements, the machine controls may involve everything from simple relays to programmable controllers. The latter may be used in conjunction with computers to perform mathematical computations for selective assembly, or data gathering for maintenance and production information. This control area has changed more than any other in the last five years.

The success of the assembly machine from a technical standpoint can be tied into the control system. Perhaps the use of programmable controllers has been overdone on some assembly machines in the last few years. Nevertheless, for involved assemblies and systems, the advantages of programmable controllers are becoming increasingly evident. The controller route calls for in-house expertise to make sure controller advantages will contribute to the success of the system. (*Figure 6*).

PROOF OF PRINCIPLE

When the success of an entire system hangs on the efficient performance fo one particular operation, it may be necessary to verify the process with a "proof of principle" demonstration. Case in point: The go ahead to proceed with design and build of an extensive system for this 3-cylinder, 6-piston compressor was given only after automatic closing of the final assembly was verified. The operation was accomplished by using a special fixture, and synchronizing a two-stage hydraulic motion with the final pin press.

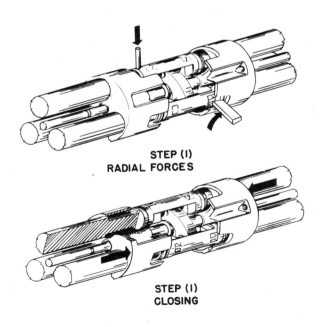

STEP (I)
RADIAL FORCES

STEP (I)
CLOSING

Figure 4. Transferline, R-4 Compressor Pump Assembly

PROOF OF PRINCIPLE

Occasionally a process is involved that calls for proof of principle in a series of separate operations. Such was the case in assembling a computer keyboard switch that contained this swing plate and connector. The fine copper wire is .006 of an inch in diameter, yet it must be fed into the station from a spool, then formed, cut to a precise length and welded to the swing plate at a rate of 1,440 completed parts per hour. When this was verified, another demonstration was required to prove that the tiny wire could be welded to a pin in the switch housing on the main line at a sustained production rate of 1,125 pph. The overall system represents an advance in capability that was not considered feasible a few years ago.

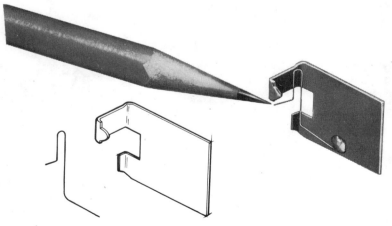

Figure 5. Powerfloat/250 Indexomatic, Keyboard Switch Assembly

PROGRAMMABLE CONTROLLERS

As automatic assembly machines become more sophisticated, mechanical control systems are giving way to computers and programmable controllers. Assemblies with a large number of parts calling for memory or test sequences with data collection lend themselves more readily to the use of controllers. Typical is the system shown here, wherin difficult assemble-and-test capability is combined with the operational convenience of a programmable controller. This installation contains a magnetic memory that monitors every operation in assembling and testing a variety of master brake cylinders. A reliable test for functional performance of the assembly, with subsequent segregation of rejects, provides another check point for quality control.

Figure 6. Powerfloat/250 Indexomatic, Master Brake Cylinder Assembly

BASIC CONTROL QUESTIONS

1. Are there any particular control systems to be specified? Does the company have experience with these control manufacturers?

2. How about costs? Control features for memory, counters, track switches, bowl switches, operator switches, etc., can be expensive. Perhaps they should be options on the machine.

3. Has complexity of the assembly been considered? Number of parts, several assembly modules, or will testing and measuring be involved? A large number of parts requiring memory or test sequences with data collection will lend themselves to controllers.

4. Does the company have a similar process operating satisfactorily with relays? If so, they should stay with the relays. Programmable controllers are usually not a cost effective way of just replacing relays.

5. Remember that relays provide a lot of muscle but little intelligence per buck ($), while programmable controllers provide a lot of intelligence but little muscle per buck ($).

6. Is the machine cycle less than 2 seconds? Seriously consider programmable controllers for increased reliability. A major automotive company, as an example, requires solid state control on all machines operating two million or more cycles.

7. Programmable controllers are easy to re-sequence (re-program). Is the process absolutely set or will changes be required? If changes can be expected, a programmable controller is advisable.

SUPPLIER SELECTION

Next on the list is selection of a supplier. First, it should be understood that there is no such thing as a cheap assembly machine. It costs money, not only in original purchase price, but in time and money spent in the buyer's plant. Supplier selection is very important. Above all, the supplier selected should have the ability to stand behind the machine he delivers. Complete involvement by the assembly builder, both technically and financially, is a must to insure a successful machine. There is a large number of competent builders in the assembly industry today. They are successful because. . .

1. They have the technical ability to understand assembly problems and build a machine to handle those problems.

2. They get involved. The builder's engineer -- the one who handles design and construction of the machine -- will put a little bit of his "life blood" in it. Also assigned to the project is a leadman or floorman who contributes considerable experience in building assembly equipment.

3. They have management ability, plus the financial stability to build a piece of equipment and stand behind it until it works, even though the problems involved may not have been all their's.

It is not advisable to ask too many assembly builders for quotations, as evaluating each supplier's individual quote can be rather complex. Everyone has his own ideas about how a specific part should be handled.

If all bidders are equal (they often are in the eyes of the Purchasing Department) all one has to do is select the lowest bidder and place the order, right? Unfortunately, everyone knows this is not true. Each one of the vendors who has been asked to quote on the assembly project will come in with a little different type of machine, as influenced by background and experience in the field of automatic assembly. There are large ones, small ones and those in between. Each one has expertise in certain areas and perhaps little or none in others. Keeping in mind the ultimate success of the machine, suppliers should be evaluated in terms of individual requirements of the user.

1. Is the assembly simple so that anyone can build it? True experience in building even an uncomplicated machine is hard to find.

2. Is the builder experienced in building a machine of the complexity required for the assembly?

3. What are the in-house capabilities of the supplier?

4. Will the builder do the engineering himself or will he farm it out?

5. Will a competent engineer be assigned to the project?

6. Do they have in-house control experience?

7. What is their track record in assembly? Check with others who have purchased from them.

8. Very important, service. Without good service no machine can be successful. Check with others to see how the prospective builder sticks with a problem.

If these questions have not been asked prior to going out for quotes, make sure they are brought up prior to order placement.

INQUIRIES AND APPROVALS

An inquiry should not simply say build a machine to assemble a widget. It should take into consideration all of the product information that has been gathered, and this information should be made available to the prospective vendors. The sequence worked out for the assembly in engineering should be detailed on the inquiry so that the prospective suppliers know exactly what is required. They may not agree completely with what was asked for, or they may disagree with the sequence, but it gives them a starting point from which recommendations can be made on how the parts can best be handled. It is a good idea to ask for ballpark quotes prior to going firm. This provides a better grasp of what is available before you finalize your thinking. It also gives one a general idea of the dollars involved.

In many companies "approval" is another word for internal selling, and selling is often necessary to get a program going. The procedures for approval to start a project differ from one company to another, mostly depending on internal accounting procedures. Final approval usually comes after the quotes are in. Payback is one answer. New projects with good projected marketing volumes go first, no matter what engineering wants. Marginal projects with some bench automation may go second.

All pertinent facts and information must be covered in your presentation to management. The time it takes for approval is the largest variable in any program implementation. In cost reduction programs re-quote time goes up as justifications change the original program.

PURCHASE ORDER PLACEMENT

The time from selection of the vendor to order placement is very important. Even if all of the various aspects that the builder must follow have been spelled out in the request for quotation, the success of this machine can very much depend on what happens between builder and buyer just prior to issuance of the purchase order. It is suggested at this point that a meeting be called between all of the interested parties to go over various aspects of the request for quotation.

A detailed review of the vendor's reply may show misunderstandings about what he has proposed to build and what is really wanted. This will involve not only how each individual part is to be handled, but the best method of handling the part in terms of mechanical or air control, in order to make the cycle times that have been requested. Even such seemingly obvious things as cycle time should be discussed in detail.

This meeting also will give the customer's people a better opportunity to get acquainted with the builder. If it appears that he has obvious weaknesses in any particular area, now is the time to talk to him about it, so the machine he builds will be exactly the one that is needed.

At this meeting questions may surface that call for compromises to assure the success of the assembly machine. Some of the things wanted on the machine may not be possible. Perhaps it is a technical problem, or perhaps what was asked for will cost more money than the results of that particular operation can justify. Remember, at this point it is easy to add and subtract from the machine, and the time spent in this pre-award analysis will be returned tenfold later on.

At this time it is recommended that top company management visit the prospective builder so that they are aware of who has been selected and why. They should become acquainted with the facilities of the builder they are going to deal with. Let company management talk to the prospective builder management so they will have confidence in who was recommended. This is no time to run a solo. Management backing is needed to insure a successful installation.

DESIGN/BUILD PHASE

As the building goes forward, prime technical responsibility should rest with two people: one person within the customer organization such as a manufacturing engineer who is technically proficient in the product, and the vendor's project engineer who is responsible for building the machine. All technical information should be funnelled through these two responsible individuals, rather than having a lot of people get into the act. If possible, the same people should be kept on the job. Changing of personnel very frequently means a change of ideas. Sometimes these new ideas are for change only.

During the building phase of the machine the manufacturing engineer may find himself called upon to do more than just look at drawings and keep track of building progress. Remember, the builder does not have expertise in the function of the product, and he will be asking questions along these lines. If he is obviously heading in a direction that is unworkable due to the function of the product it is the responsibility of the manufacturing engineer to turn him around and guide him in the right direction. His responsibility as project engineer is to build a machine that will handle and assemble the product at the rates specified. Problems with function of the part itself can be answered only by the customer. Occasionally at this point a compromise may be necessary to assure the success of individual stations on the machine. An unknown may have entered the picture that neither the company nor the builder were aware of in regard to the parts themselves.

A change in direction may be necessary. Perhaps a station has to be split up, another check or a manual operation added in order to assure success. We all know from experience that these compromises are not easy but they are nevertheless necessary. When possible, these compromises or technical changes should be made at the level of the people involved, as long as it does not disturb final output of the machine itself. People who are not familiar with the actual building of the machine may not be aware of all the discussions that have gone on before. A decision can be made on purely emotional basis that could jeopardize success of the machine once it is in the plant. I believe that all who have had experience in this area will agree that technical changes on the machine are best left to those who understand them.

ACCEPTANCE

Prior to acceptance of the machine, plant visits to the builder should be arranged for the people who will be responsible for the machine's production. They should be made to feel that they have a part in building the machine. It is sometimes advisable to have the actual production workers come to the builder's plant during the final production proving stage. These visits should not be made too early, since the builder during his initial stages of production proving or debug may be experiencing problems that could cause needless anxiety for the production people.

The big day arrives -- The builder is ready to demonstrate on his floor the agreed-upon rate of production for the new assembly machine. Who is present at this acceptance is a matter of internal requirements. Normally, the acceptance team is made up of the manufacturing engineer, project manager, engineering manager, quality control, the controls man if necessary, and production management, or any combination.

Again, it depends on internal requirements. If the builder has done his homework and has been supplied with a sufficient quantity of parts to debug, and has made his own internal run, this final acceptance should be relatively easy. (Frequently this internal run is longer and tougher than the final acceptance run.) He will demonstrate over a given time period the successful production rate of the machine, and that all requirements of the contract have been met. If you have acceptance runs in your own plant, we suggest that this same team be the key people involved.

INSTALLATION AND FOLLOW-UP

Normal installation practice in the assembly industry is for the builder to send the leadman who is intimately familiar with the machine, having been with it during its entire life on the builder's floor. He will also assist in the start-up. The builder's project engineer, who has been involved with the machine from its first days on the design board, will be present too. They are often accompanied by a controls man, particularly on some of the more complicated machines.

It is strongly recommended that, in addition to the customer's manufacturing engineer who has been following the project all along, an hourly man be assigned to the machine to assist in the set-up. This does not refer to the operator but to someone who is familiar with the workings of the plant. This will help speed up the entire installation, as nothing is more frustrating to the builder than having his personnel in an unfamiliar plant trying to figure out how to get things done. An excellent example of proper installation and consistent follow-up is shown on the last photo (*Figure 7*).

Quite a bit of ground has been covered in describing the design and construction of a successful machine, but the main ingredients of success are the ones mentioned at the beginning of this article.

It is involvement of everyone from the manufacturing engineer to top management, and the involvement of the operating people whose responsibility it is to run the machine, with the complete understanding and cooperation of both builder and user.

There has to be integrity on the part of both parties. Part of success is both wanting it to go. Without this, no machine can operate successfully or give the type of positive figure everyone is looking for on the southeast corner of the balance sheet.

FINAL INSTALLATION

The last requirement for successful automatic assembly is proper installation, operation and service. This is most often accomplished when responsibility for performance is assigned to one capable individual who cares about the machine. A good example is this installation which began turning out pen cap assemblies in 1960. At that time the machine was assigned a "father" who supervised performance for the next 16 years. It has been very usccessful, producing 200 million satisfactory assemblies for a major pen manufactuere, and it's still going strong. As in this case, one "father" can be responsible for other assembly machines, thereby gaining the benefit of his experience for additional operations.

Figure 7. Transferline, Pen Cap Assembly

CHAPTER 5

Basic Systems Concepts for Automated Assembly.

Selections of a basic system concept is the most important step in the actual process of automating an assembly operation. A broad variety of solutions is available, ranging in production rate from automobile underbodies at 60 per hour to spark-plug assembly at 24,000 per hour; ranging in size from instruments and watch assemblies to complete automobile bodies. A general outline of the available concepts is shown in *Figure 1*.

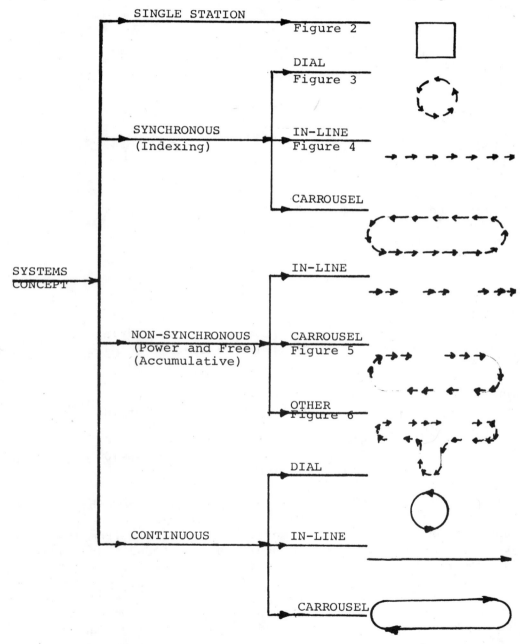

Figure 1. Basic Manufacturing Systems Concept.

This arbitrary breakdown of assembly systems into four types and ten sub-categories is meant only as a general guide to currently popular solutions to assembly problems. It by no means covers all possible methods. Combinations of these basic systems are sometimes used. It is possible that a system could be built in which the automated stations are brought to a work-piece much as a turret lathe or turret mill presents the tooling to a piece. We will not go into such variations but will confine ourselves to more commonly used solutions.

Many factors determine the optimum choice of these systems. Some of these factors are:

1. Production rate.
2. Physical size and weight of work-piece.
3. Manual operations involved.
4. Number of automated operations involved.
5. Complexity of operations.
6. Material handling and supply logistics.
7. Desired material flow.
8. Line balance of operations.
9. Tooling concept.
10. Motive power.
11. Controls.
12. Repair and off-line operations.

First we will look at examples of each of the basic concepts and then discuss the factors which govern their selection.

SINGLE STATION CONCEPT.

This consists of a single work station, (which may perform one or more work functions), to which are brought two or more parts of the assembly to be joined and fastened together or otherwise altered to complete an assembly unit.

A number of different operations may be performed providing the tooling does not get excessively complicated or difficult to maintain.

More often, the single station concept finds its greatest value in performing a given function many times per part: be it assembly, machining, inspection, marking or other such operations.

Shown in *Figure 2* is a single station machine which assembles 31 vanes, turbine or pump, (*Figure 3*) into automotive transmission converters in less than 30 seconds. Housings, or shells, arrive on conveyors from the right and are automatically loaded into the machine and unloaded when completed. The vanes are manually loaded in the magazine on the left and automatically placed into the shell or housing which is indexed in a rotary manner to present the proper slots to the machine. These machines are also incorporated into both synchronous and non-synchronous systems in which multiples of the blade loading machines are combined with tab rolling, machining, balancing, etc.

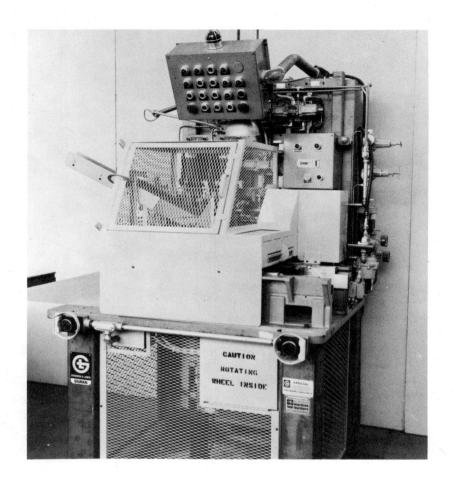

Figure 2.

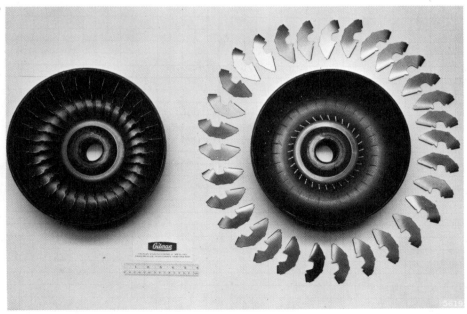

Figure 3.

Figure 4 shows another single station machine assembling folding rule parts (*Figure 5*).

Fig. 4 and 5

1,000 wooden folding rules, composed of 23 parts each, are assembled every hour on this single-station assembly machine. Built around a standard press brake, one oscillating box type of parts selector orients and feeds the tracks from above supplying 11 hinge clips to be accurately positioned in the die set. The 12 thin wooden rule segments, preprinted, are manually loaded in an oriented position to 12 magazines in the foreground. In a little over 3 seconds, the clips are loaded to the die, the segments placed into accurate positioning over the clips, the press brake crimps the clips and the finished, unfolded wood rule is ejected to a conveyor leading to a final inspection station.

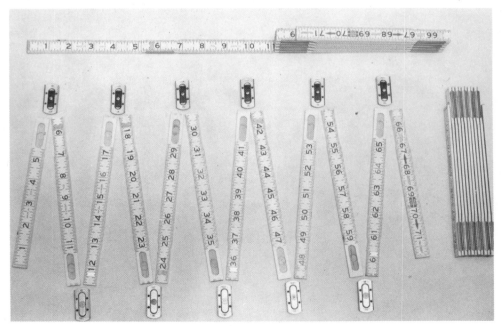

Figure 5.

Still another example of the single station assembly machine is shown in *Figure 6*. In this example, three sheet metal parts are loaded to a fixture which is positioned for a laser welding head which then, under computer control, automatically welds three transverse seams plus other areas required to make a complete assembly.

Figure 6.

Unique in concept, size and flexibility is the laser welding system as tooled for welding a three piece automobile underbody. This replaces spot welding with sealant for water-tightness, with a continuous weld.

This system is a single-station, hydraulically powered, computer controlled machine positioning parts with $\pm.005$ inches under the focussed 0.25 spot of laser beam from a 6 KW, CO continuous laser. The laser generator is located on the platform in the upper left of the photo while the associated vacuum pumps, blowers, heat exchangers, etc., are under the platform.

SYNCHRONOUS DIAL CONCEPT.

The synchronous dial consists of a circular fixture plate which contains a multiple of fixtures or nests uniformly spaced around the periphery. At pre-determined time intervals, the plate rotates a fraction of one turn, progressively presenting each fixture to each of a number of stations.

The number of fixtures, or stations, is dependent upon the number of operations required. Usually available in four, six, eight, twelve, sixteen, twenty-four, thirty-two stations, etc. Such a device may also be used with an odd number of fixtures, double-indexed, when it is desirable to have a work-piece pass a manual operational station twice in the course of its assembly.

Fixtures may be single or multiple tooled to obtain higher production from the same indexing cycle time.

The indexing cycle time may be controlled through some uniform timing device, often used to pace an operator, or the index may be controlled on a "demand-type of basis" in which all working stations must complete their function before the dial will index. In the latter case, varying time cycles of individual stations can be accommodated. In the latter case, also, it is necessary to provide any manual operator working on the line with palm buttons to initiate the index for safety purposes.

The dial index machine (See *Figures 7* and *8*) offers many advantages such as the minimum floor space required, minimum number of fixtures, high speed operation and a high level of standardization among the available bases and tooling components. A wide variety of standard drives, standard work stations, controls, etc., may be acquired either as complete machines or for in-house machine construction.

Limitations of the dial index include physical size, (most dials range from 6 to 200 inches in diameter), torque required to index fixtures and parts, available space for part loading and such auxiliary equipment as hopper feeders.

Speeds of up to 3,600+ parts per hour (more with multiple tooling), parts ranging up to 18 inches in diameter, and six to ten operations are common. Since it is a synchronous machine, all operations are limited to the slowest function. Manual operations may be included with the machine pacing the man.

Dial index units may be tied together with transfer devices or attached to other systems such as auxiliary, sub-operation units for sub-assembly, inspection, machining, etc.

Additional types of drives are shown in *Figure 9* and *Figure 10*.

Figure 11 and *Figure 12* illustrate variations in the use of dial index tables and mechanisms.

Figure 7. Synchronous dial index machine consisting of a machine base housing motor drive and operating cams, index table with rotary indexing fixture plate, one or two vertically oscillating tooling plates, hopper feeder, operating tooling and operator pushbutton control.

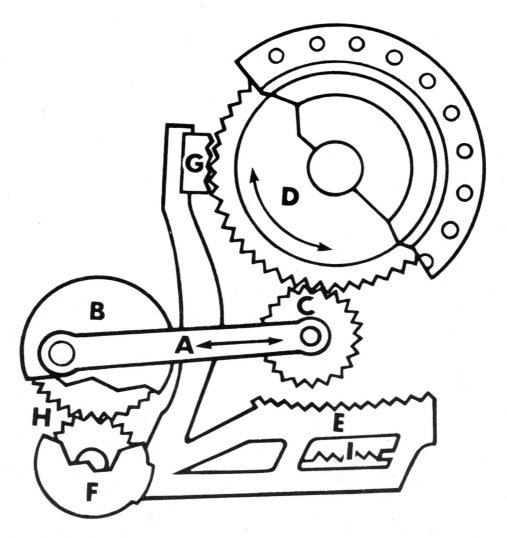

Figure 8. The indexing mechanism for the dial table in Figure 7 works as follows: When unit is operating, connecting rod A maintains a constant back-and-forth motion, driven by eccentric crank B. During dwell cycle, (illustrated), pinion C rides freely across index gear D, which is securely locked in position by spring-loaded rack G.

Indexing results when cam F simultaneously releases rack G from the index gear and causes spring-controlled rack E to engage with pinion C.

The stroke of the connecting rod assures precisely controlled indexing force and rate that moves the table into its next position! Spiroid gear H supplies minimum backlash drive force for all motions. System operates in either direction.

Overload protection is assured as the adjustable, spring-loaded detent I responds instantly to overloads or jam-ups to disengage the indexing mechanism. Since it is located between the input drive and the actual point of force application to the table, drive forces do not amplify jam-up forces.

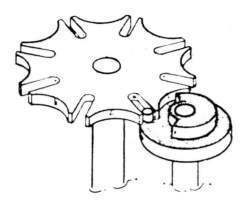

Figure 9. The Geneva Index motion is one of the oldest and most common indexing devices and is shown in its older version in which the pin, or cam follower in the rotating wheel engages and turns the Geneva plate. During the dwell cycle the three-quarter round acts as a "shot-pin" in the radius in the plate O.D. Many variations of this design are manufactured.

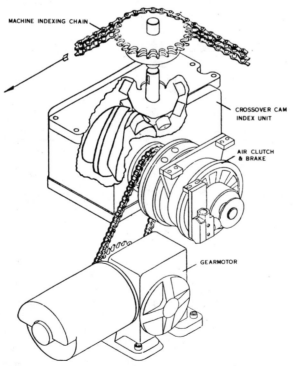

Figure 10. The Barrel-cam indexing mechanism may be used directly on a dial table or as a means of moving a chain on an in-line indexing machine as shown. Smooth operation, controlled acceleration and decelleration are among the advantages of this drive.

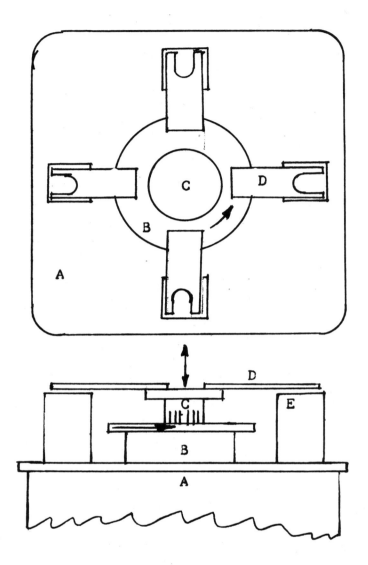

Figure 11. An unusual application of a tooling plate dial index machine is illustrated above as a "lift-and-carry" device for transporting parts from station-to-station where they must be placed down into nests or fixtures.

Mounted to the machine base (A) are four fixtures (E) and one four-station dial index table (B). The vertically oscillating tooling plate post (C) is splined to the index table plate and carries four arms (D). In operation the cam operated center post lifts the parts out of the nests and dwells. The index table (B) turns 90° rotating the fingers (D). The center post is cammed down and the parts are deposited in the next nest for further work.

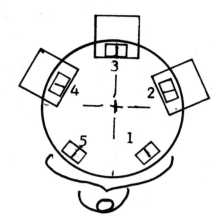

Figure 12. A unique application of a 5 station, double-indexed dial table presents the same fixture to one operator twice in the same assembly cycle. A hypothetical example of such a sequence would be:

Sta. 1 Manually unload assembly; load housing.
    3 Auto load a gear.
    5 Manually load a cover.
    2 Auto load a shaft.
    4 Auto spin rivet both shaft ends.

The same assembly on an in-house would require 3 men:

1. Manually load housing.
2. Auto load gear.
3. Manually load cover.
4. Auto load shaft.
5. Auto spin rivet.
6. Manually unload.

SYNCHRONOUS IN-LINE CONCEPT.

Synchronous in-line machines consist of a series of fixtures or holding devices attached to chain belts or steel belts or moved by fingers from one work station to the next in a straight line motion. All fixtures are moved at the same time and over the same distance resulting in all fixtures sequencially passing all stations in a straight line. Indexing intervals may be controlled by timers or they may be controlled on a "demand type basis" indexing only when all stations have completed their individual functions.

In-line indexing machines which are fixtured usually return the fixture in the conveyor base below the assembly area. Parallel return lines may be used. If the part serves as its own fixture, it may not be necessary to return any part of the machine as in a "lift and carry" or "walking beam."

In-line indexing mechanisms offer the advantages of more operations than are practical in a dial, the speed and accuracy of the index principle and the "in-one-end-out-the-other" principle of good work flow. A basic "over-and-under" slat conveyor is shown in *Figure 13*. If the work piece can serve as its own fixture, no pallet return is required. If fixtures are necessary, they may be returned underneath as in *Figures 14* and *15* or on a parallel track as in *Figure 16* which makes an automotive odometer assembly (*Figure 17*) at 1,333 per hour.

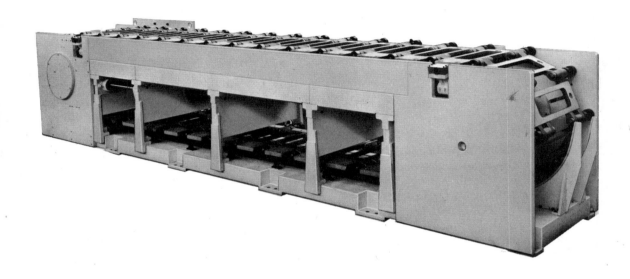

Figure 13.

Limitations of this concept include the accumulative inefficiency of many stations (usually limited to 15 to 20 working stations on hopper fed equipment) and the necessity for the slowest operation pacing the line.

Speeds up to 1,500 assemblies per hour are common with more possible through multiple tooling. Manual operations may be added with the machine pacing the man.

Physical sizes range upward to complete car bodies as shown in *Figure 18*. An entire "body in white" is assembled, critical dimensions established and approximately 600 spot welds made by nineteen robots and eighty-eight fixed guns every cycle, sixty cycles per hour.

Indexing mechanisms include air and hydraulic cylinders, Geneva motions, barrel cams, and hydro-static drives.

Figure 14. Automotive front suspension arms are assembled on this "over-and-under," fixtured, in-line indexing machine. The fixtures indexed from right to left on the top of the machine are inverted by the "ferris-wheel" mechanism at the left of the photo and are returned to the front of the machine below in the machine base. Note the cam of the machine below in the machine base. Note the cam shaft which extends along both sides of the machine below the operator walk-way to actuate the various stations.

Figure 15. Close-up of an assembly on a shot-pinned fixture. Mounting bushings are pressed into the assembly at one end and ball joint assemblies are riveted into the other end at 600 assemblies per hour.

Figure 16. Another in-line indexing machine configuration in which the fixtures are returned along a parallel path. This permits manual operations to be performed along the front of the machine while automotic stations are located along the back path. Note the cam shaft below and the memory and indicating panels above. This machine assembles automotive instrument panel odometers at 1333/hour on a 2.7 second cycle.

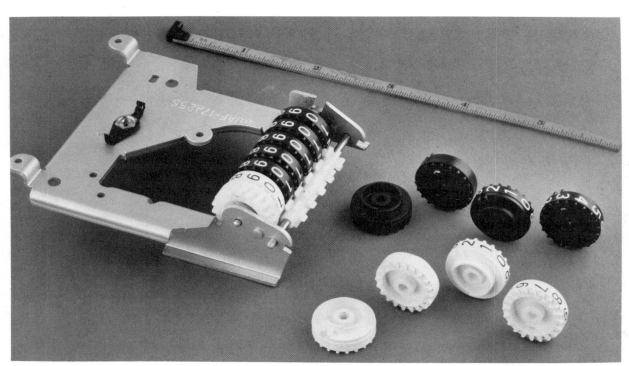

Figure 17.

Figure 18.

SYNCHRONOUS CARROUSEL CONCEPT.

Like the synchronous in-line concept, these machines consist of a series of fixtures or holding devices attached to chain or steel belts, or moved by fingers from one work station to another. However, the carrousel machine moves the work in a rectangular path, or some variation of the same, returning the pallets to their starting point, all in a horizontal plane. All parts are indexed at the same time for the same distance on either a timed or demand-type basis.

Advantages of the carrousel include utilization of all the fixtures in the system, since none are returned below; possibility of more operations in the same space; operations may be performed on all sides of the machine; and work pieces are returned to the starting point which is sometimes a plus.

Figure 19 illustrates a typical, untooled carrousel base of the chain driven type. Indexing may be accomplished by Geneva motions, barrel cams, rack and gear, or by a number of other concepts.

Figure 20 shows a tooled carrousel composed of off-the-shelf station modules.

NON-SYNCHRONOUS CONCEPTS.

"Non-synchronous" describes a system in which individual stations may operate independently and without regard to other functions in the system. Work pieces, in the case of fixtureless machines, (such as disc brake rotor

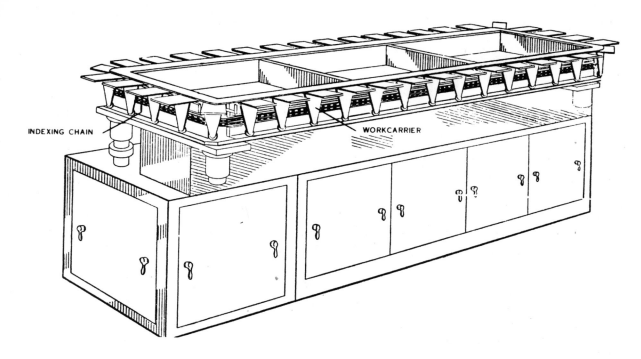

Figure 19.

Figure 20. Synchronous carrousel assembly machine built up from standard components. Indexing base, drive, transfer belt, controls and cam shaft actuations are all standard in a wide range of selection of number of stations, indexing distance, time cycle, etc. Standard stations are added as required for parts pick-and-place, transfer, pressing, etc. The illustrated machine is approximately 95% pre-engineered.

lines and cylinder block lines), or work pallets, (as on a fixtured machine), are free to move independently, powered either by belts, chains, motors or other power sources. Parts or pallets may be stopped individually at any point for manual or automatic work and may accumulate, or "back up" at these stations to allow for variations in station work time. Work pieces or pallets may be removed from the mainstream of the system on to repair loops or other non-scheduled operations and reinserted in the mainstream at the first opening without affecting any of the other parts or stations.

"Power and Free" or "Accumulative" systems, as they are sometimes known, are relatively new on the manufacturing scene as compared to indexing equipment which dates back possibly one hundred years. There are distinct advantages, as well as disadvantagees to this concept.

On the plus side is the possibility of single-tooling the faster operations while multiple-tooling the slower operations with as many stations as necessary on the same machine base. One example is shown in *Figure 21* in which an automotive master brake cylinder is assembled on a double-fixtured power and free carrousel with a basic cycle time of 6.0 seconds to produce 1200 assemblies per hour. Leak testing the part requires multiple tooling in the inspection area. A number of identical stations have the pallets "programmed" and released to permit testing of a number of parts simultaneously on the required, longer cycle time. A number of manual stations are incorporated along with the automatic assembly and functional testing, all computor controlled.

Figure 21.

Figure 22 illustrates the sequence of another master cylinder assembly and testing system, this time an in-line non-synchronous. Five manual operations, six automatic assembly operations, one universal test station and three total function test stations are included. At 450 assemblies per hour line speed, three functional test stations are required.

This highlights the second advantage of the non-synchronous system in which the accumulative characteristics of the independent pallet travel permits "averaging-out" the variations in manual operation time without penalizing the automatic operations.

The third advantage lies in the virtually unlimited number of operations which can be tied together, subject only to mechanical limitations such as drive torque. The addition of automatic repair loops (which divert parts automatically found to be defective through inspection stations) permits manual repair and automatic re-insertion into the system for a "one-in-one-out" capability.

The fourth and perhaps most important advantage of the non-synchronous system is the capability of adding manual standby stations after many--or all--of the automatic stations for use when one goes down. Safety requirements limit the use of such manual standby on synchronous systems. Redundancy in automatic stations becomes more practical in programmable or computer controlled non-synchronous systems. Obviously, considerable additional line length, floor space, and hardware cost are involved.

Many non-synchronous systems today are completely modular. The basic work transfer mechanisms are standard and independent of any operation stations. Completely independent work stations also are standard, ready to be tooled for specific operations. *Figure 23* illustrates such a station module which can be added or removed from a system without interfering with the system operation. Re-tooling for a model or part change or repair of an existing station can be accomplished through station redundancy.

On the negative side of the ledger is the much higher cost of the non-synchronous system. Several fixtures should be provided for each working station to take advantage of the accumulative characteristic; as many fixtures as required to take advantage of anticipated line variances and downtime. This results in greatly increasing line length and floor space usage. Controls costs and complexity multiply. Production rates are usually limited on non-synchronous compared to synchronous, usually down to 600-900 parts per hour. Concepts to increase production rate have usually resulted in very expensive pallets with high maintenance costs, although such devices as "accelerator" bars and indexing devices to move pallets into and out of stations minimize this penalty.

IN-LINE non-synchronous systems offer the "in-one-end-out-the-other" work flow advantages with the disadvantages of space requirements and the unusable pallets which must be returned to the starting point. (See *Figure 24*)

CARROUSEL configurations, such as *Figure 25*, bring the work back to the starting point with the concurrent material handling problems, but utilize all the pallets in less floor space.

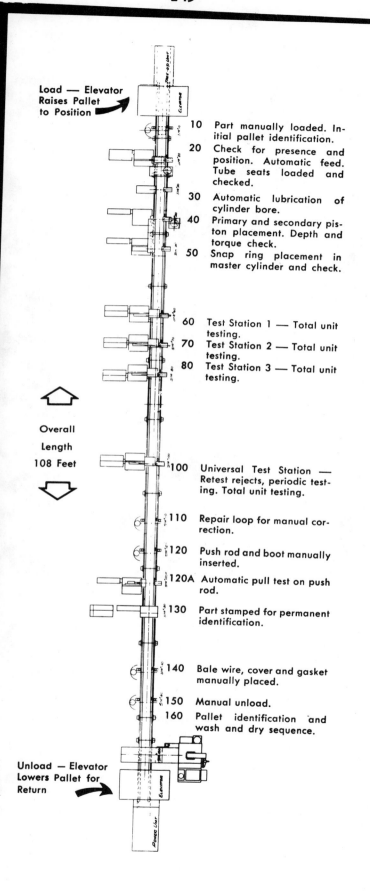

Figure 22.

Figure 23

A self-powered, self-contained modular work station designed to be used with a non-synchronous system. It is usually a complete machine tool in itself, with motor, clutch, brake, cam shaft, controls, etc., merely lacking a fixture and work piece. In the example shown, the module is mounted to the power and free conveyor with the fixtures passing through the opening on the right. Note the round rod which enters a "V" slot in the fixture while a shot pin is actuated by the cylinder on the right to accurately position and hold the fixture during the operation. In this module only a torque operation is performed, but any assembly function can be tooled - or re-tooled - and the module moved up to an existing base.

Figure 24

Figure 25

OTHER CONFIGURATIONS are almost limitless. *Figure 26* shows a "T" shaped power and free base ready for the addition of pallets and work stations. "L" shapes, "U" shapes or any other irregular configuration are available.

MOTIVE POWER is applied to the parts or pallets in non-synchronous systems in a number of different ways. The most common of these are:

1. Electric motors on-board each pallet obtaining low-voltage current from bus bars below the pallet and either stalling out when accumulating or switched off by cam means. *Figure 27.*

2. Clutches on board each pallet which lock-up or un-lock a pallet sprocket in engagement with a continuously running chain below the pallet. *Figure 28.*

3. Drive rollers on each pallet cammed to 0-35° angle in contact with a continuously running round shaft below the pallet permit a variable speed of the pallet and possible controlled acceleration and deceleration. *Figure 29.*

4. Sliding friction between the pallet and drive chains offers inexpensive construction, low maintenance, rapid acceleration, and opens up the space below the pallet for work stations to operate from below. *Figure 30.*

Figure 26.

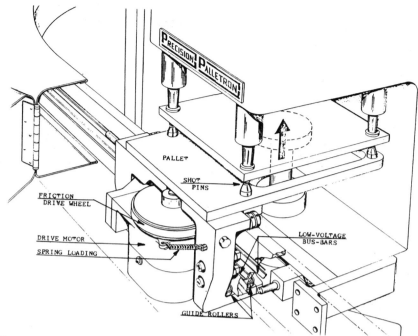

Figure 27. Motorized pallet obtains power from low-voltage bus-bars. Electric motor may be stalled when stopped at a station or by a "back-up" of pallets waiting for a station, or it may be mechanically disconnected from the power source by camming brushes away from bus bars. This system permits "switching" pallets from track to track or placing them in temporary storage or repair loops.

Figure 28. This photo shows a pallet with a spring clutch which either "free-wheels" or "locks up" in engagement with a continuously moving drive chain through a sprocket depending upon whether the clutching lever is free or depressed by contact with a station stop or adjacent pallet.

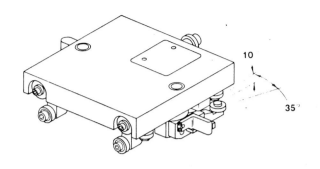

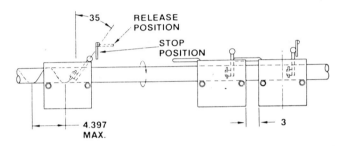

Figure 29. *A variable-speed drive consisting or a rotating shaft the length of the machine engaging a roller attached to each pallet. With the roller set at right angles to the shaft, no pallet motion is impaired. With the roller turned at an angle, forward or reverse motion is imparted to the pallet depending upon the direction in which the roller is turned and the rate of motion governed by the angle. Thus a variable speed, acceleration or deceleration is possible. Since the rotating shaft cannot go around corners, some auxiliary mechanism is required to move the pallets to the return side of the machine base.*

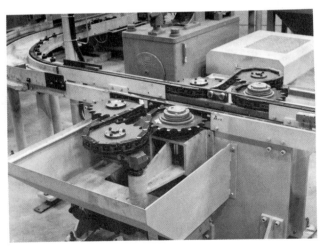

Figure 30. *Typical sliding friction pallet or part conveyor. In this case separate drives are used for each of the two chains and chain take-up is built into the drive. Pallets are guided around curves and idler rollers span the gap where the chain enters the drive. The plastic topped chain shown has proven to last years with two-shift operation at rates to sixty feet per minute. Tops are replaceable molded inserts.*

5. Anti-friction roller idlers carried on a pair of chains engage the lower edges of the pallets to transport parts smoothly. Built-in friction in the rollers can be controlled to offer acceleration at a desired rate while keeping the stopping forces (and safety for operators within bounds. (*Figure 31*). *Figure 32* sketches a typical section of such a conveyor. For any but the very lightest operations, pallets and/or parts are usually lifted off the rollers at working stations. This facilitates shot-pinning for location absorbing press forces, etc.

Many variations of the anti-friction drive have been built. For example, expediting the load and unload of working stations is accomplished with the addition of "accelerator bars." When the station function is completed, and the pallet is lowered back on to the rollers, hardened bars are brought into contact with the rollers at the station. The effect is to cause the rollers to turn and double the speed of the pallet out of the station and bringing the next pallet in. The same feature may be obtained using short, 2-pallet transfer bars to take one pallet away and bring in another.

Figure 31. Typical "manual" station on a frictionless power and free conveyor. Chains on either side carry bearing rollers with sufficient friction to immediately move free pallets, yet with very little resistance will "idle" along under the pallet. Note the clear space below the pallet available for operations from underneath. Note also the air operated stop pin for automatic stop of each pallet and manual release by palm button, (which could be foot treadle or automatic timer). Note also the memory pin, or "flag" in the upper corner of the pallet for use in identifying rejects or other desired information.

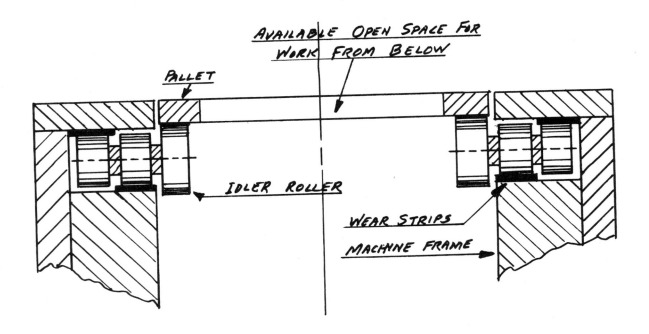

Figure 32.

Another variation on the anti-friction drive is illustrated in Figure 33. This provides positive movement of the pallet at zero to double the chain speed.

Thus, the anti-friction non-synchronous drive can be arranged to offer full power and free at chain speed, locked-in continuous operation at chain speed plus/minus desired variation, or full indexing in areas provided with transfer bars.

Which of the many non-synchronous concepts best suits the individual application will depend upon many factors inherent to the specific application. Generally speaking, the most satisfactory results will be obtained from the simplest, most maintenance-free design consistent with the production requirements. The "state of the art" is evolving rapidly and perhaps the best is yet to come.

CONTINUOUS SYSTEM CONCEPT.

The continuous concept consists of work pieces or work fixtures attached to a chain, belt or other conveyor driving mode which runs without stopping; on a continuous basis. Any work performed whether manual or automatic, is done "on-the-fly" by stations or operators who move with the work piece while performing a function.

Figure 33. "Power Pace" section of frictionless power and free conveyor features a silent chain contacting idler bearings under pressure from below. If the silent chain and the conveyor chain run at the same speed, the pallet will be "locked-up" to the same speed such that an operator is unable to stop the pallet. If the silent chain runs faster or slower than the conveyor chain, the pallet will be "locked-in" to a slower or faster speed than the conveyor chain. (Holding the silent chain stationary will double the pallet speed. Running the silent chain at double the conveyor chain speed would hold the pallet stationary.)

Continuous systems are probably the most widely used types of assembly lines in industry today, primarily for manual assembly in manufacturing operations. Final assembly lines for automobiles, engines (*Figure 34*), transmissions, refrigerators, etc., are almost all continuous though manually tended. Automated continuous lines are widely used in the food, tobacco and beverage industries. See *Figure 35*. Scattered examples of heavy hard goods continuous systems which are automated include some piston machining, engine hot-test and portions of cylinder head assembly. *Figures 36* and *37*.

Several factors limit the development of continuous automated assembly systems. One, station cost is substantially higher than for other concepts. Not only must a station have all the functions involved in a stationary unit, but the entire station must synchronize and move with the work piece. This obviously creates additional expense in construction, engineering and de-bug. Two, such cost would be best justified over many systems rather than one of a kind. Three, the additional mechanism required and the potential for additional malfunctions and failures, compounds the cost problem. Unless enough stations and systems are required with sufficient hourly production requirements per station, the development of such units for hard-goods production is not justified,

Figure 34. A continuous "J" hook assembly line for automobile engines. This photo actually shows a "power and free" which is used essentially on a continuous basis during manual operations and permits banking at intervals. Bracket attached to the part is usually pivoted to present the engine's four sides to the operators and frequently the J-hook is also pivoted for positioning anywhere in 360° rotation.

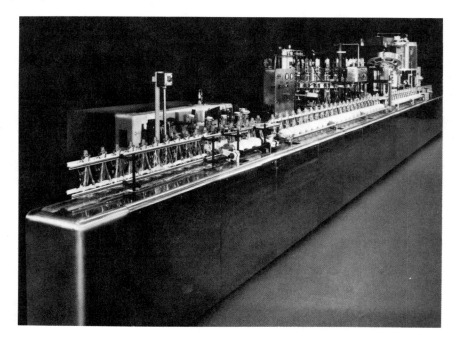

Figure 35. Continuous line consisting of three modules for bottling syrup. Each module removes the bottles from the main line and carries them around a circular table. Module one is an air-operated cleaner. Module two is a filler. Module three is a capper. Production rate is 350-400 parts per minute.

Figure 36 shows a continuous rotary assembly system which produces in excess of 100 assemblies/minute or 6,000 assemblies/hour. This system consists of 6 components as diagrammed in Figure 37.

Item A is an indexing dial rotating about a horizontal axis into which 2 parts are automatically fed and assembled, inverted and discharged into B which is a continuously rotating transfer wheel.

Item C receives the sub-assembly, adds a powder, tamps the powder, repeats this process four times and discharges the parts into the second rotating transfer wheel D where it is again inverted.

Unit E receives the part, automatically loads three more parts plus an adhesive and includes a screw torquing function. Unit F is a final rotating transfer wheel which unloads the completed assembly to a chute.

All the above functions are cam controlled and have inspection operations incorporated. Systems of this complexity and speed are unusual in the hard-goods manufacturing areas. Usually, many identical systems are required to justify the engineering, development and de-bugging costs necessary to achieving the 70-80% efficiency this system reaches.

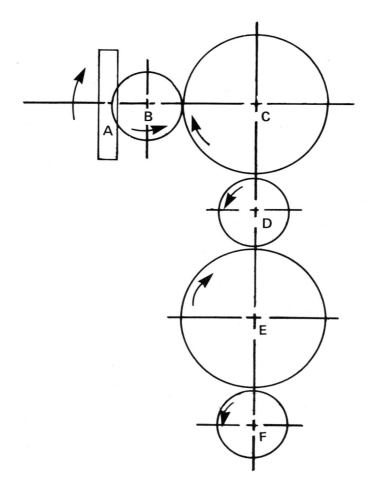

Figure 37.

Advantages of the continuous concept include the very high volume of production possible and the elimination of "stop-and-go" features of the synchronous and non-synchronous systems.

A good example of potential for continuous concept might be an automotive spark plug which is required in enormous volumes. Even so, a continuous system developed for 400 spark plugs per minute (which is entirely practical) results in a potential 48,000,000 spark plugs per year on a one-shift basis. Very few products accumulate this type of volume without excessive change-over for models, sizes, etc. Frequently, it becomes more feasible to have more machines, each at a lower rate, with little or no change-over required.

Continuous manufacturing systems concepts are available now. They only await the specific demand to be implemented.

FACTORS IN SYSTEM CONCEPT SELECTION.

The following are some of the factors involved in the choice of a system. Seldom will any one factor make the final determination. Most often several, or all of the factors, will have an effect on the ultimate choice. None of the factors discussed elsewhere in this book, are included below. The assumption is made that product design has been optimized, that human resources have been considered, and that only the following factors remain.

1. Production rate. *Figure 38* charts the usual and approximate production rates applicable to the various concepts for single tooled systems only. Multiple tooling will multiply the quantities obtainable.

    A production rate of 750 assemblies per hour might conceivably be obtained from any of the systems. Physical size, complexity of operations, number of operations, etc., would all then have to be taken into consideration to determine the most efficient system.

    A single station concept is suggested for quantities up to 1,000 parts per hour, provided the machines does not get too complex with too many operation, or too difficult to maintain.

    The synchronous machines usually operate up to 3,600 assemblies per hour with the number of operations and complexity, as well as the mixture of manual labor determining on whether it goes on a dial, in-line or carrousel.

    The non-synchronous systems are usually limited to under a thousand an hour, single tooled, with other factors determining when they are preferable.

    The continuous systems are applicable in the lower volume rates to larger assemblies such as engines, transmissions, complete automobiles, etc. where the cost and difficulty of automating the continuous line is more readily justified. For completely automated systems of the continuous type, we look, usually, to such items as food and beverage industries, high-speed requirements such as spark plug assembly, watch assembly, etc.

    As a matter of convenience, *Figure 39* is offered to assist in determining net production from systems of varying total efficiencies at varying production rates and cycle times.

2. Physical size and weight of work piece.

    To one degree or another, automation has been applied to a range of assemblies from micro-electronic circuit boards to complete locomotives and aircraft. The problems involved in the selection of system concept range from the accuracy of positioning required for instrument type of assemblies up to the physical equipment required to move such an item as an aircraft.

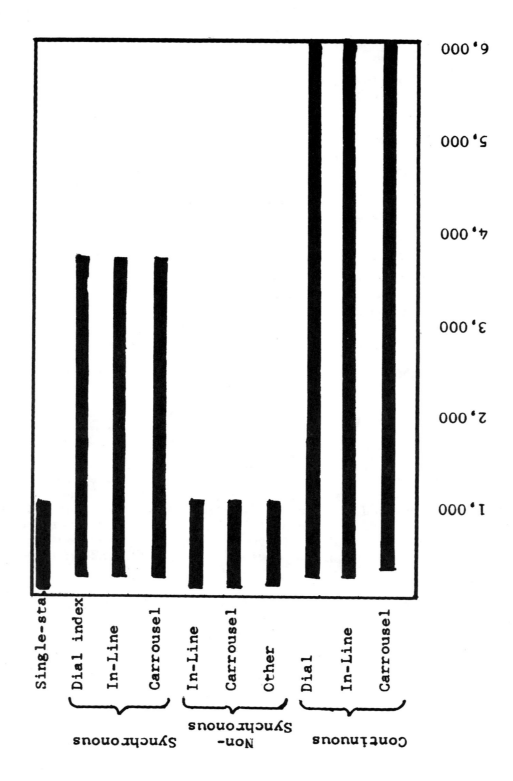

Figure 38. Approximate production rate in assemblies/hour for the principal basic concepts, single-tooled.

P R O D U C T I O N   P E R   H O U R

Cycle Time in Seconds	GROSS 100%	NET @ 85%	NET @ 80%	NET @ 75%	NET @ 70%	Cycle Time in Seconds	GROSS 100%	NET @ 85%	NET @ 80%	NET @ 75%	NET @ 70%
.5	7200	6120	5760	5400	5040	3 1/2	1029	875	823	772	720
.6	6000	5100	4800	4500	4200	3.6	1000	850	800	750	700
.7	5140	4369	4112	3855	3598	3.7	972	826	777	729	680
3/4	4800	4080	3840	3600	3360	3 3/4	960	816	768	720	672
.8	4500	3825	3600	3375	3150	3.8	947	805	757	710	663
.9	4000	3400	3200	3000	2800	3.9	923	784	738	692	646
1.00	3600	3060	2880	2700	2520	4.00	900	765	720	675	630
1.1	3272	2781	2617	2454	2290	4.1	878	746	702	658	614
1.2	3000	2550	2400	2250	2100	4.2	857	728	685	642	599
1 1/4	2880	2448	2304	2160	2016	4 1/4	847	719	677	635	592
1.3	2769	2353	2215	2076	1938	4.3	837	711	669	627	585
1.4	2571	2185	2056	1928	1799	4.4	818	695	654	613	572
1 1/2	2400	2040	1920	1800	1680	4 1/2	800	680	640	600	560
1.6	2250	1912	1800	1687	1575	4.6	782	664	625	586	547
1.7	2117	1800	1694	1588	1482	4.7	765	650	612	573	535
1 3/4	2057	1748	1645	1542	1440	4 3/4	757	643	605	567	529
1.8	2000	1700	1600	1500	1400	4.8	750	637	600	562	525
1.9	1894	1610	1515	1421	1326	4.9	734	623	587	550	513
2.0	1800	1530	1440	1350	1260	5.00	720	612	576	540	504
2.1	1714	1457	1371	1286	1200	5.1	705	599	564	528	493
2.2	1636	1391	1309	1227	1145	5.2	692	588	553	519	484
2 1/4	1600	1360	1280	1200	1120	5 1/4	685	582	548	513	479
2.3	1565	1330	1252	1174	1096	5.3	679	577	543	509	475
2.4	1500	1275	1200	1125	1050	5.4	666	566	532	499	466
2 1/2	1440	1224	1152	1080	1008	5 1/2	654	555	523	490	457
2.6	1385	1177	1108	1039	970	5.6	642	545	513	481	449
2.7	1333	1133	1066	1000	933	5.7	631	536	504	473	441
2 3/4	1309	1112	1047	981	916	5 3/4	626	532	500	469	438
2.8	1285	1092	1028	964	900	5.8	620	527	496	465	434
2.9	1241	1055	993	931	869	5.9	610	518	488	457	427
3.00	1200	1020	960	900	840	6.0	600	510	480	450	420
3.1	1161	987	929	871	813						
3.2	1125	956	900	844	788						
3 1/4	1107	941	886	830	775						
3.3	1091	927	873	818	764						
3.4	1059	900	847	794	741						

Figure 39.

One example of such a choice might be the assembly of an automobile "body in white." Up until this point in time, the body has been assembled in single station units known as "bucks" in which the various sheet metal pieces are assembled and welded together in enormous machines involving many hundreds of spot weld guns, each of which might be "dressed" at least once a shift and sometimes hourly. This results in extensive downtime and labor. In some cases operators may crawl into a complete machine and dress points individually with a file. A more recent solution to the problem has been the indexing system in which the operations are spread out over 10, 12 or more stations resulting in greater accessibility. This example of the comparison between an indexing in-line and a single station has resulted in considerable controversy which only time, technical break-throughs and economic improvements will resolve.

Another example of a comparison between two basic concepts is that of the continuous line versus synchronous and non-synchronous. Automobile engines and transmissions have traditionally been manually assembled on continuous conveyors, usually "J" hooks hanging from a chain conveyor. The difficulty of synchronizing a work-station with a moving conveyor of low precision has inhibited the development of automation for such lines. On the other hand, non-synchronous and synchronous systems should be carefully compared with continuous concepts at such times new assembly lines are being considered rather than being "locked in" to a traditional concept.

Very small assemblies frequently pose the problem of the required precision location. Single station machines offer the easiest solution where there are few operations with the synchronous and non-synchronous machines, in that order, considered for greater number of operations and mixture of manual requirements.

3. Manual operations involved.

    Where production rates, product design, and economics are compatable, it is always desirable to eliminate manual operations completely. Where this cannot be done, or if not economical to do, consideration must be given to the number of manual operations required, their complexity, their probable efficiency, etc.

    For a single station concept this may consist merely of loading and unloading a part.

    On a synchronous machine, a number of manual operations may be involved. The dial concept permits one or two, the in-line offers space for a number of operators, and the carrousel offers still greater opportunity for mixture of men and automation.

    The non-synchronous offers a special inducement for the mixture of manual operations with automatic, in that the variations in manual operating time from part to part may be averaged out for a minimum effort of operator pacing of automatic operations. Furthermore, the non-synchronous will permit having several operators performing the same function while the rest of the automatic line operates on single stations.

The continuous line conventionally has been almost entirely manual on the larger assemblies at lower production rates. The smaller parts with the higher production rates are almost fully automatic.

4. Number of automated operations involved.

Figure 40 suggests an average level of number of automated operations involved by concept.

The single station concept is limited in space and accessibility to a very few automatic operations.

The synchronous dial also has a distinct space and accessibility of limitation. Fewer or greater numbers of automatic operations are possible depending upon station efficiency.

As the number of automated operations goes up, the required efficiency per station assumes critical proportions. *Figure 41* charts the resultant efficiency for various numbers of stations by the average individual station efficiency. It now becomes obvious that individual station efficiencies must average out to the high 90% level to permit more than a half-dozen or so stations to be locked together.

The synchronous systems are, therefore, generally limited to 20 functions or less, depending upon the complexity of the functions being performed. A number of techniques have been devised to boost the efficiency of the synchronous systems, such as repair loops, auditing and repair stations, etc. However, the disadvantages of accumulative inefficiencies and the fact that one station down shuts down an entire system, combine to limit the number of automated operations feasible on one synchronous system.

The non-synchronous system came into being as a partial solution to: (a) mixture of manual operations with automatic and the averaging out effect of the non-synchronous system, (b) possibility of multiple tooling some operations while single tooling the faster operations, (c) possibility of removing parts or pallets in process for repair and reinsertion at any point in the system. Theoretically the accumulative effect of station inefficiencies, resulting in a declining net total efficiency, can be recovered at intervals with only total scrap parts being lost. Depending upon the types of operations involved, virtually any number of operations can be combined in a non-synchronous system.

The continuous system will have the same general limitations as the synchronous inasmuch as all stations are effectively locked together with little chance of recovering lost pieces.

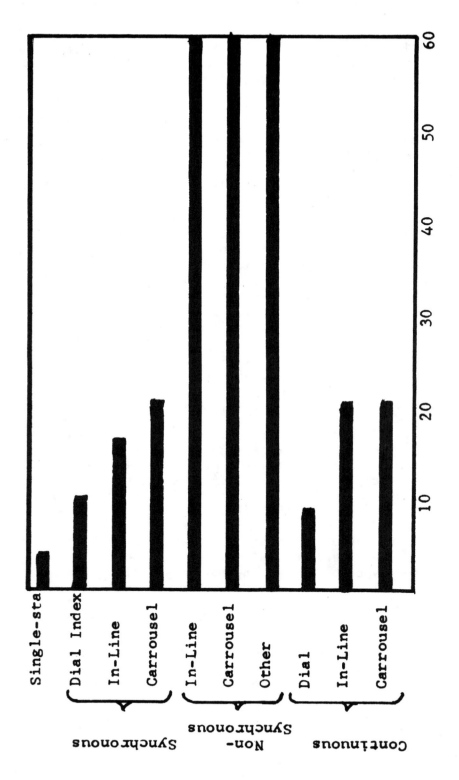

Figure 40. *Approximate number of automated operations which may be used with the principle basic concepts.*

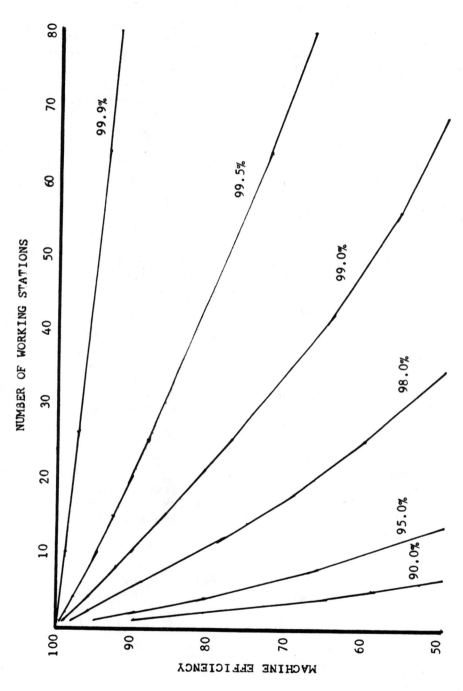

Figure 41 graphs machine efficiency vs. number of working stations at various station efficiencies. This chart assumes a synchronous system in which all stations are "locked" together; one station down—all stations down. Further, the station efficiency curve is based on the product of all the stations' individual efficiencies. Efficiency losses will include the assembly of bad parts due to the machine failure, variable station time, waiting for previous stations, etc.

Since most assembly machines are quoted at something approaching 80% machine efficiency, it becomes obvious that a system with 11 working stations will require a cumulative average station efficiency of 98%. A machine with 22 working stations would require an average station efficiency of 99%.

5. Complexity of operations.

   This factor is a highly individual one depending strictly upon the part being assembled and the functions which must be performed. As previously described, the automobile body in white assembly, may be engineered into a single station body buck concept, into a synchronous line, into a non-synchronous line, or into a continuous line. The laser under-body welding system shown in *Figure 6* was engineered into a single station system to take maximum advantage of a highly expensive laser source. Higher production might indicate more than one station with "beam splitting" the laser beam and sharing an expensive source with more than one station. By the same token, multiple test stations requiring computer integration of data may readily be placed on the same synchronous or non-synchronous system, sharing a common computer.

   Complexity, like the number of automated operations involved, must be carefully weighed against efficiency of the various concepts on an individual basis.

6. Material handling and supply logistics.

   Space and equipment for material handling is sometimes greater than that required for the assembly itself. A typical synchronous assembly involving a dozen hopper fed components of fairly large size, may use up 500 square feet of space for the assembly machine, and 2,000 square feet of space for the material handling and storage prior to and subsequent to the machine itself. The single station laser under-body assembly in *Figure 6*, operating at 60 assemblies per hour, might require as much material handling space as the machine itself.

   A synchronous dial with five or six hopper fed stations encircling it, along with bulk auxiliary hoppers, may become a maze of equipment difficult to supply and maintain.

   The synchronous, non-synchronous in-line equipment, offers the opportunity to spread out such auxiliary units to permit more orderly handling of material and accessibility for operators and maintenance men. Hopper decks, monorail material supply, conveyor storage towers, etc. all are used to simplify and minimize space required for material handling and open up accessibility to the equipment for service.

   Continuous systems will consume incoming material at an even greater rate and should be carefully engineered to insure that log jams do not occur.

7. Desired material flow.

   It is usually desirable to have a flow of assembly material in a straight line from one point in the factory to another. The single station concept offers no problem here.

   The dial concept and the carrousel concepts of synchronous, non-synchronous and continuous, all tend to bring the assembly back to the starting point or, alternatively, leave unused portions of the equipment to obtain a straight line flow.

The in-line concept and "other" configurations offer possibilities of straight line flow from point A to point B.

8. Line balance of operations.

   All single station, synchronous and continuous systems effectively lock together all operations at a constant speed, that of the slower operation involved. This can occasionally be avoided on a synchronous system by having two or more identical operations working on alternate parts brought down a synchronous transfer.

   Other than this two to one or three to one balancing of operations, the non-synchronous systems offer the greatest opportunity to accomplish a line balance of operations.

9. Tooling concept.

   System concept selection will depend, to some extend, upon the desired tooling concept. Whether the part must be fixtured or whether the part can serve as its own fixture will have to be considered in all systems. Fixtures cost original money, they have to be maintained, and they are a source of possible malfunctions, so if they can be dispensed with--either through product design changes to the part, or if the part is adaptable to the concept--there are distinct advantages in doing so.

   If the part is fixtured there may be advantages to coding the pallets mechanically or magnetically. Computer control and memories offer the advantages of keeping track of which fixtures produce the most repair parts, memorize the location of the previous operations by fixture number, and permit off-line and repair operations under controlled conditions.

   Single station and most synchronous and continuous systems have all working stations locked into the cycle. Non-synchronous systems permit the use of completely selfcontained, independent work function stations which may be removed, replaced or retooled without effecting the system operation.

   Multiple tooling and multiple fixturing are possible on all systems. Non-synchronous permit a mixture.

   The assembly in question must be carefully reviewed in the light of the most desirable tooling concepts prior to selection of the basic system.

10. Motive power.

    Basic systems are powered by electric motor, pneumatic motors and hydraulic motors. Electric motors are the most common and probably the most efficient type of drives.

Pneumatic drives are probably the least expensive. Hydraulic drives probably offer the greatest power and force in a given space and cost.

Any of the three may be used directly or through mechanical linkages.

Hydraulic equipment may be engineered to the "hydro cam" concept with a minimum of mechanical connection.

Both pneumatic and hydraulic power are adaptable to explosive atmospheres.

Air power is a relatively expensive source. Hydraulic drives are subject to leakages and temperature variations.

Selection of the motive power is frequently a personal choice dictated by familiarity.

11. Controls.

The usual choice of controls falls between mechanical, electrical, pneumatic and hydraulic.

Mechanical controls are usually in the form of cams which in themselves can eliminate a multiplicity of cylinders, valves, relays and the other usual control paraphernalia subject to malfunction and failure. Cam controls are usually more expensive in the beginning, but have indefinite life.

Electrical controls may be conventional relay type, solid-state type, programmable controller type, or computer control. The degree of complexity of the control desired, the probability of sequence or part changes in the future, and frequently the familiarity with the equipment by the involved personnel will determine the proper approach. Pneumatic controls are the least expensive and are adaptable to explosive atmospheres. Complete logic systems are available. They have the disadvantage of requiring a clean, dry source of air for their operation. Hydraulic controls are capable of operation in explosive atmospheres and generally control more power.

Many times, combinations of the above controls work out to be the optimum answer.

12. Repair and off-line operations.

If the assembly system case-in-point indicates an expected high level of repair operations, in-process repair loops, available in synchronous or non-synchronous systems, offer desirable solutions.

The desirability of off-line operations should be considered when comparing synchronous and non-synchronous concepts.

These are some of the factors involved in the systems selection which will effect an intelligent choice. It is important that they all be considered prior to committing the project to any one of them. It may be necessary to compare more than one concept far down the road before the optimum choice becomes obvious. Traditional methods of manufacturing a given product should not be permitted to influence such considerations if genuine progress is to be made.

PRODUCTIVITY

Productivity is a much used--and misused--term for the comparative output of any processing method whether it be physical, mechanical, mental or whatever. An existing or proposed process must be compared to some previous standard, norm or level.

Industrial production productivity should eventually relate to profits if stability, strength and growth are to be measured. In small organizations the owner or manager will undoubtedly work in such an overall mode. In major corporations where the complexity of such calculations are beyond the level of the manufacturing engineer, the "bean counters," or accountants establish more easily measured standards for the day-to-day control and guidance.

Some of the more common "productivity" measurements are:

    Sales dollars/employee
    Value added/employee
    Sales dollars/man-hour
    Sales dollars/labor dollar
    Output units/potential
    Production hours or minutes/pc or hundred pieces
        (see conversion chart, *Figure 42*)
    Per cent of standards established
    Per cent of optimum production

The last four of these ratios are most commonly used by the manufacturing engineer. He takes the most expensive or least efficient of the operations under his care, (and the two cases are not necessarily synonymous), and works up the list. Management attempts to go with the projects which offer the greatest return-on-investment up to the limit of the available funds for capital investment, (or the diminishing returns). Since the manufacturing engineer usually works to accounting guidelines as to labor cost and R.I.V., we will look more at means of measuring the hardware performance.

"Machine Efficiency" is another much used--and misused--term. It means something different to the equipment user and the equipment builder. For obvious reasons, the machine tool builder looks for a means of measurement of his quality of engineering, manufacturing and de-bugging, (the factors under his control), which will form a basis of acceptance by the buyer and a measurement of his satisfactory completion of a purchasing contract. A conventional equipment quotation will specify acceptance on the builders floor when meeting such conditions as:

    gross production rate over a specified time
    net production rate over a specified time

PRODUCTION RATE CONVERSION CHART

PCS/2000 HR. YEAR MILLIONS	PCS/8 HOUR DAY	PCS PER HOUR	PCS PER MIN.	CYCLE TIME SEC.	MIN. PER PC.	HRS. PER PCS.	PCS/2000 HR. YEAR MILLIONS	PCS/8 HOUR DAY	PCS PER HOUR	PCS PER MIN.	CYCLE TIME SEC.	MIN. PER PC.	HRS. PER PCS.
14,400	57,600	7,200	120.0	0.5	0.0083	.0138	1,846	7,384	923	15.4	3.9	0.0650	.1083
12,000	48,000	6,000	100.0	0.6	0.0010	.0167	1,800	7,200	900	15.0	4.0	0.0667	.1111
10,280	41,120	5,140	85.7	0.7	0.0017	.0195	1,756	7,024	878	14.6	4.1	0.0683	.1139
9,000	36,000	4,500	75.0	0.8	0.0133	.0222	1,714	6,856	857	14.3	4.2	0.0700	.1167
8,000	32,000	4,000	66.6	0.9	0.0150	.0250	1,674	6,696	837	14.0	4.3	0.0717	.1195
7,200	28,800	3,600	60.0	1.0	0.0167	.0278	1,636	6,544	818	13.6	4.4	0.0733	.1222
6,544	26,176	3,272	54.5	1.1	0.0183	.0306	1,600	6,400	800	13.3	4.5	0.0750	.1250
6,000	24,000	3,000	50.0	1.2	0.0200	.0333	1,564	6,256	782	13.0	4.6	0.0766	.1279
5,538	22,152	2,769	46.2	1.3	0.0217	.0361	1,530	6,120	765	12.8	4.7	0.0783	.1307
5,142	20,568	2,571	42.9	1.4	0.0233	.0389	1,500	6,000	750	12.5	4.8	0.0800	.1333
4,800	19,200	2,400	40.0	1.5	0.0250	.0417	1,468	5,872	734	12.2	4.9	0.0817	.1362
4,500	18,000	2,250	37.5	1.6	0.0267	.0444	1,440	5,760	720	12.0	5.0	0.0833	.1389
4,234	16,936	2,117	35.3	1.7	0.0283	.0472	1,410	5,640	705	11.8	5.1	0.0850	.1418
4,000	16,000	2,000	33.3	1.8	0.0300	.0500	1,384	5,536	692	11.5	5.2	0.0867	.1445
3,788	15,152	1,894	31.6	1.9	0.0317	.0528	1,358	5,432	679	11.3	5.3	0.0883	.1473
3,600	14,400	1,800	30.0	2.0	0.0333	.0555	1,332	5,328	666	11.1	5.4	0.0900	.1502
3,428	13,712	1,714	28.6	2.1	0.0350	.0583	1,308	5,232	654	10.9	5.5	0.0917	.1529
3,272	13,088	1,636	27.3	2.2	0.0367	.0611	1,284	5,136	642	10.7	5.6	0.0933	.1558
3,130	12,520	1,565	26.0	2.3	0.0383	.0639	1,262	5,048	631	10.5	5.7	0.0950	.1585
3,000	12,000	1,500	25.0	2.4	0.0400	.0666	1,240	4,960	620	10.3	5.8	0.0967	.1613
2,880	11,520	1,440	24.0	2.5	0.0417	.0694	1,220	4,880	610	10.2	5.9	0.0983	.1639
2,770	11,080	1,385	23.1	2.6	0.0433	.0722	1,200	4,800	600	10.0	6.0	0.1000	.1667
2,666	10,664	1,333	22.2	2.7	0.0450	.0750	1,028	4,112	514	8.6	7.0	0.1167	.1946
2,570	10,280	1,285	21.4	2.8	0.0467	.0778	0.900	3,600	450	7.5	8.0	0.1333	.2222
2,482	9,928	1,241	20.7	2.9	0.0483	.0806	0.800	3,200	400	6.7	9.0	0.1500	.2500
2,400	9,600	1,200	20.0	3.0	0.0500	.0833	0.720	2,880	360	6.0	10.0	0.1667	.2778
2,322	9,288	1,161	19.4	3.1	0.0517	.0861	0.360	1,440	180	3.0	20.0	0.3333	.5556
2,250	9,000	1,125	18.8	3.2	0.0533	.0888	0.240	960	120	2.0	30.0	0.5000	.8333
2,182	8,728	1,091	18.2	3.3	0.0550	.0917	0.180	720	90	1.5	40.0	0.6667	1.1111
2,118	8,472	1,059	17.7	3.4	0.0567	.0944	0.144	576	72	1.2	50.0	0.8333	1.1389
2,058	8,232	1,029	17.2	3.5	0.0583	.0972	0.120	480	60	1.0	60.0	1.0000	.1667
2,000	8,000	1,000	16.7	3.6	0.0600	.1000	0.102	408	51	0.9	70.0	1.1667	.1961
1,944	7,776	972	16.2	3.7	0.0617	.1029	0.090	360	45	0.8	80.0	1.3333	.2222
1,894	7,576	947	15.8	3.8	0.0633	.1056	0.080	320	40	0.7	90.0	1.500	.2500

(These gross production figures must be multiplied by anticipated efficiency to obtain net production)

Figure 42.

functions performed
blueprint specifications to be met

Uusually such a demonstration is "clocked" and an accurate log kept of all conditions. Unnatural conditions or conditions not under the control of the builder are deleted from these demonstrations, such as:

parts not to blueprint
material handling losses where suppliers plant does not have production equipment

Such attempts to isolate "machine efficiency" from "productivity" are necessary to a clean-cut definition of an equipment purchasing contract, but the user has much more to consider. See *Figures 43* and *44*.

The manually operated line usually has the highest labor cost and the lowest equipment cost. The "machine efficiency" is at the mercy of the operators. The conveyors between machines act as surge banks leveling out the variable operator time. Since an operator usually is not as fast as an automatic load/unload and is subject to fatigue, the line rate is lower.

The synchronous "locked-in" line actually drops in machine efficiency from the manual line if we mean parts produced divided by potential output since the individual machine inefficiencies are cumulative. In practice, however, the synchronous line usually runs faster and the net parts produced are greater than the manual line. Even taking into account the higher equipment costs, true "productivity" in terms of dollars/part is usually higher.

Adding surge and repair loops to the synchronous line permits "recovery" of much of the efficiency losses in the individual stations. Short loops of a few parts can account for variations in manual cycle time or variable automatic cycles. Large loops can allow time for correction of malfunctions such as hopper feeder jams, track cleaning, etc. For extreme stoppages, such as press die changes or welder tip dressing, "storage towers" may offer up to an hour or more surge capacity. The additional mechanisms to remove parts from and replace them on the synchronous conveyor do cost more, but the efficiency recovered should much more than replace the efficiency lost when going from manual to synchronous.

Instead of attachments to a synchronous line to increase its net efficiency, the non-synchronous, "power and free" system starts out with engineered banks designed to absorb "less-than-storage-capacity" stoppages. Careful analysis of anticipated malfunctions will indicate the banks required. Alternate paths of part and/or pallet flow offer optional product design possibilities without setup changes. Repair loops at pre-engineered points offer maximum recovery of lost parts.

The continuous system shown also offers surge storage capability, though usually more difficult to utilize at the speeds at which continuous systems operate.

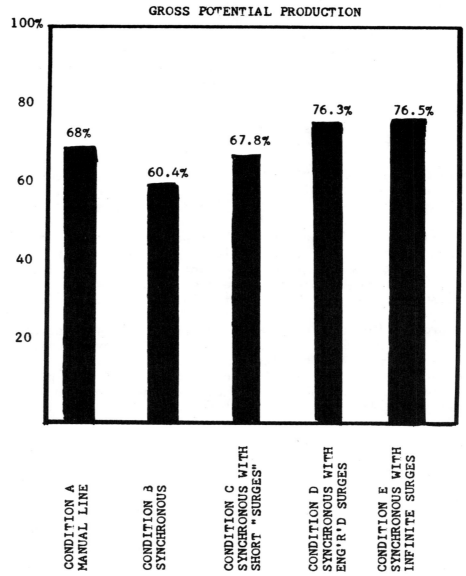

Figure 43 charts the results of a study made by a major manufacturer of the "productivity output" of an 8 operation, 98 station, synchronous, transfer-line system. While this is a machining line, the general conclusions apply to assembly as well. Data for <u>Condition A</u> is estimated from similar lines, consisting of manually operated machines with conveyors between. <u>Condition B</u> represents a completely "locked-in" line with no storage or repair facilities between machines. <u>Condition C</u> indicates the improvement when very short "surge" banks, representing approximately 3 minutes running time, are inserted after every operation. <u>Condition D</u> is the improvement to be expected if the "surge" banks are engineered according to the needs of each operation; more for frequent stoppages and less or none for better operations. <u>Condition E</u> has been developed from a computer model based on unlimited "surge" banks after all machines. The balance of the production losses are accounted for if operator personal time, scrap, equipment failure, maintenance time loss, etc.

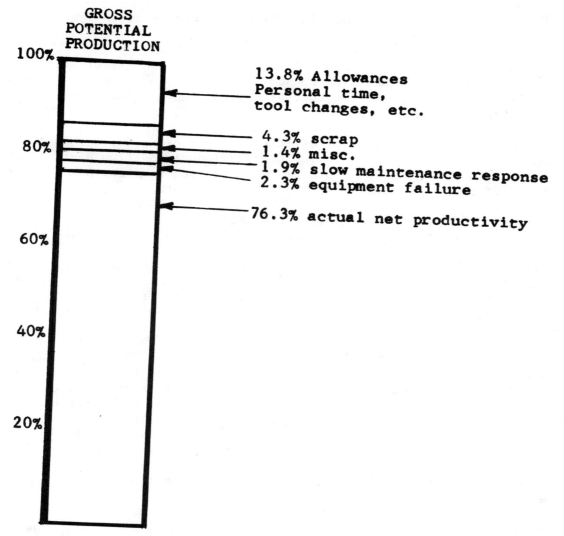

Figure 44 represents a breakdown of the productivity losses from the previous example. These statistics are actual and the result of an intensive in-plant study.

The personal and misc. allowances will vary from plant to plant and are usually beyond the manufacturing engineers' baliwick.

Scrap losses should be lower on assembly equipment than on the described machining line.

Equipment failure, translated into malfunctions on assembly equipment, will probably be higher but will be handled by the operator. Machine breakdowns requiring a maintenance man will probably result in much greater losses due to "slow maintenance response."

The net "productivity" of 76.3% is very good for a machining line. (The same study indicated a range of 46% to 88% for machining systems of all types.) Assembly machines in the same environment as the above study should range in excess of 65%.

Maximizing system efficiency in a line with many stations can be approached in several ways:

1. De-bug to a much higher level of individual station efficiencies. This is a very expensive and time-consuming procedure. The perfect machine has yet to be built and the cost of approaching perfection may be unjustified.

2. Design for a higher level of efficiency. For examples; use cam operation instead of air or hydraulic cylinder; positive parts placement; memory systems instead of machine stoppages; overtravel protection for stations; and automatic station lockouts.

3. Engineer-in devices to minimize the cumulative effect of individual station efficiencies such as repair loops, surge banks, etc.

Assuming the first two possibilities have been exhausted or an optimum level has been reached, the third direction offers some interesting directions. Depending upon the results of an analysis of efficiency losses, "surge and repair" loops permit a substantial recovery of efficiency "losses."

Figure 45 plots the efficiency curve of an average 95% station efficiency over 4 stations resulting in a little over 80% net for the four. If all but 1% of these efficiency losses can be recovered through surge and repair loops, the system could result in 99% average station efficiency as shown by the dotted line. These recoverable losses might be:

1. Variations in cycle times at any one station, auto or manual.

2. Short-time stoppage which can be cleared in less time than the system would use up the bank ahead of the station.

3. Repairable rejects which can be taken care of during the bank time.

Non-recoverable losses might include:

1. Scrap made by the station or previous stations.

2. Equipment failure requiring more time to fix than is available in the bank.

Probably few systems would require surge and repair loops after every station. In many systems it would be advisable to use a computer model to arrive at the optimum combination of surge capabilities and repair facilities. The number and placement of loops depends upon the relative efficiency and location of problem stations in the system. The size of the loops will depend upon the production rate and probable downtime of the problem stations. These computations result in the "engineered loops" previously mentioned.

The use of "surge and repair" loops in synchronous systems approximates the advantages of the true "power and free," or non-synchronous systems.

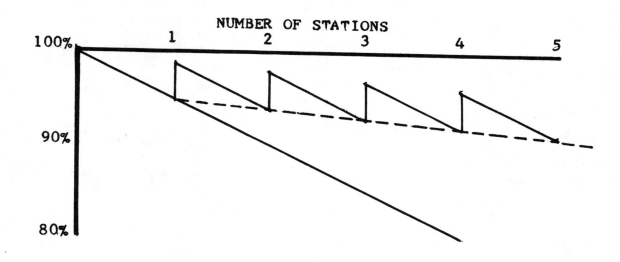

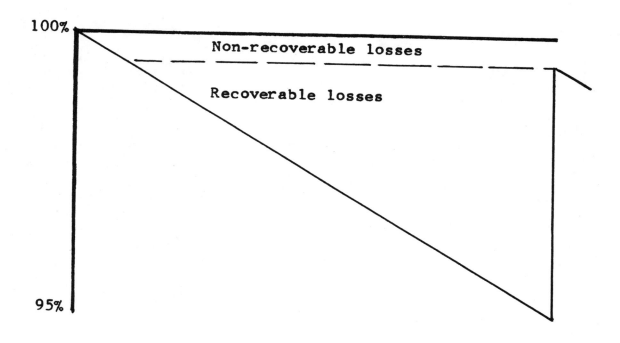

Figure 45.

Maximizing "machine efficiency" is the function of the machine designer with all the help he can get from the product engineer, the process engineer, and the previous experience of the user. It is, however, only a small part of maximizing "productivity."

SUMMARY

Basic manufacturing assembly concepts have undergone an evolution--if not revolution--over the past fifty years. It is a moot point as to whether labor costs forced automation, whether automation produced labor benefits, whether productivity was the "chicken or the egg." It is a fact, however, that today's manufacturing engineer has a large array of possibilities open to him in planning his company's productivity moves. Some he can affect; some he cannot; all must be attacked.

There are no "panaceas," no pat answers for every situation. There is only the necessity for being aware of all the potential answers and the common sense to match needs with availabilities. It is hoped that some of the thoughts presented in this chapter, some of the comparisons such as *Figure 46*, will be of some help in establishing points of reference for the reader.

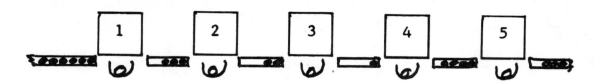

Manually operated line with roller conveyors

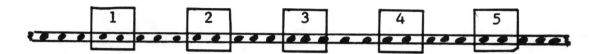

Synchronous "locked-in" line; indexing conveyor

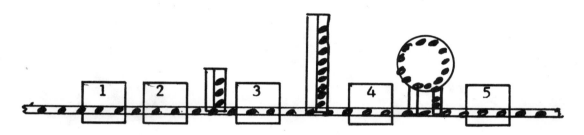

Synchronous line with "surge-and-repair" loops

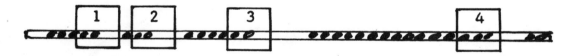

Non-synchronous line with engineered banks

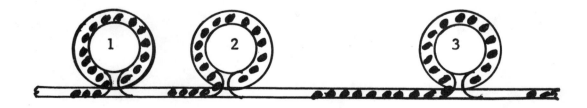

Continuous line with engineered banks

Figure 46.

CHAPTER 6

Parts Feeding

The ideal automated assembly system would receive all the raw materials and components at one end and would deliver a finished product out the other end with no manual input and 100% efficiency.

Since the ideal system is virtually impossible, and the economics of justification are usually prohibitive, all the possible options should be considered in some logical sequence of efficiency and desirability. Some components may be handled by any of several methods and the advantages/disadvantages of each must be weighed in making a selection. Manual input of some components may be mandatory due to the complexity or fragility of the part. The economics of a mechanization may prove that one operator can manually load more parts than several automatic devices and thus make automation impractical. Technical inability to handle the size, shape or other characteristics of the part may limit the choice.

This chapter will deal with the contemporary methods of supplying parts to the assembly system in the general sequence shown in *Figure 1*. However, due to the variations in definitions of terms used throughout the industry, this chapter will use the following terminology:

HOPPERS: Bulk storage devices holding relatively large quantities of material, liquid, granular or discrete parts without regard to orientation.

FEEDERS: Power or gravity devices capable of dispensing materials in random manner from a limited storage capacity.

AUXILIARY FEEDERS: Power or gravity devices designed to re-fill feeders or parts selectors, usually on demand. Typical examples are hoppers with vibratory bottoms or hoppers with elevator belts, either one operating on signal from level sensors, to maintain a parts level in selectors.

MAGAZINES: Holding and/or dispensing devices loaded with oriented discrete parts. May be attached to the assembly machine and manually loaded or they may be filled at a previous operation and loaded to the assembly machine as required.

SELECTORS: Devices which agitate quantities of discrete parts and select- or permit to pass into tracks - only those parts in the desired orientation.

ORIENTORS: Devices which examine discrete parts presented to them and turn, rotate or otherwise alter the orientation to the desired position. Primarily used when the selector is incapable of detecting orientation or when a greater volume is required than the selector can supply from the randomly presented volume.

ESCAPEMENTS: Devices which regulate the release of parts or quantities of material in controlled amounts at desired rates.

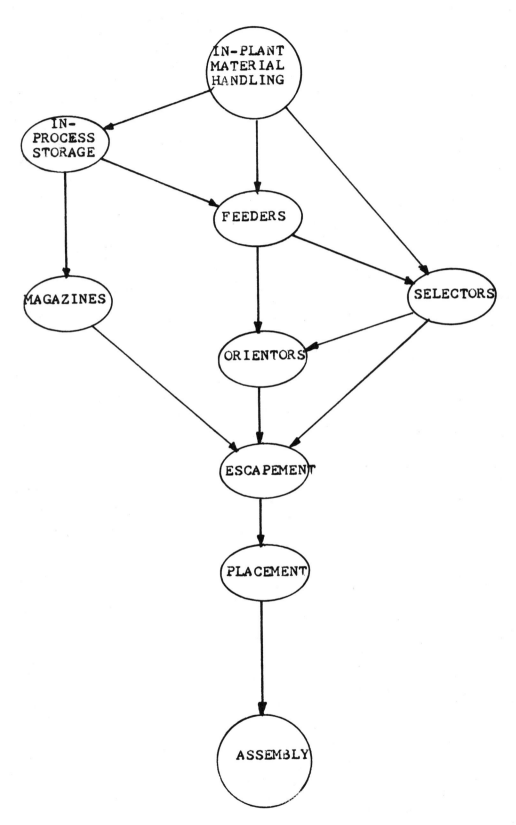

Figure 1. Elements of incoming parts handling systems.

Some parts handling devices may incorporate all of the above functions in a single unit. Again, some parts handling systems may have separate units for each of the above functions. It is helpful, however, to be able to identify the separate functions when planning a system.

Typical handling methods fall in the following categories. Inasmuch as every additional handling introduces additional opportunities for malfunctions and efficiency losses, this is generally a descending order of desirability.

1. Bulk handling

2. Manufacture on-line

3. Manufacture off-line, semi-complete

4. Manufacture off-line, retain orientation

5. Parts selectors

6. Parts orientors

7. Escapements

1. <u>Bulk handling</u> and the supply of components from "hoppers" is simple and usually trouble-free.

    <u>Liquids</u> are readily metered by volume as shown in *Figure 2* at very high rates of speed in the hundreds of units per minute.

    <u>Powders</u> may be dispensed by volume as in *Figure 3* or by weight as in *Figure 4*. Humidity may tend to cause powders to clog and block up narrow passages requiring a controlled atmosphere, vibrators, heat lamps, etc.

    <u>Gaskets</u> are "formed-in-place" from hoppers of liquid silicone compound or other material, *Figure 5*, and in very complex patterns, *Figure 6*.

    <u>Adhesives</u> are automatically silk-screened on to work pieces in fully automated systems. "Pot life" or the length of time the adhesive takes to cure as in epoxies, becomes a crucial factor and may require adhesives which must be heat or radiation cured to eliminate the need for too frequent cleaning of the equipment.

2. <u>Manufacture on-line</u> eliminates the need for any feeding equipment apart from the basic machine involved.

    <u>Stamping</u> of components from coil stock is one of the most common examples. *(Figure 7)* Parts which tend to tangle or are difficult to feed may well be handled this way. On the negative side of the ledger is the fact that most assembly machines run slower than presses and the press utilization can be very low. Presses usually take up considerable space. Press down-time for die change, etc.,

Figure 2. Automatic, continuous bottle filling at several hundred bottles per minute. Proper volume of fluid is metered, (or escaped), by piston stroke and pumped into bottle. Photo courtesy Pneumatic Scale Corp.

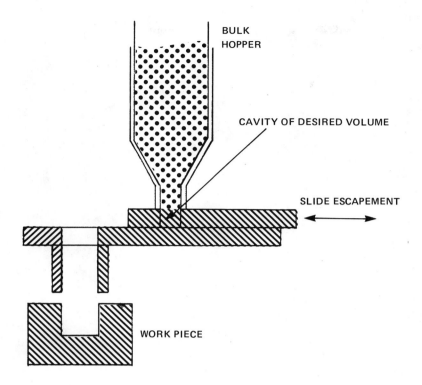

Figure 3.

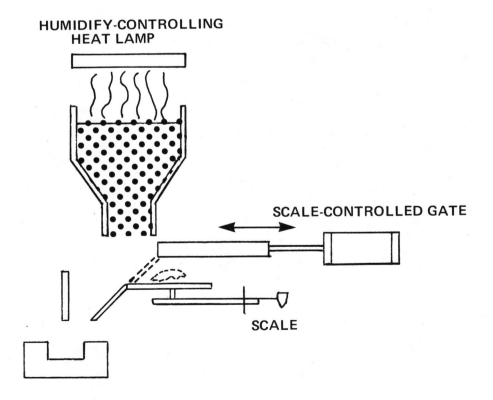

Figure 4.

Figure 5.

Figure 6.

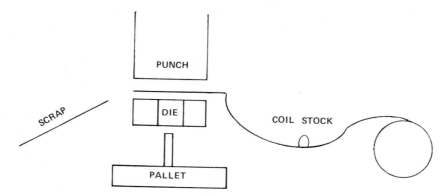

Figure 7.

may reduce the attractiveness of the method. On the other hand, floppy parts such as gaskets may well be die-cut at assembly, eliminating some messy feeding problems.

<u>Spring-winding</u> is sometimes the only answer to handling the automatic load of parts that tangle badly. Springs requiring ground ends or heat treatment pose some special disadvantages which may make automation impractical.

<u>Plastic injection molding</u> has been employed on automatic assembly lines to eliminate handling problems.

3. <u>Manufacture off-line, semi-complete</u>, is a common way of minimizing handling problems with electric terminals, gaskets, rubber parts, etc. Parts are left "in-the-web" and fed to the assembly machine in coils or "sticks". Parts may then have an added forming operation if required or a cut-off directly into the assembly or fixture.

4. <u>Manufacture off-line and retain orientation</u> is next best in the order of desirability. This usually takes the form of "magazined" parts directly from the prior operation.

<u>Electric motor laminations</u> are stamped out on progressive dies and blanked into stacks, retaining the orientation. These stacks are then transferred in trays to the assembly operation.

<u>"O" rings</u> may be purchased on cardboard tubes ready for placement in the assembly machine station.

<u>"E" rings and snap rings</u> are available on plastic or cardboard rods with the gaps oriented for efficient assembly with a minimum of malfunctions.

Gaskets may be bought in stacks, tied together with string or twine in an oriented fashion.

<u>Screws and electrical terminals</u> are frequently shipped held in an oriented position on adhesive tapes.

5. <u>Parts selectors</u> are available in many types for automatically supplying discrete parts which must be individually selected from bulk oriented for use in the assembly system. Many basic types are

illustrated on the following pages, some of them hundreds of years old and still used for specific applications such as coin sorting. By far the most popular today is the vibratory feeder, due primarily to its extreme flexibility and the fact that standard bases and bowls can be readily tooled for a wide variety of applications.

Virtually all feeders have one characteristic in common - they work on gravity. The efficiency, rate of feed, etc., are all related to the rate at which the part will fall and the tendency of the part to come to rest in a predictable manner.

Research resulting in a "Handbook of Feeding and Orienting Techniques for Small Parts" has been conducted by Professors G. Boothroyd, C. R. Poli, and L. E. Murch of the University of Massachusetts. A coding system for parts of various configurations has been developed resulting in a logical organization of characteristics such as shown in *Figure 8* and *Figure 9*.

Figures 10 through 15 illustrate various types of mechanical parts selectors with related data from the handbook tables.

Figures 16 through 21 illustrate additional types of nonvibratory parts selector concepts.

Vibratory selector concept is described with *Figure 22* as a linear feeder and *Figure 23* as a circular bowl construction. The means of selection of part characteristics is virtually infinite in variety. A few examples are shown in *Figure 24*. The design, building and de-bugging of selectors is a fine art rather than a science. Asked how he went about designing a vibratory parts selector, one engineer said he "would pretend he was a part and figure out what he would do in the bowl".

MISCELLANEOUS FEEDERS

Special Feeders for Surface Protection

These feeders are used when it is possible to automatically feed parts which would be damaged by the action of the parts hitting or rubbing against each other as occurs in vibratory feeders. Ground surfaces of valve parts are an example. Another requirement for this type of feeder are parts for consumer products which, if scratched, destroy the appearance of the product. There are basically two concepts of commercial feeders produced to meet these requirements. Both are rectangular in shape rather than circular. One type consists of a series of belts. The parts are gently circulated, oriented, and conveyed by these belts. The other type of feeder is a vibratory type of feeder, but a surface of soft brush bristles is in contact with the parts. In both types the number of parts in the feeder at any one time is minimized and usually require buik feeders. This type feeder tends to discharge the parts with their long axis horizontal. This may require additional tooling to re-orient the part to a position where it can be used by the assembly machine.

Miscellaneous Moving Feeders

These type feeders are limited to headed parts which will hang. They

Figure 8. Coding system for rotational parts.

FIRST DIGIT

0	rotational	L/D < 0.8
1	rotational	0.8 ≤ L/D ≤ 1.5
2	rotational	L/D > 1.5
3		
4		
5		
6		
7		
8		
9		

SECOND DIGIT

0	ALPHA symmetric		
1	BETA symmetric steps or chamfers on external surfaces	part can be supported by head with center of mass below supporting surfaces	
2	BETA symmetric steps or chamfers on external surfaces	part cannot be supported by head (center of mass too high)	
3	BETA symmetric steps or chamfers on external surfaces	features small for orientation purposes	
4	no BETA symmetric steps or chamfers on external sufaces	BETA symmetric grooves, BETA symmetric recesses or holes	on cylindrical surfaces only
5	no BETA symmetric steps or chamfers on external sufaces	BETA symmetric grooves, BETA symmetric recesses or holes	on end surfaces only
6	no BETA symmetric steps or chamfers on external sufaces	BETA symmetric grooves, BETA symmetric recesses or holes	on both cylindrical and end surfaces
7	no BETA symmetric steps or chamfers on external sufaces	internal features with no corresponding external features	
8	no BETA symmetric steps or chamfers on external sufaces	Either BETA assymetrical features or geometrical features too small for orientation purposes	
9	no BETA symmetric steps or chamfers on external sufaces	features other than geometrical features	

THIRD DIGIT

0	BETA symmetric	to be fed end-to-end
1	BETA symmetric	to be fed side-by-side
2	external projections are not BETA symmetric	on cylindrical surfaces only
3	external projections are not BETA symmetric	on end surfaces only
4	external projections are not BETA symmetric	on both cylindrical and end surfaces
5	external projections are not BETA symmetric	features small for orientation purposes
6	no external projections	keyways, recesses, holes or flats
7	no external projections	slots, recesses, holes or flats on external end surfaces only
8	no external projections	features too small or other geometrical features
9	no external projections	features other than geometrical features

3996

An ALPHA symmetric part does not need to be oriented end-to-end.

A BETA symmetric feature is a feature that can be represented by a solid of revolution.

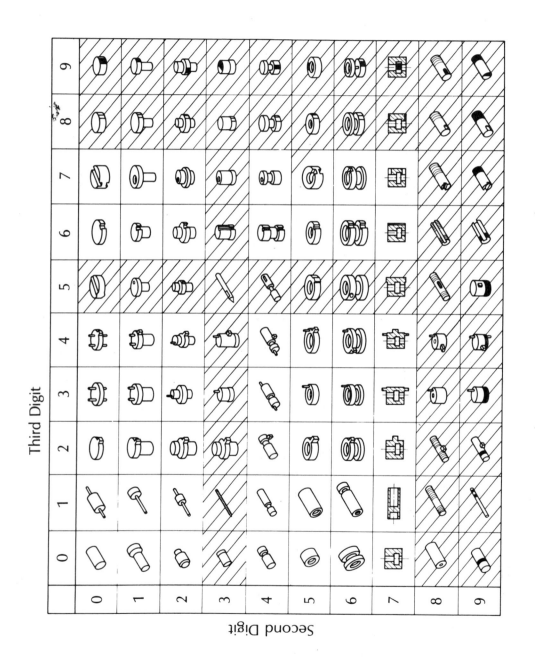

Figure 9. Coding system for rotational parts -- examples.

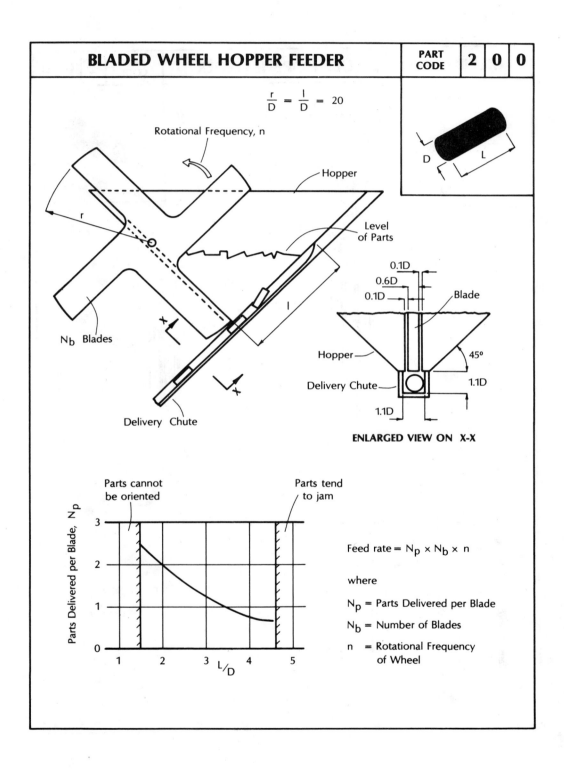

Figure 10.

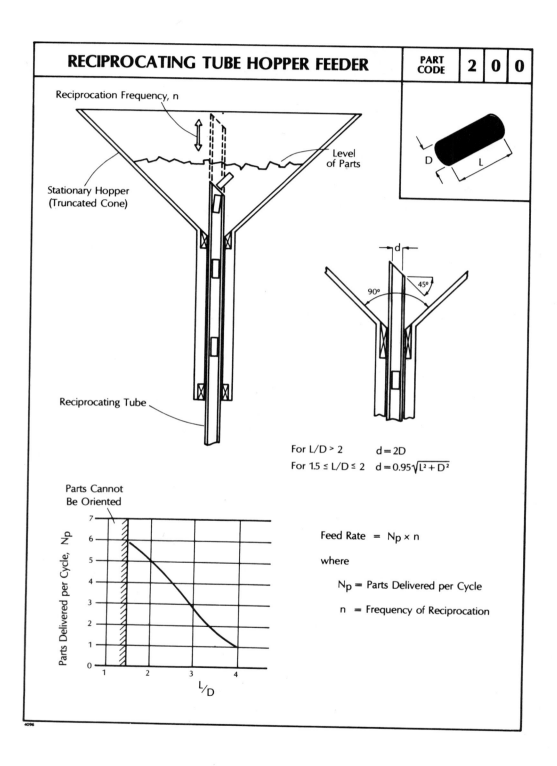

Figure 11.

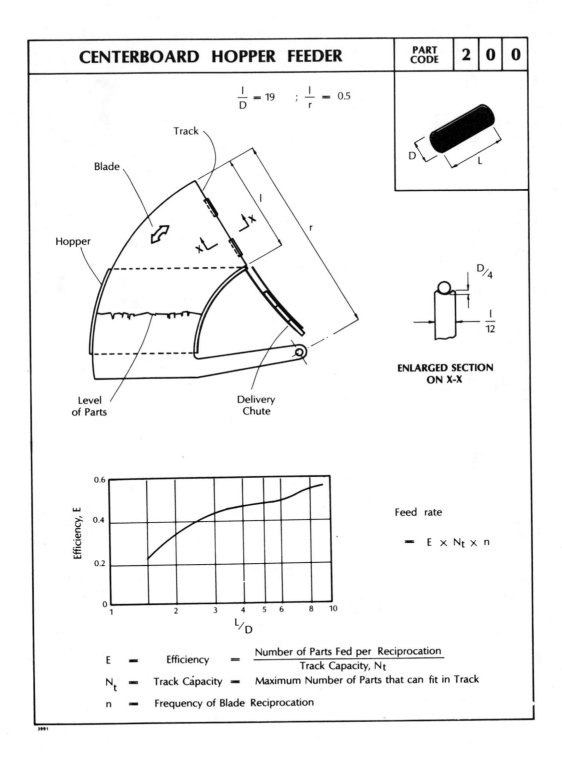

Figure 12.

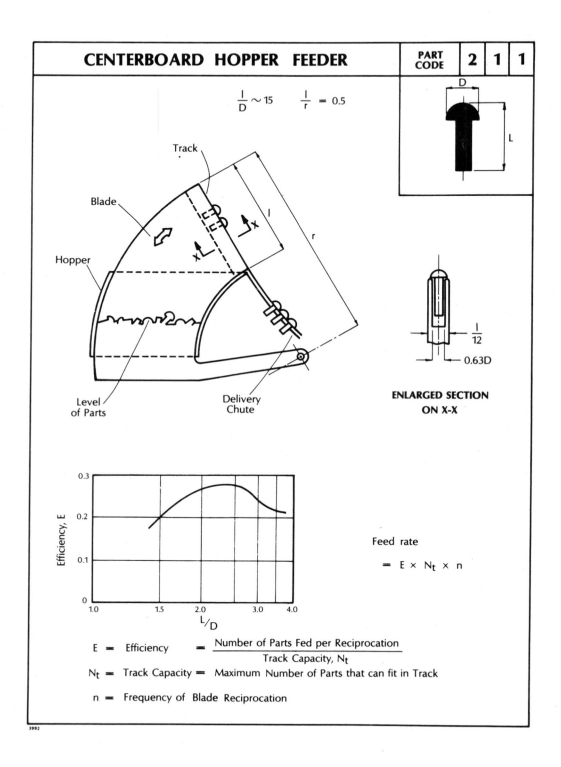

Figure 13.

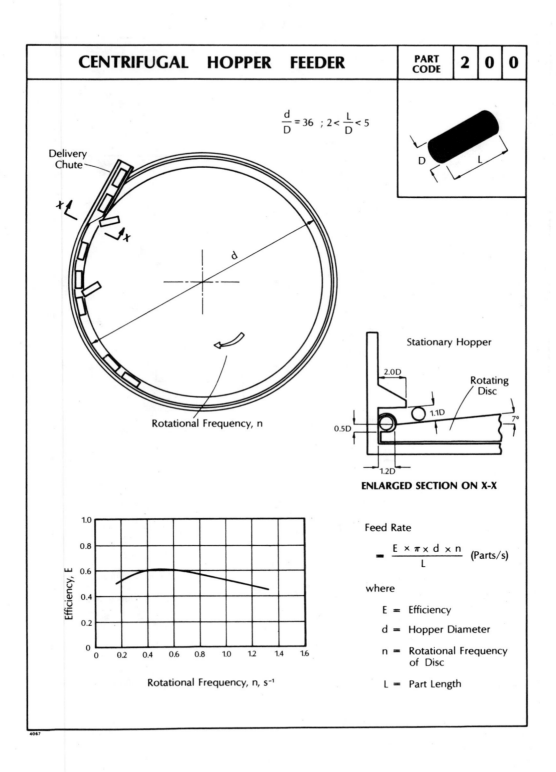

Figure 14.

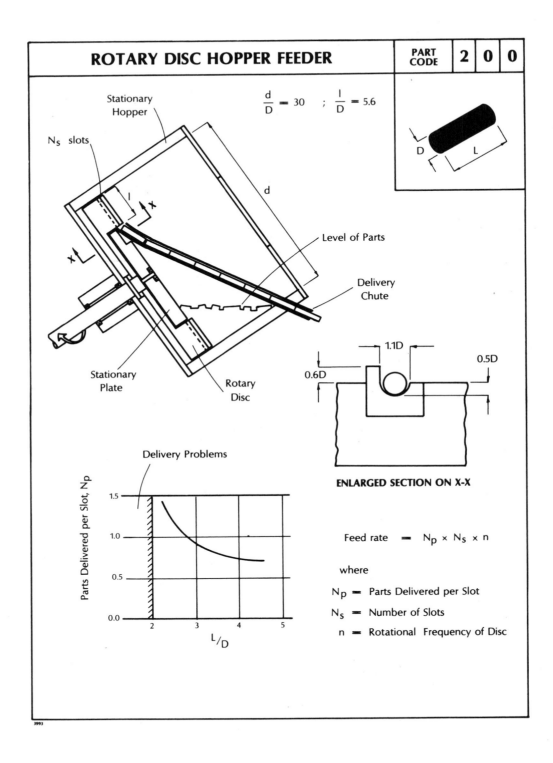

Figure 15.

Figure 16. Rotating Drum Selector

The rotating drum selector is designed around a bin having a stationary base and centerpost with a rotating sidewall with profiles machined in its lower surface. As the drum rotates, paddles attached to its supporting web agitate the mass of parts to cause a continual realignment of part positions.

Parts properly oriented will slip into the profiles, or recesses in the drum and be carried upward about one-quarter of a turn at which point they may discharge into a track or chute.

Headed parts which are to be fed head down will have a natural tendency to position themselves accordingly in this feeder. Caps, cups or other parts having sufficiently different shapes from face to face are readily oriented at high speeds in this feeder. A minimum of tumbling is possible since parts do not travel all the way upward in the drum rotation.

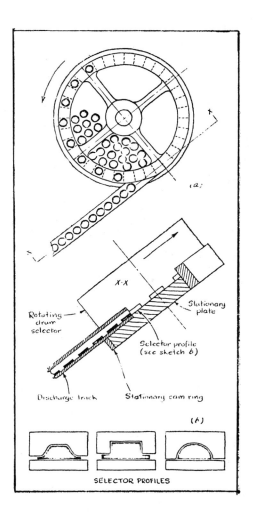

Figure 17. Pin Selector

The pin selector might really be called a variation on the selector ring type since it uses a rotating ring with selector slots to pick up the parts. The addition of pins cammed in and out of the profiles as the ring rotates permit retaining of cups or caps properly in orientation while other parts are rejected to the bin.

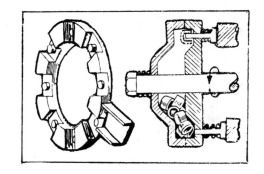

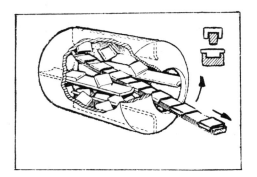

Figure 18. Barrel Parts Selector

Selection of properly oriented parts is made as the parts fall on to the discharge track by means of the shape or profile of the track. The function of the barrel is to store, elevate, drop on to the track and to agitate maloriented parts to cause them to fall differently each time.

Stampings such as cups, caps, clips and frequently complex forms, may be fed from this type. Large parts and high rates of feed may be obtained due to the length of the discharge track which may be flooded with parts with the passing of each flight inside the barrel.

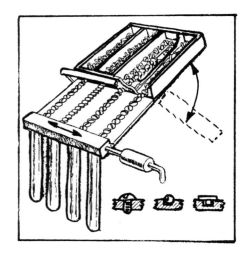

Figure 19. Oscillating Box Part Selector

Sometimes known as a "nail-hopper" this device depends upon parts falling into slots, grooves or profiles in the box bottom as it oscillates. At the top of its stroke the box profiles align with the discharge track permitting properly oriented parts to pass through. A gate or profile at this point rejects improperly oriented parts. As the box moves downward, parts tend to reorient themselves through a slight tumbling action.

Multiple tracks are frequently found in this type of feeder. Some screwdriver feeders will supply as many as ten different screws to ten different tracks by means of a partitioned box.

Figure 20. Rotating Agitator Parts Selector

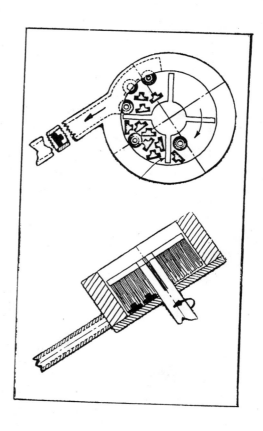

The action of a rotating brush, rubber wiper or other flexible medium tends to sweep parts around the periphery of this feeder. Properly oriented parts will pass into the discharge track. Screws were once fed in this type of unit where milled slots in the base plate accepted the screw stems and they were brushed into the discharge slots hanging by their heads. Also used for feeding eyelets and other small stampings.

Figure 21. Elevator Parts Selectors

Very large parts or large quantities of small parts develop the need for elevator selectors. A bonus factor of this type of feeder is the advantage of loading the bin close to floor level while having the discharge of the feeder at considerable, and sometimes adjustable, height.

Par sizes which may be selected in standard units will range up to 24" in length and as much as 15" in diameter, although metal parts of these maximum dimensions would be too heavy in most cases.

Typical Elevator Selector Details

Figure (a) illustrates the basic arrangement in which bulk parts are placed in a hopper or bin of 1 - 30 cubic foot capacity. A belt or chain conveyor equipped with "flights" picks up one or more parts per flight and elevates them to a discharge point, either to one side or over the top. Part orientation or selection may take place during this transport or in the track into which they are discharged.

Figure (b) shows angled flights for a side discharge.

Figure (c) describes the straight flights used in an over the top type of discharge.

Figure (d) and *Figure (e)* compare the two types in feeding rings.

Figure (f) illustrates a means of feeding and orienting rubber brushings which have a length/diameter ratio of one.

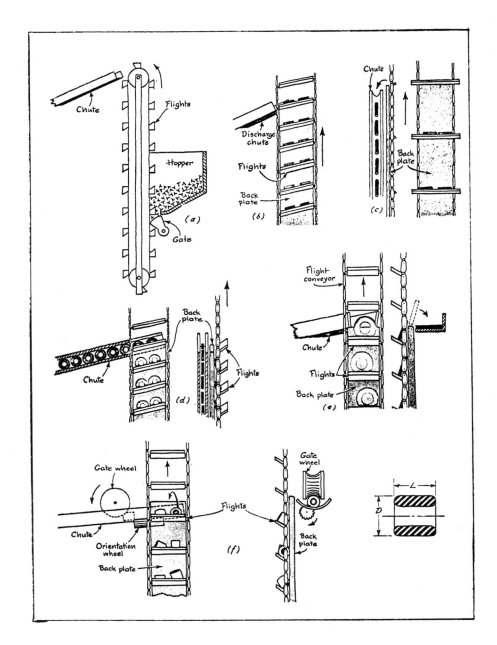

(Figure 21. Cont.)

Production rates depend to a large extent upon the part sizes and number of selections per part. In general, elevator selectors will feed at the same rates as other selectors, all of them being dependent upon gravity. Multiple tracks and multiple pieces per track, however, can offer very high outputs due to the physical size of the feeder.

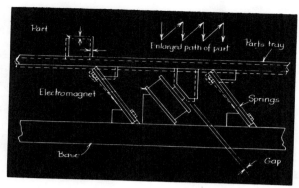

Figure 22. *Vibratory Parts Selectors*

All vibratory units operate on a resonant, vibrating reed principle.

A track, or tray, carrying loose, discrete parts is supported on springs above a fixed, massive base.

The mass inertia of the moving weight is balanced against a given spring recovery rate to produce a resonant frequency equal to the frequency of the applied power. In most cases a half-wave rectified 60 cycle A. C. power source is used and a "natural frequency" of 60 cycles per second is desired.

Properly "synchronized", very little power is required to keep such a system oscillating. This is much the same as applying minute impulses to a clock pendulum to keep it in motion.

The mounting springs are set at an angle, variable on some makes, which causes the track to move downward and backward when power is applied to the bowl by means of electromagnet, air cylinder, hydraulic cylinder or mechanical linkage. This motion is designed to be faster than the parts will fall by gravity and they drop on to the track at a point somewhat further along the track. When the power is removed, the springs move the track forward and upward carrying the parts along through gravity and friction.

Repeating this sequence at 60 cycles per second with an "amplitude" of horizontal movement of - for example, .100 inches would theoretically result in advancing the parts some 6" per second or 360" per minute. Slippage, friction, vertical amplitudes greater than the parts will fall in 1/60 second, and other factors will reduce this feed rate according to their effect in specific cases.

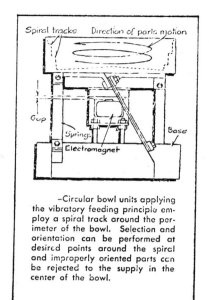

-Circular bowl units applying the vibratory feeding principle employ a spiral track around the perimeter of the bowl. Selection and orientation can be performed at desired points around the spiral and improperly oriented parts can be rejected to the supply in the center of the bowl.

Figure 23.

Bowl feeders are the most common vibratory selector type. A spiral ramp around the inside of a bowl feeds parts forward and upward. At any point various devices such as "plows", gaps, profiles, air jets, etc., may be employed to pass parts with the desired orientation and to reject into the bowl parts improperly oriented.

"Linear" feeders are also offered as standard items. These may be adaptations of bulk feeding units such as have been long used in the sand and gravel industry or they may be engineered feed tracks capable of transporting oriented parts any distance without losing their orientation. In some cases such "transport rails" have been used to perform the orientation as well.

Many variations of the vibratory principle are available. The most common type uses the central electromagnet with permanently mounted springs at fixed angles, and operates at 60 cycles.

Similar feeders operating at 120 cycles per second are available.

Air-operated vibratory units permitting variable frequencies as well as variable amplitudes are also made.

Adjustable angle springs are a feature of one supplier who mounts the electromagnet (or air drive) in several "blocks" permitting a variation in both vertical and horizontal components of the vibration.

Constant amplitude vibratory feeders vary the frequency, length of time, or length of stroke according to signals from transducers mounted to the bowl.

Wiper rejects more than one thickness of parts, returning them to the bowl. Contours may be used to pass given profiles only.

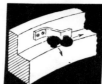
Spacer forces parts with head up over the track edge past their center of gravity, returning them to the bowl.

Air jet wiper blows improperly oriented parts back into the bowl.

Cutout passes parts with sharp, open end down, but lets rounded cup end fall back into the bowl.

Cutout with a hood holds parts standing up as they pass over gap, while parts with horizontal axis drop back into bowl.

Rails over cutout retain parts with head up, permit parts with head down to drop back into the bowl.

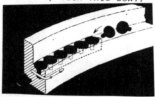
Two orientation methods for use with long, headed parts. Slot lets parts hang by heads. At discharge tube, arm trips parts to fall head first if desired.

Figure 24. Shown are a few of the arrangements which may be used on the track of a vibratory feeder to select or orient parts for position. Selection of the particular method for a specific part must be based on consideration of the characteristics of that part.

are of two types. One type consists of a bulk hopper with an elevator of some style which rises out of the bulk hopper. As the elevator moves, some of the product in the bulk hopper is elevated. The discharge from the elevator is tooled so that the parts end up usually in gravity rails diameter to diameter, hanging by their heads. A variation of this concept rotates the whole bulk container through 180 deg. arc and the parts fall onto the rails. This type of system is usually used for small screws.

Another special type of feeder combines and interfaces a bulk storage hopper with the outer circumference of a rotating wheel. As the parts pass around the wheel they pass selection devices which orient them. They can then pass into the various configurations of inline feeders previously discussed. This type of feeder is claimed to be quieter than equivalent size vibratory feeders.

Bowl Coatings

A number of coatings can be applied to feeder bowls to control noise or abrasion and to provide a clean wear-resistant resilient surface upon which the parts will ride during the feeding operations. The common coatings are PVC, polyurethane, epoxy and chrome plating.

Noise Control

Large parts in large vibratory bowls can be objectionably noisy and in many cases exceed OSHA requirements. Solutions have included using the least noisy type of feeder i.e. belt feeders vx. vibratory feeders, or by enclosures. Enclosures include enclosing the feeder packages in a solid soundproof box or alternately in cylindrical enclosures with a see-through plastic dome top. The difficulty is knowing if the sound enclosures will be required before the feeders and machine are built and operational. Some machines are totally enclosed rather than attempting to cover the offending bowls.

To assist you in arriving at the best selector choice at the least cost in time and money, *Figure 25* itemizes many of the facts which should be determined prior to asking for a quote.

Also of interest is the nomogram in *Figure 26* which is typical of estimating aids used by hopper, feeder and selector manufacturers.

6. Parts Orienters are usually required when the part design does not lend itself to any "natural selection". If the part does not tend to fall or assume a predictable position with respect to the desired orientation, some external device, or orienter must be employed.

   Figure 27 suggests one condition requiring a device to examine a shaft, determine where the cross-hole is located and turn it to the desired location, in this case end-for-end, by rotating the track section holding the shaft. Hole sensor may be optical, pneumatic, eddy current detector, etc.

   Figure 28 shows the same shaft, oriented end-for-end, and requiring a radial orientation to line up with a pin hole in the mating part. In this case the part may be inserted into a head containing a ball detent which is rotated more than 180° ending up with the shaft always in the required radial position.

   Figure 29 offers 5 more examples of parts probably requiring special orientation devices. Shaft in "a" might be bottomed on a limit switch which is made if the solid end is down and vise versa. Sphere in "b" might be randomly rotated until hole locks on a ball detent. Double-ended stud in "c" may require sensing with a screw thread gage to determine which end is forward. Washer with oil groove on one side might require an optical or surface contour check. The collar with the key-way might also use a ball detent rotated in its I.D.

   The possible solutions to orientation devices are as many as the applications. If the product design must be as-is, and if gravity or other forces will not cause a natural selection, special orientation devices must be employed.

Figure 25.

PARTS SELECTOR CHECK LIST

Many of the factors involved in selecting a parts selector are listed below. Complete information and material samples should be available before requesting quotations or making a decision.

1. PART NAME — Proper terminology will always simplify follow-up and reduce errors.

2. PART NUMBER — Inquiries and quotations will be safer when specific.

3. PART DRAWINGS — For tolerances, deviations from samples, etc.

4. PART SAMPLES — Particularly in the field of parts selectors, samples are essential to accurate rate and pricing estimates.

5. PART MATERIALS — Specify unusual conditions such as fragility, surface finish, hard or soft, weights, flash or burr conditions to be expected.

6. PART HANDLING — Will they arrive in pans or tubs, floor level or overhead, dry, wet, oily, tangled, etc.

7. PART STORAGE — Required quantity or volume of parts which the selector must store in terms of pieces, time or volume.

8. PART ORIENTATION REQUIRED — Supply a sketch indicating the position in which the part is to be delivered and whether it should be end-to-end, O.D. to O.D., side-by-side, hanging by heads, head up or head down, stems first or last, slots or flats oriented, etc.

9. DISCHARGE TRACKS — Specify whether required or not, single or multiples, power or gravity, vertical or horizontal, length, discharge height and location.

10. ESCAPEMENTS — Specify whether required or not, single or multiple track, single or multiple parts/cycle, fixed or adjustable for number of parts/cycle, and the preferred type of actuation such as air, electric, hydraulic or mechanical.

11. AVAILABLE POWER — Specify types of electrical, pneumatic, hydraulic or mechanical power available.

12. AMBIENT CONDITIONS — Dust, dirt, humidity, oil in the air, vibration or any other unusual condition which might affect the operaion of the equipment.

13. SELECTOR LIFE — Unusual requirements, either very short or very long life of feeder.

14. TIME LIMITATIONS — Required quotation date and delivery date should be sufficient for adequate experimentation prior to quote and de-bug after build.

15. JUSTIFICATION — If known, and if within your purchasing policy, you may expedite and obtain the optimum recommendation from your supplier by offering some idea as to what equipment cost is justified.

16. PERSONAL PREFERENCE — Indicate preferred types of equipment which may best fit into your plant operation.

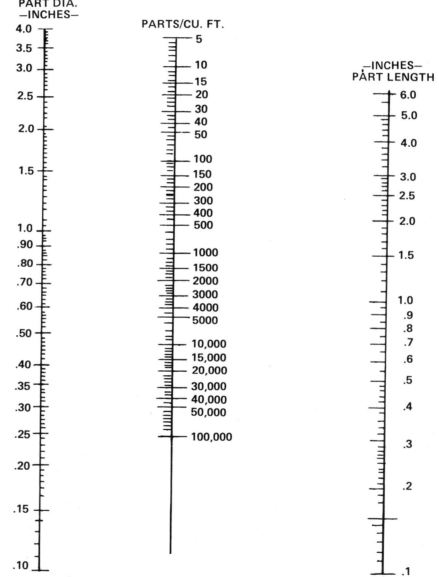

Figure 26. To determine the number of shafts to be expected in a random storage mode for given diameters and lengths, connect a line between the diameter on the left and the length on the right. The parts per cubic foot has been empirically determined.

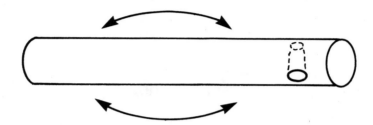

Figure 27.

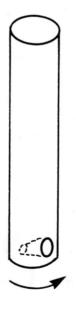

Figure 28.

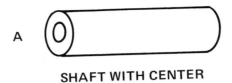

A

SHAFT WITH CENTER

B

SPHERE WITH HOLE

C

COARSE PITCH   FINE PITCH

DOUBLE-CIRCLED STUD

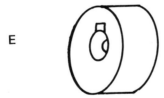

D

WASHER WITH OIL GROOVE 1 SIDE

E

COLLAR WITH KEY-WAY

Figure 29.

7. <u>Escapements</u> Industrial escapements are frequently not consciously recognized as such. Fixtures on dial tables may perform the function of escaping parts from magazines or feeder tracks; punches on punch presses may escape parts from chutes as the punches travel to perform their forming or fastening function. In order to identify the actual escaping function this article will classify escapements according to types such as ratchet, slide, drum, gate, displacement, and jaw. This cross-section of basic types can be combined and adapted indefinitely to individual needs of the moment, limited only by the ingenuity of the applier. All of the illustrated es-

capements are in current use in one form or another.

ESCAPEMENT TYPES

Ratchet: Ratchet escapements, *Figure 30*, have been used for many centuries. Their most common application is probably found in the timepiece, where the commodity released is shaft rotation resulting from stored spring energy. The release may be in accordance with the timed impulses from a pendulum, whose natural frequency is controlled by its design. The modern watch contains an oscillating balance wheel whose inertia is carefully adjusted to provide the proper timing. This balance wheel controls a ratchet to regulate rotation of a ratchet wheel and shaft. If the ratchet wheel is replaced with disks, gear blanks or other parts to be escaped, the ratchet will then release one or more parts into fixtures for machining or assembly, or into containers for packaging. Though several true ratchets are shown, such escapements may be operated by cams which insert and remove fingers, or may be spring-loaded fingers pushed in by cylinders or solenoids, etc.

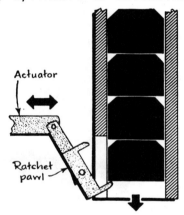

(a)—Ratchet setup to release one stamping for each cycle of the actuating device.

Figure 30. Ratchet escapements are similar to the familiar method of regulating timepieces. True ratchets may be replaced by separate fingers or pins individually actuated in proper sequence. Such escapements can be used to escape items singularly or in groups, in seriers from a single track or in parallel from multiple tracks.

(b)—An adjustable ratchet escapement which can be set within a small range for variation in quantities of parts. This device offers the advantage of working for multiples of various size pieces as in packaging work.

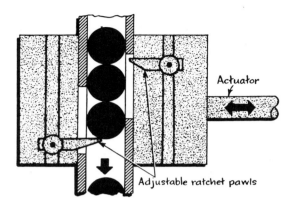

(c)—This is a compromise between a slide and a ratchet escapement which is adjustable for a wide range of part sizes or quantities. While more expensive than (b), the range and ease of setting up this device make it quite practical.

(d)—Ratchet escapements can of course be designed to control release of units from several tracks simultaneously with one actuator and one timing impulse.

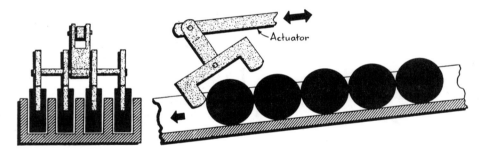

(Figure 30. Cont.)

The author would suggest that the most logical applications for the ratchet escapement exist where a straight lineal motion of parts is desired. Where the escaped commodity is other than physical parts as in a timepiece, the motion may be rotary.

Slide: Slide escapements, *Figure 31*, usually involve a change in direction of motion of the commodity to be escaped. This direction change may only be a "jog" in the material flow, but without the use of some ratchet device a slide escapement must change flow direction to operate. Multiple-track slide escapements feed more than one track, but each has its own supply track. Such escapements are quite common on screwdrivers, ball bearing assembly machines, etc. A multiple-escapement slide feeds more than one track from a single supply track. In the feeding of stacks of parts or sheets, slide escapements are limited by burrs, variations in coatings, cleanliness, etc., but this is one use of the slide method.

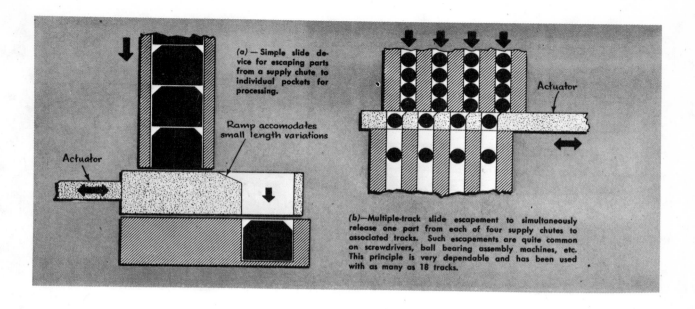

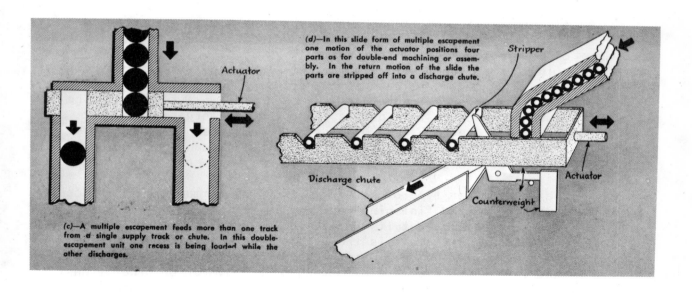

Figure 31. Although parts can move through ratchet escapements in straight lines, slide escapements involve a change in direction of motion of the material escaped.

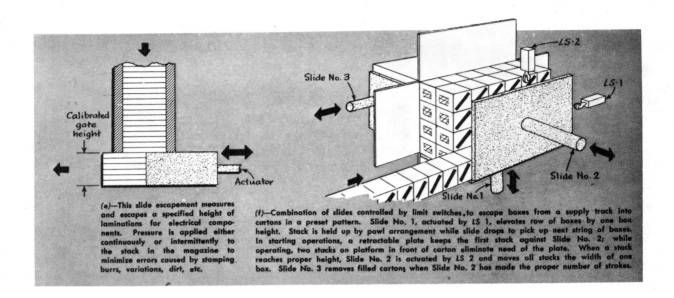

(Figure 31. Cont.)

Drum: Turnstiles at the local baseball park are a very familiar form of the drum escapement, *Figure 32*. In this form it separates the patrons and spaces them for counting purposes. Although we all dislike the bottleneck at the ball park, the escapement does control and regulate the commodity being released. A similar type of drum escapement is that commonly known as a star wheel. In the coin counter application shown, the star wheel itself does not actually time the release of the coins. The rotating drive wheel feeds the coins as rapidly as the mechanism will function. The star wheel serves to separate the coins and actuate a counter. Speeds of more than 20,000 counts per hour are possible. This arrangement is of added value since it will count only coins of the proper diameter and passes slugs or improperly segregated parts.

The drum-spider type of escapement such as commonly used on double-end grinding and finishing machines is very close to the basic slide escapement, but its rotary motion requires its mention here. A true drum escapement is the rotary feeder such as used in regulating the flow of bulk materials. Another form of drum escapement is the use of fixtures moving on a conveyor or other transfer means between load and unload stations. This simple form of the drum escapement deserves consideration wherever large items are involved.

Most of the examples shown are cyclic devices. However, the unit for dispensing liquid to bottles or other containers, *Figure 32*, is continuous in operation. Bulk materials or discrete units can be escaped in a similar manner. In this illustration the escapement head and the containers to be filled are in continuous motion. Part of the escapement rotation is used for charging the escapement cells and part for discharging the escapement cells into the

containers. Cam actuated valves, sliding valve action or many
other methods may be used for starting and stopping flow from the
resevoir to the escapement and from the escapement to the containers. Cyclic operations can be continuously performed also on
vertical drums or even straight-line units comparable to a flying
shear which performs a cyclic action on a product having a continuous linear motion.

Many types of high speed escapements have been devised which are
close to being drum types. Most of them involve a slicing action
in removing one or more parts from a track or tube and are usually
continuous in operation. The rotary slicer moves units around an
arc; the worm wheel moves units in a straight line parallel to its
axis.

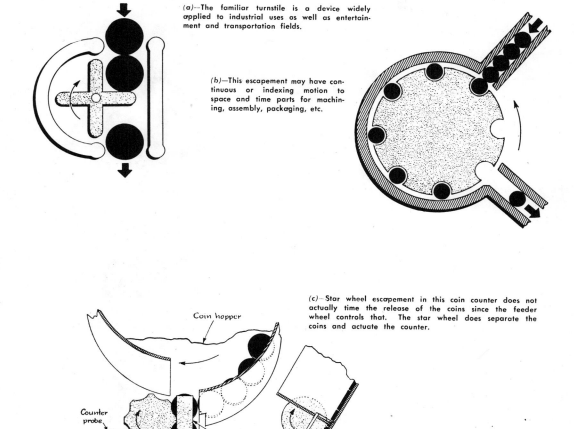

Figure 32. *These drum escapements and closely related types of high speed devices utilize some type of rotary motion in their operation. They can be used whether materials are to flow in a straight line or change direction.*

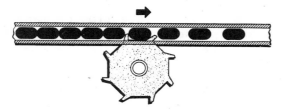

(d)—In feeding items from a conveyor into a processing machine it is frequently very important that the items be escaped to the machine at proper intervals as this star wheel is doing.

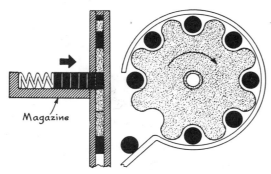

(e)—Magazine-fed, drum-spider type of escapement is very close to the basic slide escapement but uses a rotary motion. Such a device is commonly used on double-end grinding and finishing machines, where spring pressure may be used to force parts from the magazine into openings in the spider.

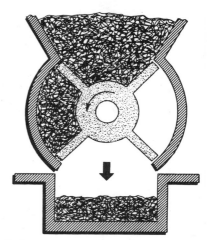

(f)—Drum escapement adapted for bulk materials is the common rotary feeder. Motion may be continuous or intermittent indexing as required.

(Figure 32. Cont.)

Gate: The most common type of gate escapement, *Figure 33*, is probably the trap door in the bottom of a hopper controlled by a timer, scale, float or other metering device. Another very common use of gates is as flippers or diverters to segregate out of limits units from acceptable units.

Displacement: A liquid escapement which might be classified as a slide, but can be called a displacement type of escapement, *Figure 34*, is formed from a piston and cylinder fitted with suitable check valves. Standard units of this type are available for very accurate dispensing of liquids and semisolids such as greases.

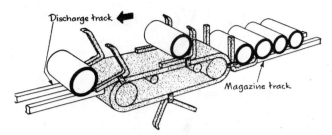

(g)—Form of drum escapement of particular use where large objects are involved can be based on fixtures carried by a transfer device.

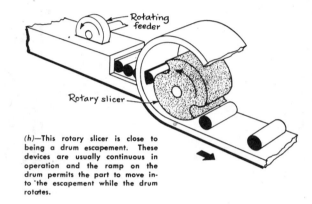

(h)—This rotary slicer is close to being a drum escapement. These devices are usually continuous in operation and the ramp on the drum permits the part to move into the escapement while the drum rotates.

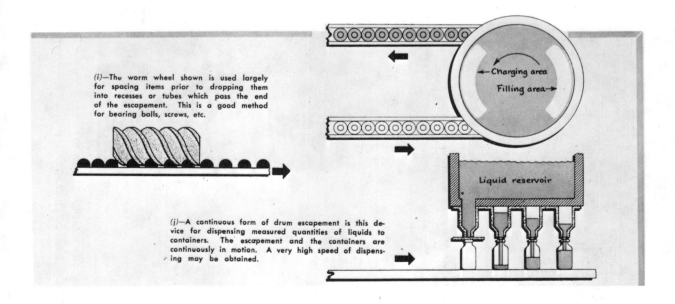

(i)—The worm wheel shown is used largely for spacing items prior to dropping them into recesses or tubes which pass the end of the escapement. This is a good method for bearing balls, screws, etc.

(j)—A continuous form of drum escapement is this device for dispensing measured quantities of liquids to containers. The escapement and the containers are continuously in motion. A very high speed of dispensing may be obtained.

(Figure 32. Cont.)

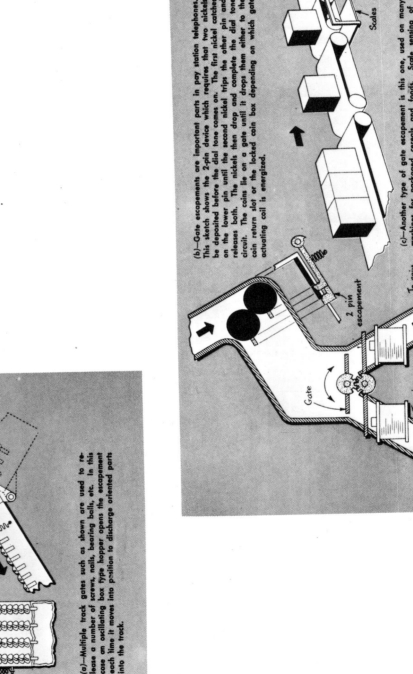

Figure 33. Many types of gate escapements are used, though the most common is probably the trap door in the bottom of a chute or hopper.

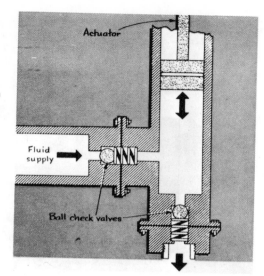

Figure 34. This liquid escapement might be classified as a slide, but is here called a displacement type. It is merely a piston and cylinder with two check valves. It fills with a given quantity of liquid on one stroke and dispenses this load on the return stroke. Action is rapid and easily controlled.

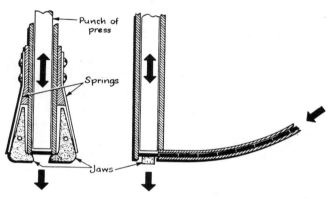

(a)—In this application of a jaw-type escapement, the punch of a press removes one part from the track where it is held by spring-loaded jaws. Very accurate location is practical where part dimensions are consistent.

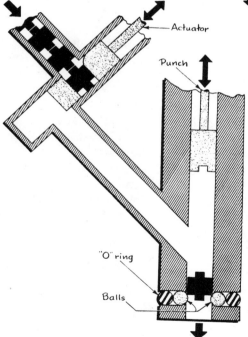

(b)—Using balls for stop-pins, this jaw-type escapement which holds parts under a punch is applied in conjunction with a slide escapement so that only one part is under the punch at any particular time.

Figure 35. Jaw-type escapements can be used to accurately position and hold parts until other machine elements release them.

Jaw: Jaw type escapements, *Figure 35*, are often used in conjunction with machine elements which perform other functions -- such as with the punches of punch presses. These escapements can be used where accurate location is required.

Other Considerations in Applying

Means of actuating these escapements is a distinct area by itself which this article does not attempt to cover. Logically, however, the most positive method is the best, whether it be a direct mechanical connection to another part of the machine or a pneumatic, hydraulic or electric actuator. In any case, the simpler the actuator principle used, the fewer parts and motions involved, the more satisfactory will be the operation. The greatest disadvantage in mechanical actuation is that it can be actuated only by cycling of the complete machine. Many times it is preferable to use an actuator which can be remotely and manually re-cycled should the magazine or feeder jam.

Most machine builders and all hopper and parts-feeder manufacturers supply one or more types of escapements, but do not offer them separately from their other products. This is due to the fact that practically every escapement is tailor-made for a given job, and an escapement alone is seldom required by the customer. However, escapements similar to these basic types are of increasing importance and are sure to be applied more frequently as automatic processing becomes more prevalent.

SUMMARY

If you have come to the conclusion from this chapter that most of the parts handling in an automatic assembly system is special design, you would be absolutely correct.

Basic concepts have surfaced over the last hundred years, but their individual application still is an art requiring the most careful individual attention. Parts selectors, for example, may be put on drawing paper in detail, but the craftsman doing the de-bugging "dresses off" a little here and there, touches up an area with a stone, rounds off an edge, and makes the difference between efficient operation and constant trouble. This chapter barely touches on the problems or the solutions, but, hopefully, offers a better insight into some of the pitfalls and some of the problems of parts handling.

CHAPTER 7

POSITIONING DEVICES

Positioning devices transfer the selected, oriented, tracked and/or magazined work piece from the escapement into a nest, fixture, pallet or the previously assembled parts. They are found in many combinations, specials, standards and universal devices.

The vast majority of placements devices are special; designed to order for specific applications. This is probably because standard devices have not yet been offered in sufficient variety and flexibility to satisfy even a very tiny portion of the requirements. We will look at a few special applications, a few standard offerings and examine contemporary universal devices in more depth, since them seem to be the answer of the future.

SPECIAL DEVICES

The simplest device is probably the "drag-off" type of mechanism illustrated in *Fig. 1*. Selected, oriented parts are presented to a spring-loaded escapement. A fixture pin, or some part of the assembly on the fixture, engages the hole in the part and drags it off the escapement into the fixture as the fixture moves under it. This method may be used on high-speed continuous machines very efficiently or it may supply parts to an indexing machine during transfer. While this device does not offer "positive parts placement" or parts presence checks, it does permit high speeds and elimination of more complex, expensive placement devices. This is a combination of tracking, escapement and positioning in one unit. With proper debugging, relatively simple parts may be handled at high speeds and efficiencies. Later stations, or machine areas may be used to check for presence, possibly on the same continuous basis or during indexing.

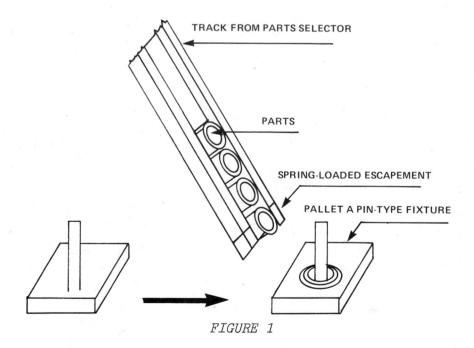

FIGURE 1

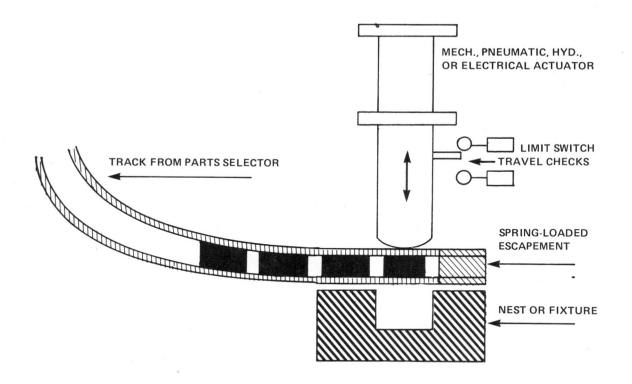

FIGURE 2

Probably the most common placement device is shown in *Fig. 2*. Selected and/or oriented parts are tracked to a spring-loaded escepement. A "pusher", operating at 90° to the parts travel in the track, pushes the part through the escapement into a fixture or assembly. The pusher acts to hold back the next part while loading the first one. Such pushers may be cam-operated, air-cylinder, hydraulic cylinder, or solenoid operated. Limit switched actuated by a finder on the pusher can serve to signal completion of both down and up strokes plus checking for the presence of the loaded part. This method would be used for parts which track well in O.D. to O.D. configuration such as washers.

Parts which are best fed end-to-end may be handled in the manner shown in *Fig. 3*. In this case, close-wound, ground-end springs are either selector-fed or manually-loaded to a tube-type magazine. A shuttle escapes the bottom spring, while holding back the rest of the stack, and moves it into location under the pusheer. In this example, the spring must be loaded into a cup in the fixture nest without letting it tip over or otherwise "losing" it as it is positioned. A spring-loaded probe, or pilot on the positioner descends. Next the pusher itself pushes the spring through some form of spring-loaded escapement, (not shown), and into the cup. Again, travel checks are shown which can serve to detect part presence as well as signal the station when it has completed strokes both ways. Actuation may be cam, cylinder or solenoid.

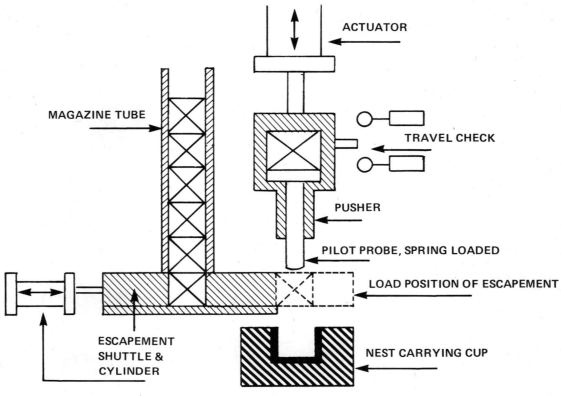

FIGURE 3

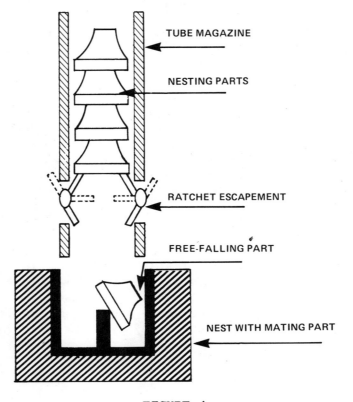

FIGURE 4

Fig. 4 is presented as a "horrible example" of parts positioning. Nesting parts are magazine loaded in the nested position, leading to possible wedging together of pieces where they should have been tracked O.D. to O.D. A ratchet-type escapement presumably releases one part per actuation, but there is no positive means of causing this, and no means of checking that it actually happened. The part is permitted to free-fall opening possibilities of tipping, wedging, missing the pin on which it is meant to assemble and no means of preventing a machine jam-up of the loading fails.

There are certain basic rules for efficient parts positioning which should be observed wherever possible:

1. Positive parts placement is always desirable as opposed to gravity fall.

2. Actuation by a consistent and controllable means. Mechanical or cam action is most consistent and readily visible. Solenoid or electric device, along with hydraulic or pneumatic cylinders must be adjustable, cushioned and monitored.

3. Parts presence checks and checks for completion of all strokes increase efficiency, prevent jam-ups and minimize scrap or faulty assemblies.

4. Overtravel protection for the positioning device will minimize equipment and parts damage as well as eliminating the necessity for stopping the assembly machine every time a bad part or mislocated part gets in the system.

Once consideration is given to good concept practices, there is no limit but ingenuity to the varieties of positioning devices. If it is not necessary to design a special device, if the "wheel does not have to be re-invented" for a given application, consider using a standard device discussed next.

Standard Placement Devices

Standard mechanisms requiring only the addition of end tooling, such as jaws, actuators, etc., are currently available and more coming on the market every year. They range from small pick-and-place units with motions of 2-3", *Fig. 5*, up to completely self-contained assembly stations capable of handling hundreds of pounds through distances of several feet, *Fig. 6*.

The elements of a flexible standard placement device would include:

1. variety of motions, and combinations of motions. *Fig. 7* suggests that a given unit be toolable for vertical/straight line (a), horizontal straight line (b), both horizontal and vertical straight line (c) and (d), inverted "U" path at right angles to line (e), and inverted "U" parallel to line (f). Any of the above should be capable of combination with standard shuttle motions, auxiliary cylinder-operated jaws and other accessories.

2. actuation should be available as a self-contained package or separate input such as from a tooling plate or cam shaft. *Fig. 5* shows a unit with mounting for tooling plate actuation, but a cam shaft lever,

cylinder or electric motor will be practical.

3. adjustability - or controllability - of stroke.

4. overtravel protection for the mechanism in the event of jams or overloading.

5. rugged, simple design for long-life repeatability & maintenance.

Figures 8 and 9 illustrate standard devices having these features and many more. Among the many units on the market are: A. G. Russell, "Uniplace", Ferguson Linear Parts Handler, Ferguson Rotary Linear Parts Handler and Stelron "Vari-Pak".

All of the above are single-purpose placement devices. They are engineered and tooled to perform one series of functions until such time as they are re-tooled for a different job. If the problem you have at hand should require a variation in the motions or functions from one machine cycle to the next, or if the total number of motions required is great, the universal positioning devices discussed next may be your answer.

A	B	C	D	E	F
2.12	.62	3.00			
2.75	1.25	3.00	2.25	.62	4.0
2.75	1.25	4.00			

Available Tooling Strokes

FIGURE 5. *Typical self-contained, cam-operated "Pick and Place" unit designed for use with dial index bases having vertically oscillating tooling plates to supply the power through the bracket at the top. The cylinder acts as an airspring for overload protection. 2-4" input strokes result in the possible patterns shown above. By Gilman Engineering.*

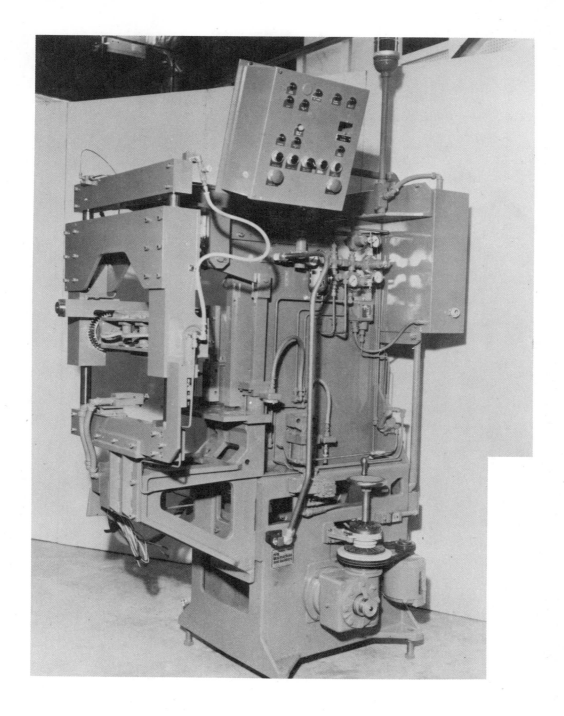

Figure 6. A large, self-contained work station which has its own drive motor, clutch, brake, cam shaft with room for three cams, a variety of motions, and complete controls. Such a unit may be operated as a complete, separate machine or attached to a synchronous or non-synchronous machine. As shown, it is designed to have work pallets on a power and free conveyor pass through the front of the machine. Pallet shot pins are part of the station. Such a unit may be arranged for positioning of parts for assembly, gaging, functional testing or any of a wide variety of operations. By Gilman Engineering.

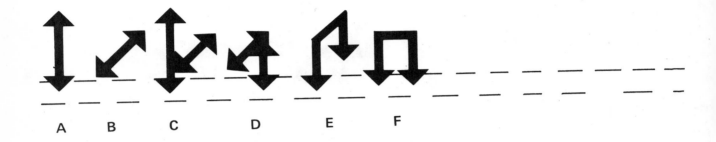

FIGURE 7. *Typical Standard Placement Device Motions.*

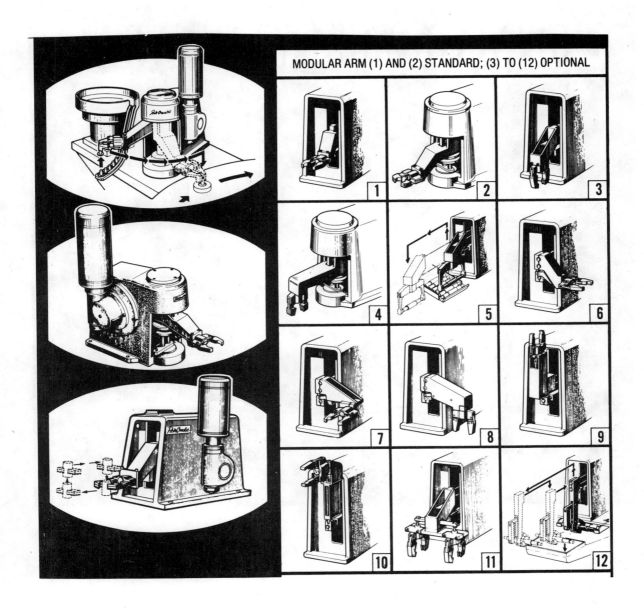

FIGURE 8. Standard, off-the-shelf pick and place device available in 5 models and sizes for use with electric motor, hydraulic or air motor drives. By PickOmatic Systems.

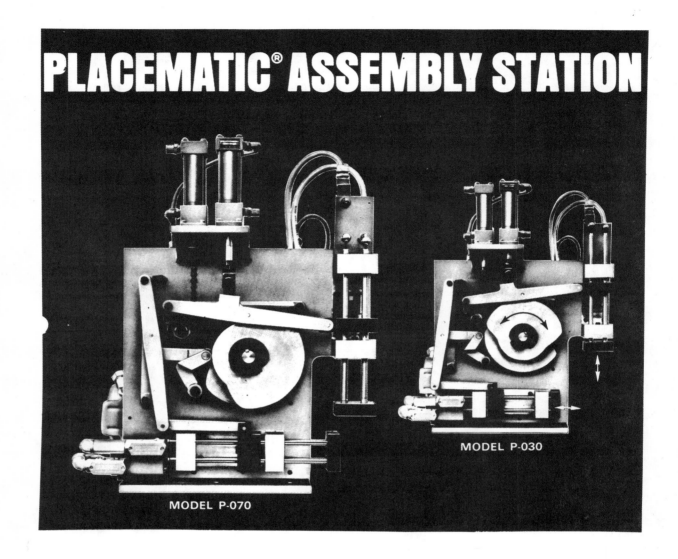

FIGURE 9. Self-contained, electric motor driven assembly station may be single or double tooled, floor or table mounted and is available in two model sizes with programmable strokes, horizontal and vertical cam motion of any configuration within 3" or 7" stroke distance. Overtravel protection is provided on both horizontal and vertical movements during forward or back stroke. Other options include travel checks, motion lockout and various motor sizes and ratios. By Gilman Engineering.

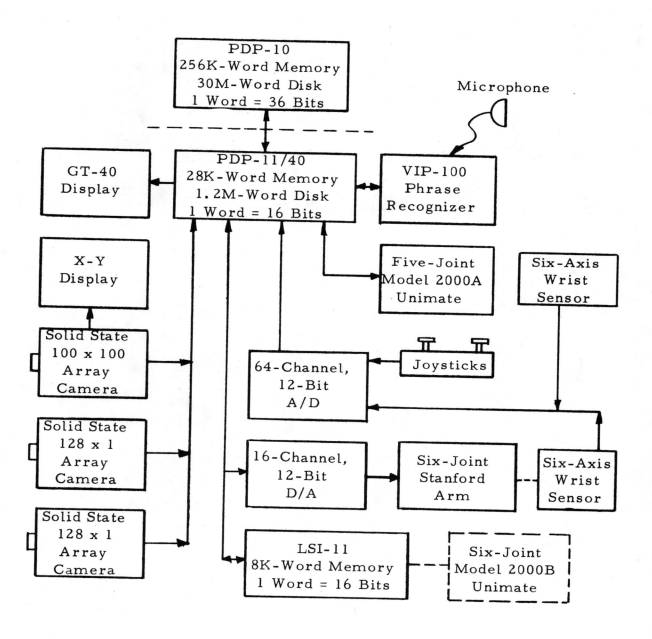

Figure 10- System Components Under PDP-11/40 Control

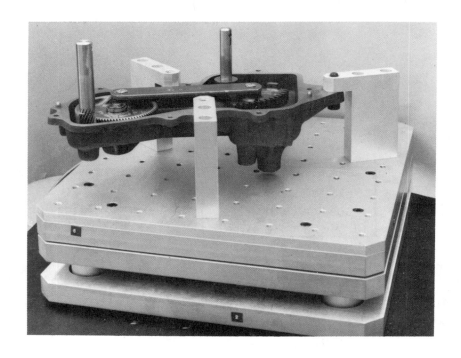

Photograph of Gearbox Base Attached to Pedestal Force Sensor

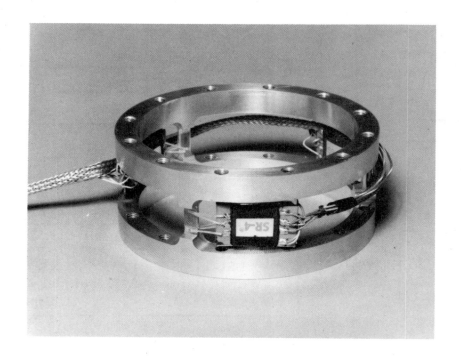

Photograph of Prototype Wrist Force Sensor

Figure 11

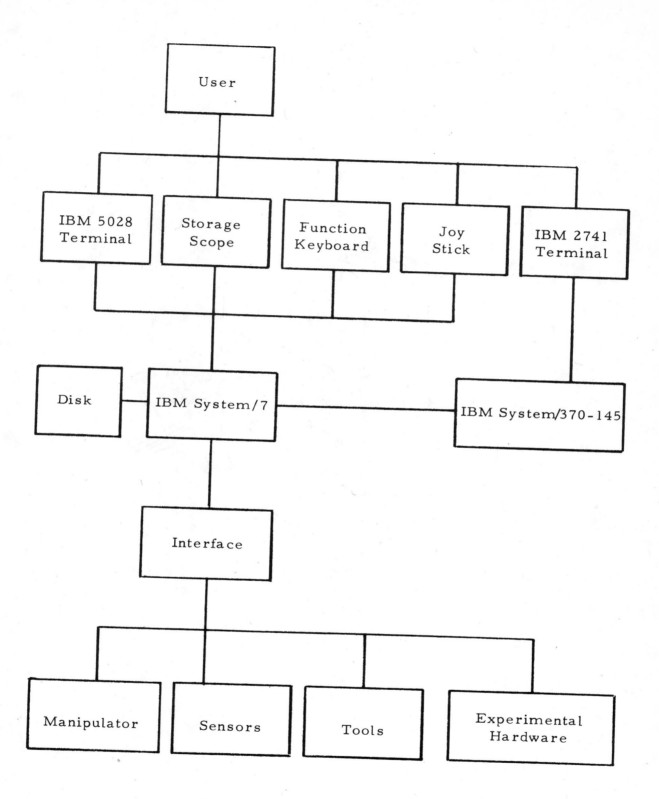

Figure 12

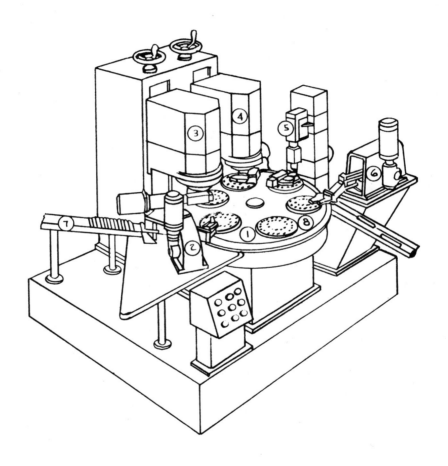

Figure 13

Rotary Index Machine

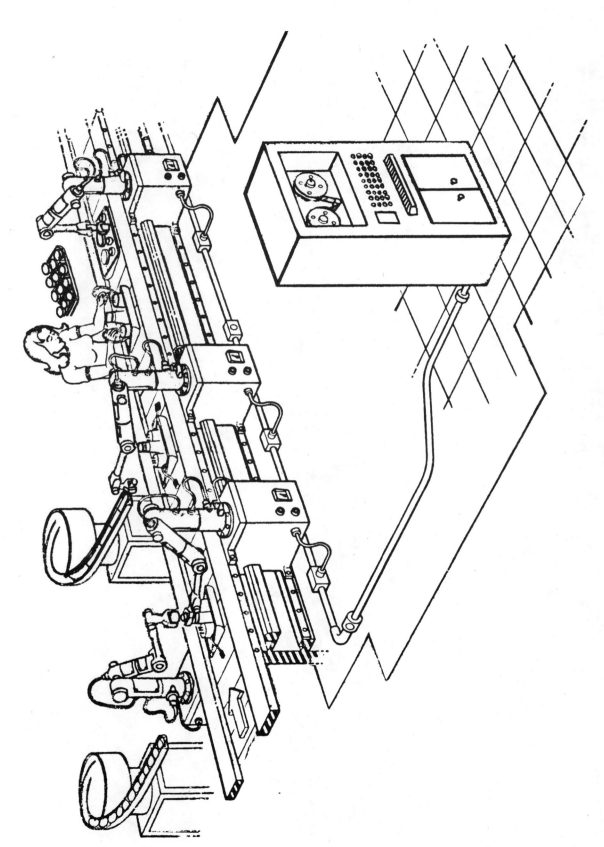

Figure 14

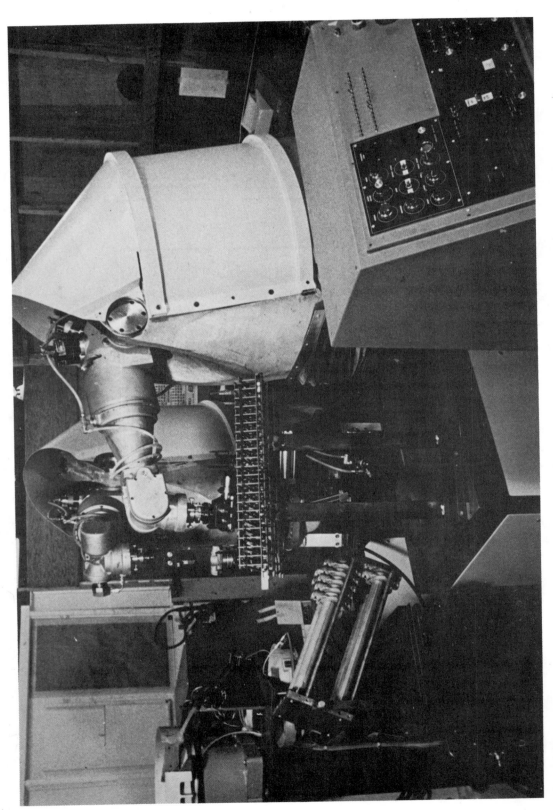

Figure 15

Prototype Two-armed Unimate Assembly Station

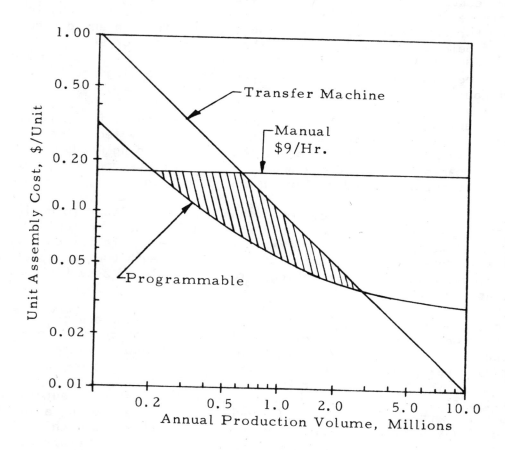

Figure 16

Regions of Economic Advantage for Three Assembly Methods.

Universal Positioning Devices

Many names have been applied to the universal positioning devices we will discuss. "Automatic Positioning Equipment" was used by some (although the acronym "APE" was resented by those whom it replaced). Robotics has been increasingly accepted as a general term and one we will use. The term "Robot" has been loosely used to indicate any mechanical device replacing manpower at the one extreme, to only "thinking", multiple feedback, sensory-equipped devices at the other extreme. One dictionary definition calls for "a mechanical man; a dull mechanical person"! For our purposes, we will use the term to mean any device capable of being programmed for more than one motion, or series of motions, and having the capability of using more than one program upon signal or feedback information. This eliminates the single-purpose piece of equipment.

"Inexorable" is the conclusion. Robotics has been largely limited to transfer and machine loading applications; but, today technology advances are rapidly being matched to the more demanding needs of assembly. Meanwhile the socio-economic environment becomes ever more hospitable to sophisticated and expensive automation.

The time, however, is ripe. More than 40% of the labor force in manufacturing is employed on assembly work. They expect not only "equal pay for equal work" but also a differential to compensate for debilitating conditions. Moreover, the roboticist today has a much broader spectrum of technology to draw upon in making a robot capable of assembly operations.

No one has pulled it all together yet, but the signs are good. Hereinafter we will look at some of the remaining problems and some of the pertinent work being done by both research laboratories and industrial development laboratories.

First, it is necessary to put conventional hard assembly automation into focus. Certainly thousands of products are already being assembled either on fully automatic machines or on highly mechanized machinery requiring little human intervention. Just as certainly more products will succumb to hard automation in the future as this technology continues its development and as automation houses enjoy the market place leverage of escalating labor rates.

It is also certain that the preponderance of __batch__ production assembly operations will not enjoy conventional automation, for both technical and economic reasons. As Professor Boothroyd, University of Massachusetts, puts it:

> "It is also generally a special purpose one-off machine and therefore cannot be considered for assembly of products other than those satisfying all of the following requirements:

(i) A volume of at least one million per year.

(ii) A steady volume of production.

(iii) A market life of at least three years.

(iv) A size of the order of between 0.5 to 20 inches with individual parts to be automatically assembled generally between 0.05 and 5 inches in their maximum dimensions.

(v) Consisting of parts which do not deform significantly under their own weight or will not break when dropped from a height of about 3 inches onto a hard surface."

In addition, hard automation is usually eschewed when parts quality requires the adaptability of a human operator, or where assembly cannot be done in "pancake" fashion.

How will robotics carve a healthy niche between hard automation and human labor? By adding these qualities to the automation bag of tricks:

1. <u>Programmability</u>

    Part manipulators which can be readily reprogrammed need not be expensed over the life of a single product or sub-assembly. (When Unimates were first introduced, wary customers insisted on short range payback usually under two years. With programmability proven and life of over 40000 hours proven, customers now depreciate over eight years as they might with any general purpose machine tool.)

2. <u>Adaptability</u>

    The issue here is how well will the assembly robot be able to emulate a human operator in coping with job variables. In due course we expect robots to have:

    a. Rudimentary vision that might permit inspection and rejection of substandard parts. It might also enable the robot to guide mating parts into juxtaposition. And, it might enable the robot to both identify and determine orientation of randomly placed parts. Ultimately the robot eye would enable the robot to pick disoriented parts out of the tub.

    b. Tactile sensing could be useful in both recognition and orientation, but its most powerful role will be in measuring and reacting to contact interaction. Is the peg going into the hole? Is the nut aligned with the bolt? Should a misfitting part be discarded?

Interest in the problem (and opportunity) is intense throughout the industrialized world. Contributions will undoubtedly be made by offshore investigators; but, at this writing, it is possible to draw upon examples of U.S. work to pretty much cover the field. There is no published foreign activity that outstrips American R & D.

Programmability in its most basic sense is rather old hat. Unimation has for the past 15 years offered commercially a programmable industrial robot and more than 1000 are in the field. Record-playback generation of programs is perfectly adequate for simple tasks such as operating a die cast machine but assembly is a much more intricate process and programming a multi part assembly by leading the robot arm through all manipulations could be most arduous.

Certain subroutines can be helpful. Unimation (and others) have generated programs that enable a robot arm to move in rectilinear coordinates or in coordinates of a robot's hand even though the arm is a polar coordinate structure. A very useful subroutine is palletizing. After record playback teaching of pick up for one part in a pallet, the subroutine generates program for pick up of parts in all other locations.

There are more elegant subroutines such as trajectory optimization and "guarded moves", as formulated in the IBM computer Sciences Department, the arm moves according to motion commands until sensor signals announce an environment change which in turn evolves a "branch" command to take alternate action. Stanford Research Institute has demonstrated voice command of a robot wherein a programmer can verbally demand actions such as straight line motions in any direction or curvilinear motion around an arbitrary center.

It comes down to this - programmability is pretty much in hand. Adaptability is another matter.

Before striving for human-like adaptability in a robot, it is probably cost-effective to make the assembly process itself more adaptable. In this regard, the work of Professor Boothroyd and his colleagues at the University of Massachusetts is worthy of note. He serves both robotic and hard automation interests. First of all, he suggests that parts be designed to ease the assembly process.

Next, Boothroyd suggests we deal with assembly operations using "Group Technology" concepts to take some of the "black art" out of automatic feeding and orienting devices. Under NSF sponsorship, the University of Massachusetts' team will create a handbook of feeding and orienting techniques based on a group technology approach.

The handbook comprises six sections:

"1. A coding system for small parts.

2. A compendium of feeding and orienting techniques arranged under part code numbers.

3. A catalogue of various non-vibratory feeders.

4. A description of vibratory feeders.

5. A catalogue of orienting devices used in vibratory feeders.

6. Data on the natural resting aspects of parts."

The National Science Foundation sponsors a number of programs in addition to Boothroyd's which are directed toward flexible assembly automation. Most appropriate of these to this presentation are the continuing efforts of Stanford Research Institute and C.S. Draper Labs. Both of these organizations are devoting the bulk of their work to adaptability issues.

Figure 10 is a block diagram of the SRI hardware being used to expand robot capability. Evidently experiments can include visual scene analysis, tactile sensing as well as the voice communication already mentioned. SRI has demonstrated ability to locate, identify and acquire disoriented parts on a moving conveyor. They have also developed a tactile sensing algorithm that enabled their computer controlled Unimate to perform complicated packaging chores that depended upon force sensing.

Draper Labs is working to get a handle on the assembly process. Of prime importance is the investigation of the "Part mating phenomenon". This involves systematic modelling and experimental verification of the events which occur as parts interact during assembly. Out of understanding comes some design of sensor inputs that would provide tactile adaptability. *Figure 11* shows two different interactive force sensing systems. The top picture is a pedestal strain gauge instrumented to deliver data on forces and torques occuring during assembly interactions. The bottom picture shows a sensor that delivers complimentary data through measuring forces and torques that occur in a robot's wrist. The sensor data must be interpreted and then used in some closed-loop format to control robot articulations so as to bring about part mating.

Industry is not sitting idle. In the first place, both SRI and Draper have attracted industrial associates to their programs enlisted from the ranks of prospective users and prospective builders of programmable assembly machines. Exemplary, but not exhaustive, of industrial development work is that carried out by IBM, by General Motors Manufacturing Development Laboratory and by a cooperative Ford Motor Co. and Unimation Inc. development.

As might be expected, IBM is interested in whether or not computer controlled assembly will be a viable activity. The exploration at the Thomas J. Watson Research Center is under the direction of Dr. Peter M. Will.

Figure 12 is a block diagram of the hardware IBM is using in its explorations. All key elements are on hand, manipulative power, sensory power, and computer power. Sound experiments have demonstrated the operating system and low level control language sufficient to make some rather complex assemblies. To date, no claim has been made that hardware is either in performance or cost appropriate to field installations. The ongoing research will be directed toward high level languages for part description, orientation, trajectory and inspection specification plus procedures for the programming of two cooperative manipulators.

Figures 13, 14 and 15 are related. *Figure 13* is a convenient jumping off point, a recognizable rotary index assembly machine. Next is *Figure 14* which should also be recognizable. This is an in-line indexing assembly machine which mix-matches automatic stations with human operator stations. The drawing and the concept are owing to General Motors. What is new is the use of programmable automation stations. General Motors calls this system, which is being built up experimentally, P.U.M.A. - Programmable Universal Machine for Assembly.

The General Motors objective is to demonstrate that such a line can be used to assemble a range of automotive subassemblies with but modest retooling and easy reprogramming. At the outset, adaptability is not expected. The system makes only the programmability step forward. This just might open up a good range of automotive assembly activities which heretofore have remained manual.

At first glance, *Figure 15* may look to be retrograde, somewhere between *13 & 14* but Unimation Inc. and Ford Motor Co. hope that some distinction will be discernable. This is a single assembly station in which a complete multipart assembly job is done in station by a two armed robot. The design objective is to perform the assembly 1.5 times as fast as a human operator. Initially this assembly machine will have little adaptability. The work table will be compliant and programs will provide for hand-hand coordination. All else will be simply programmable including rapid hand changes to effect different assembly activities.

On the other hand, the system design has faith in the eventual success of sensor adaptability and it will be ready to adopt SRI, Draper, or any other adaptive technology that proves to be practical.

Comparing the in-line system of *Figure 14* and the in-station concept of *Figure 15*, one can conclude that the in-station robot arms must have more manipulative power than the in-line robot arms. In-line assembly may involve but one or two operations per station and therefore four or fewer articulations may suffice for each robot arm. For truly flexible in-station automation, six articulation arms are probably essential.

There is no dramatic schism between these two concepts. Both can be viable and both must pass economic filters that are highly dependent upon the cost effectiveness of the robot arms employed. The fundamental issue is unit assembly cost and that is dependent upon the product of the cost of an assembly station and the part assembly time. A fast sophisticated station may outstrip a slow simple station.

Draper Labs in the person of Paul M. Lynch has examined the regions of economic advantage for three assembly methods, manual, hard automation, programmable automation. While the data is variously quantified by Lynch, the generalized plot of *Figure 16* tells the story. The log of unit assembly cost is plotted versus the log of annual production. The shaded belly is the region of cost effectiveness for programmable assembly. That belly will evidently grow as manual labor costs rise and as programmable assembly stations are speeded up or reduced in cost as the result of quantity production.

There is one concluding prediction for an audience of assembly experts. That is, robotics will not erode the base of hard assembly automation. Robotic or programmable assembly will find its market in the vast arena of manual assembly that cannot now be served by builders of hard automation. Furthermore, the use of programmable assembly machines can only bode well for the producers of orienting equipment, feeders, conveyors and such. These automation components will continue to be essential and cost effective.

Robot Applications Guide

An industrial robot is a programmable manipulator capable of performing a variety of work tasks such as spotwelding, tool handling, painting or parts transfer. It is versatile and flexible and can be retooled and reprogrammed easily to perform a new task when the old task changes or goes away. It is reliable and, with normal maintenance, will experience less than 2% downtime. Robots come in a variety of sizes and shapes, with physical capabilities covering a wide range. All of the robots presently available have some common characteristics. They are all basically one-armed, blind idiots. They have a limited memory, cannot think and cannot see or hear. Consequently, their application requires some rigid constraints and some careful planning.

All of the robots included in this discussion are programmed or "taught" by physically guiding them through their intended task and recording positions in space along the way. The robots then perform their task by "playing back" these recorded positions. Since the positions taught are related to the robot's own coordinate system, it follows that the parts on which the robot works must be located in the same place every time or else the robot must know where the part is at all times. This is one of the primary requirements of any application and is often the most costly phase of the project.

Some, but not all, of the robots are "intrinsically safe" and can be used in explosive hazard applications, such as paint booths. All of the robots can be used in the normal production environment and usually require only electrical power and cooling water for services. Although normally floor floor mounted, many of the robots can be inverted and mounted above a line or fixture or wall-mounted alongside a work station. This feature should be kept in mind when reviewing potential applications; one overhead mounted robot might work where two floor mounted robots would otherwise be required.

Robots are being successfully applied to numerous spotwelding operations, some parts transfer operations and to prime spray small parts. In all of these cases, the parts are fixtured or located and are stationary while the robot works. With the recently introduced Trallfa and Cincinnati Milacron 6CH robots, some moving line operations have become feasible. Typical welding operations include subassembly welding wherein an operator loads and unloads the fixture(s) and the robot handles the welding gun and respot line welding where the body is indexed into a stop station and located while the robot welds. Typical parts handling involves the transfer of an assembly from a stop station on a build-up line to the hooks on an overhead delivery conveyor. The line tracking capability of some robots will permit either one or both lines to be moving during pickup and transfer of the part. Operations which are not feasible at present are those where the parts are not consistently located from job to job, such as spotwelding on a moving respot line, parts transfer from racks or dunnage into a fixture or tool handling requiring more than one tool.

In a typical spotwelding operation, a robot is capable of operating at about the same speed as a man, except when welding through access holes, where a robot is slower than a man. In a transfer operation, the robot is generally faster than a man, since the load-carrying capability of the robot makes the use of a hoist unnecessary. Painting speeds with a robot are comparable to or faster than manual painting operations. As a rough rule of thumb, an application where one robot can replace two men per day, without extensive facility changes, will generally pay off well enough to warrant detailed investigation.

As with any automated equipment installation, robot applications must be backed up to prevent production loss in the event of machine or peripheral equipment downtime. In many cases, this backup capability can be attained by leaving the manual system which the robot replaces intact. For example, in a transfer operation, the manually operated hoist and bridge may be left in place. Of course, availability of manpower to operate the standby equipment must be considered. In other cases it may be possible to provide a "float" of completed assemblies after the robot operations.

The key to minimizing robot downtime is a well trained maintenance force, provided with adequate tools and spare parts. Maintenance training and programming instruction are provided by the robot manufacturers, either at their facilities or on-site. Since the mechanical and hydraulic systems are relatively straightforward, the major training requirement is for electricians. Past experience has shown that it is best to assign the primary maintenance and programming responsibility to the electricians, with other skills assisting, as required.

Initial complements of spare parts and test equipment should be provided. Purchase the cassette tape recorder, if offered with the particular robot, for permanently recording programs. This permanent record can then be used to quickly restore programs in memory, as required. Project cost analyses should include items for training, spares, testers and recorders.

Experienced assistance is available from the robot manufacturers covering all phases of a robot application from initial survey through installation and programming. However, the primary responsibility for implementation and continued operation lies with the plant. Also, technical advice provided by the robot manufacturers (especially related to performance of the robot) must be considered carefully, since the vendors have a natural tendency to amplify the capabilities of their equipment.

The following table compares some of the characteristics of the four robots which are presently available; Cincinnati Milacron, Trallfa, Unimate and Versatran.

FIGURE 17

	CINCINNATI MILACRON 6CH	TRALLFA	UNIMATE 2000 & 4000	VERSATRAN SERIES F
Drive System	Hydraulic	Hydraulic	Hydraulic	Hydraulic
Control System	Minicomputer	Microprocessor	Solid State	Microprocessor
Memory System	Core Memory	Tape or Disc	Plated Wire	Semi-Conductor
Memory Capacity (Max.)	900 Points	4 to 128 Minutes	1024 Points	1300 Points
Point-to-Point Control	Yes	Yes	Yes	Yes
Continuous Path Control	Yes	Yes	2000 Only	No
Line Tracking Ability	All Axes	All Axes	One Axis	One Axis
Load Capacity (Max.)	300 lbs.	10 lbs.	350 lbs.	1100 lbs.
Speed (Max.)	50 in/sec	67 in/sec	50 in/sec	36 in/sec
Accuracy	± .050 in.	± .040 in.	± .080 in.	± .050 in.
Interlocks (External)	24 in, 16 out	6 in, 5 out	6 in, 6 out	32 in or out
Wall or Overhead Mount	Yes	Yes	Yes	Yes
Number of Axes	3 to 6	5 to 6	5 to 6	3 to 7
Intrinsically Safe	No (in process)	Yes (FM approved)	No	No
Services Required	Electrical-460V 11 KVA Water - 10gpm	Electrical-460V 4.5 KW	Electrical-460V 30 KVA	Electrical-460V Water - 2gpm

TECHNICAL DATA - CINCINNATI MILACRON 6CH ARM

Cincinnati Milacron offers only one basic model of its robot. This is a 6-axis machine with separate hydraulic power supply, electrical distribution panel, control console and mechanical unit as shown below.

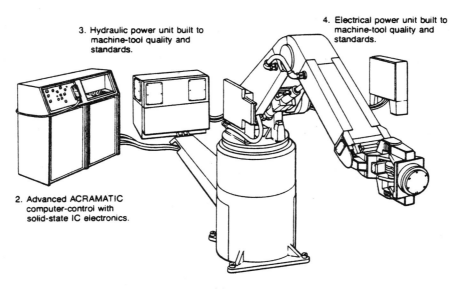

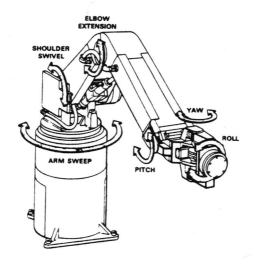

FIGURE 18

Specifications of the Cincinnati Milacron robot are as follows:

Floor space   *Net weight*
6CH Arm . . . . 6 sq. ft. (0.5 sqm)   4,000 lb (17.6 kN)
Hydraulic power supply. . . . . .17 sq. ft. (1.5 sqm)   1,200 lb (5.3 kN)
Electrical power unit . . . . . . 3.4 sq. ft. (0.3 sqm)   700 lb (3.1 kN)
ACRAMATIC computer control. . . .10 sq. ft. (0.9 sqm)   1,100 lb (4.8 kN)

Load capacity
Load 10" (254mm) from tool mounting plate . . . . 125 lbs (5.5 hN)
Load at tool mounting plate, max. velocity. . . . . 175 lbs (7.7 hN)
Load at tool mounting plate, reduced velocity,
   depends on arm and wrist attitude. . . . . 300 lbs (13.2 hN)

Positioning accuracy, axid drive
Accuracy to any programmed point . . . . ±0.050-in. (±1.27 mm)
Drive for each of 6 axes . . . . .Direct, electrohydraulic

Jointed-arm motions, range, velocity
Max. horizontal sweep . . . . 240°
Max. horizontal reach . . . . . . 97-in (2464 mm) to tool mounting plate
Min. to max. reach, floor to ceiling. . . . . .0 to 154-in (3912 mm)
Max. velocity with full load. . . . . 50 ips (1270 mmps)
Arm motions. . . . . Three; arm sweep, shoulder elevation, elbow extension

Wrist motions
Pitch . . . . 190°
Roll. . . . . 240°
Yaw . . . . 180°

Power requirements. . . . . .230/460 volts, 3 phase, 60 Hz, 11 KVA
Environmental temperature . . . . 40 to 105°F (5 to 40°C)
                     (Opt conditioner for temperatures above 150° F)
Minicomputer memory capacity. . .400 points std (additional storage option-
                                                   ally available)

The standard control has the capacity to store 400 program points. These may be entered either by teaching or loaded by the tape reader. An additional block of 500 points may be added by acquiring an additional memory board when the unit is purchased or later.

Each program point is a complete set of data of the arm's status and contains the following information:

1. Position data required to locate the tool center point (TCP) in the working space.

2. One of eight selectable velocities at which the TCP will approach the position. 1, 2, 5, 10, 20, 30, 40, 50 inches per second are normally used, but any eight may be specified.

3. One of the following functions to occur at the position:

   a. <u>Continue</u>. The TCP passes through the position without stopping.

   b. <u>Wait</u>. The arm stops and waits for one of 24 selectable contact closures.

   c. <u>Output</u>. Upon reaching the position, one of 16 selectable state defined contact closures is activated (2 amp. max.).

   d. <u>Time Delay</u>. The arm pauses at position for 1 to 240-1/4 sec. increments (1 min. max.).

   e. <u>Tool #1</u>. At position an <u>output</u> is given followed by a <u>wait</u>. Wiring is on the arm.

   f. <u>Tool #2</u>. A second pair of <u>output</u> and <u>wait</u> functions.

   g. <u>Branch</u> (sub-routine). At position, a selector is made of one of 24 branch programs depending on contact closure.

   h. <u>Tracking</u> (optional). At position the arm waits for one of 15 selectable branches to be performed in synchronization with a moving line.

4. The point identification in the main, a branch, or a tracking section of the program and its sequence in that section.

This robot normally operates in a straight-line mode, that is, the tool at the end of the arm moves in a straight line between programmed points. The control system automatically calculates the relative motions required for each segment of the trunk and arm to move the tool in the straight path. This greatly simplifies programming. Each segment may also be moved individually when programming, if so desired. The robot can also simulate continuous path movement by moving in straight lines through a series of closely spaced points without stopping.

The Cincinnati Milacron robot also has the ability to perform work on a moving object and to transfer parts to or from a moving line. An external device which continuously monitors line speed is required.

The Cincinnati Milacron robot is not "intrinsically safe" and cannot be used in hazardous (explosive) environments. The manufacturer is, however, in the process of obtaining such certification.

A layout showing component sizes and the work area of the robot is shown on the following page.

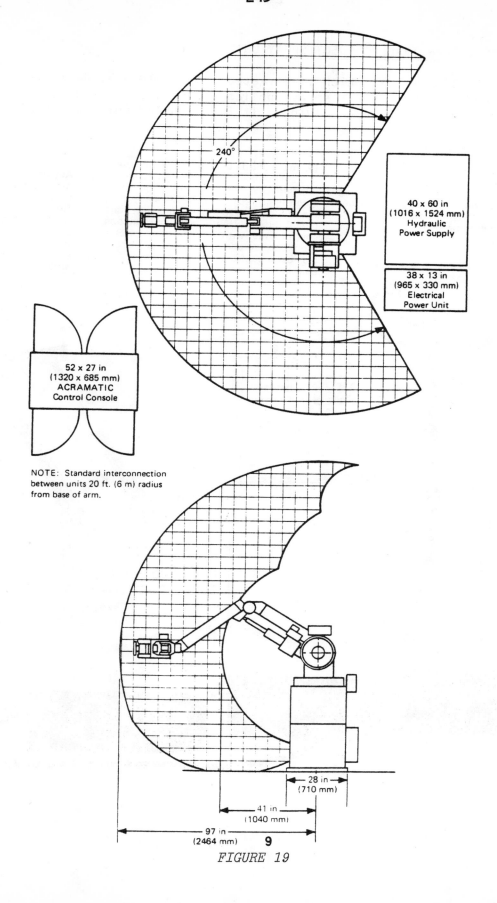

FIGURE 19

TECHNICAL DATA - TRALLFA ROBOT SPRAY UNIT

The Trallfa robot, which is marketed and serviced by DeVilbiss, is intended primarily for spray application of paint, deadeners and adhesives and for arc welding. It is a modular unit, as shown below:

❶ The Hydraulic Power Supply, containing the oil reservoir, electric motor, pump, valve manifold and the air heat exchanger.

❷ A standard bracket for attachment of a spray gun is mounted on one of the turning motors.

❸ The Control Unit containing the main junction box, the electronics rack with the «plug-in» type circuit boards, the control panel with the recording and play-back controls and the tapereader.

❹ The Remote Control Panel containing necessary lamps and push buttons for start/stop and function controls as well as an emergency stop button.

❺ Hydraulic Servo Actuators for positioning of the manipulator arm. Three linear and two turning actuators allows the arm to emulate the movements of the human sprayer.

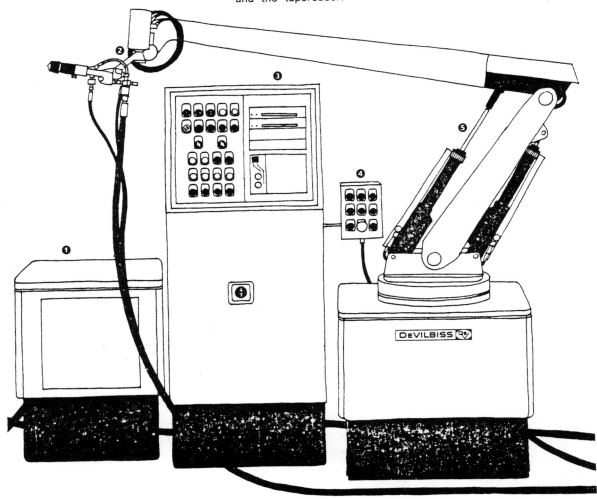

FIGURE 20

Specification of the Trallfa robot are as follows:

WEIGHTS

Hydraulic power pack
356 lbs/160 kg

Control Cabinet
244 lbs/110 kg

Manipulator
1000 lbs/450 kg

WRIST ROTATION

Horizontal 210°
Vertical   210°

MAXIMUM ARM SPEED

5'7" per sec/1.7m per sec

ELECTRICAL CURRENT

4.5 KW 230/460V
60 cycle or 220/380V
50 cycle 3-phase

GUN TRAVEL LIMITS

See diagram

PROGRAMMING CAPABILITY

105 seconds max. duration
(SCT Cassette)
300 seconds max. duration
(RPS Deck)
Up to 15 separate 20 second
programs per RPS Deck
Program selection from 4 decks
can be made at random with RPS

SYSTEM OPTIONS:

For extended flexibility within
the automatic system, the following
optional configurations have been
created for the TRALLFA ROBOT:

Intrinsically Safe Version with
approval from F.M., BASEEFA, P.T.B.
and STATENCY PROVNINGSANSTALT
Electrostatic Spray Gun Capability
Early Program Reset
Program Interrupt
Sixth Motion
2 additional Output Functions
Central Computer Control
Floppy Disc Memory

The Trallfa robot has a continuous path control, using either magnetic tape or a microprocessor and floppy disc memory. It has capability to work on a moving line, with an external feedback device to monitor line speed. It is available as an intrinsically safe (F.M. approved) unit and can be used for electrostatic spraying. A 6-axis machine, with microprocessor control, floppy disc memory, electrostatic spraying capability, is intrinsically safe and capable of working on a moving line. No special recorders or test equipment is required.

A layout showing size and work area of the 5-axis robot is shown below:

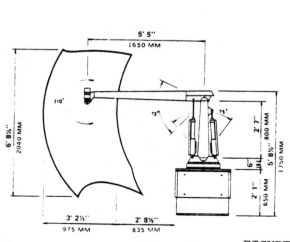

FIGURE 21

TECHNICAL DATA - UNIMATE

Unimation builds a family of robots in two series, 2000, *Fig. 22*, and 4000, *Fig. 23*, in both 5-axis and 6-axis models. The 2000 series robots are available in either standard or extended-reach models and normal or heavy-duty versions. A "space saver" model of the 2000 series is available, which can be mounted in any attitude. A "RIG" model with continuous path control, which can be used for MIG welding and flame cutting is also available in the 2000 series. The 4000 series is available in either standard or heavy-duty versions in the 5-axis machine. The 2000 series robots (except for the space saver) are a single unit configuration. The control console on the 4000 series is separate.

The Unimate robots have a limited line tracking ability. This system, which requires an external feedback device to monitor line speed, allows the robot to move one axis in synchronization with the line.

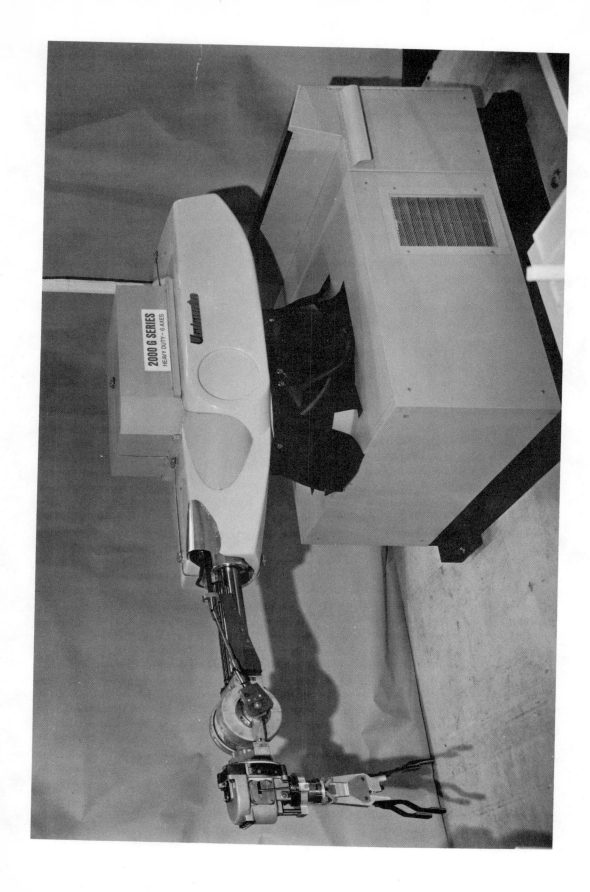

Figure 22 Unimate series 2000 robot

Figure 23 Unimate series 4000 robot

Specifications of the various Unimate robots are as follows:

Model 2000-2100 Specifications

Load Capacity:	Up to 100 lbs. (Subject to application analysis)
Hand Characteristics:	Mounting surface for tools or gripper: 2⅞ inch diameter bolt circle or per your requirements.
Memory Options:	128 steps 256 steps 512 steps 1024 steps Flexible arrangements available.
Programming:	Record playback using "Teach Control." Point to point.
Positioning Accuracy:	0.050 in.

Operating Data:

Max. radial velocity:	30 in./sec.
Max. radial stroke:	Model 2000—41 in. (Model 2100—53 in.)
Max. vertical stroke:	Model 2000—57° (Model 2100—52°)
Max. vertical velocity:	50 in./sec.
Max. rotation:	220°
Max. rotational velocity:	110°/sec.
Max. wrist bend velocity:	110°/sec.
Max. wrist bend:	220°
Max. wrist swivel:	300°
Max. wrist swivel velocity:	110°/sec.
Max. wrist yaw:	200°
Max. wrist yaw velocity:	110°/sec.

Wrist Load Capacity

Bend:	1000 in.-lbs.
Yaw:	600 in.-lbs.
Swivel:	800 in.-lbs.
Interlocks:	Up to 12 channels of incoming and outgoing command signals.
Optional Accessories:	Standard accessories as required, custom hand and finger tooling, protective boots.
Environmental Conditions:	Ambient temperature 40°F to 120°F. Humidity. 0-90%.
Power requirements:	220/440 volts, 3 phase, 60 Hz. 11.5 KVA. 90 PSI air at 2.5 CFM when required.
Weight:	2800 lbs.

Space Saver Model 2000-2100 Specifications

Load Capacity, Hand Characteristics, Memory Option, Programming, Positioning Accuracy, Wrist Load Capacity, Interlocks, Optional Accessories, Environmental Conditions, Power requirements and Operating data are the same as for basic Model 2000.

Model 4000 Specifications

Hand Characteristics:	Mounting surface for tools or gripper: 3⅝ inch diameter bolt circle or per your requirements.

Memory Options, Programming, Interlocks, Optional Accessories and Environmental Conditions are the same as for basic Model 2000.

Load Capacity:	Up to 500 lbs. (Subject to application analysis)
Positioning Accuracy:	0.080 inches

Operating Data:

Max. radial velocity:	30 in./sec.
Max. radial stroke:	52 in.
Max. vertical stroke:	50°
Max. vertical velocity:	35°/sec.
Max. rotation:	200°
Max. rotational velocity:	65°/sec.
Max. wrist bend velocity:	110°/sec.
Max. wrist bend:	230°
Max. wrist swivel:	300°
Max. wrist swivel velocity:	110°/sec.
Max. wrist yaw:	200°
Max. wrist yaw velocity:	110°/sec.

Wrist Load Capacity

Bend:	3500 in.-lbs.
Yaw:	2800 in.-lbs.
Swivel:	2300 in.-lbs.
Power requirements:	220/440 volts, 3 phase, 60 Hz, 30 KVA, 90 PSI air at 2.5 CFM when required.
Weight:	5000 lbs.

Unimation RIG specifications

Mode 1 — Point-to-point

Programming:	Record-Playback using Teach Control
Positioning Accuracy:	± 1 millimeter (± 0.040 in.)
Max. radial velocity:	765mm/sec. (30 in./sec.)
Max. vertical velocity:	1270mm/sec. (50 in./sec.)
Max. rotational velocity:	110°/sec.
Max. wrist bend velocity:	110°/sec.
Max. wrist yaw velocity:	110°/sec.
Max. wrist swivel velocity:	110°/sec.

Mode 2 — Continuous path with Velocity Control

Programming:	Record-Playback using special marking tape.
Working Speed:	102 — 3250mm/min. (4 to 128 in. min.) resultant electrode tip velocity.
Speed Regulation:	± 2%
Accuracy:	± 1mm (± 0.040 inches)°
Automatic Speed Selection:	4 independently adjustable channel.
Environmental Conditions:	Ambient temperature 40°F to 120°F. Humidity. 0-90%.
Power requirements:	220/440 volts, 3 phase, 60 Hz. 11.5 KVA. 90 PSI air at 2.5 CFM when required.
Weight:	2800 lbs.

°Depending upon application details.

Wrist Load Capacity of Heavy Duty Models:

	2000 5 Axis	2000 6 Axis	4000 5 Axis
Bend	2000 in.lbs.	2000 in.lbs.	11,000 in.lbs.
Yaw	1200 " "	1400 " "	2800 " "
Swivel		800 " "	

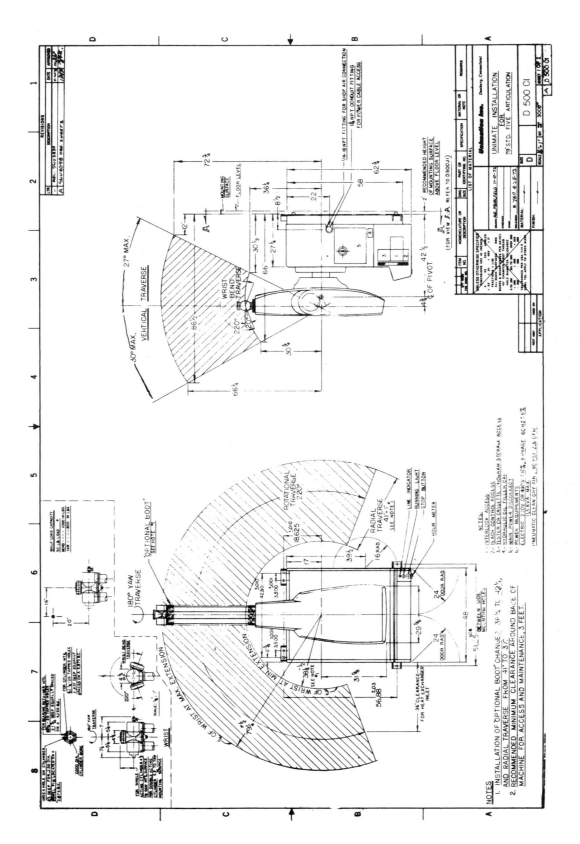

Figure 24 Unimate 2000 work area

TECHNICAL DATA - AMF VERSATRAN

AMF Versatran offers three models of its basic robot in the new F series. All three models, light, medium and heavy use the same control and hydraulic power unit. Up to seven axes of motion are available, as shown below:

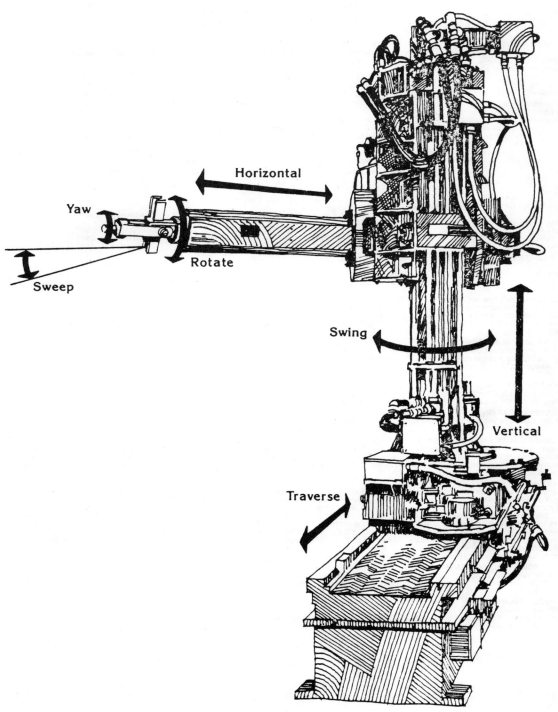

FIGURE 25

Specifications of the Series F Versatran robots are as follows:

VERSATRAN SPECIFICATIONS
Series F Mechanical Units—Model 600 Control

PRIMARY ARM MOTIONS—(Hydraulic Actuation, Servo Control, Analog Feedback)

Model	Stroke			Velocity per Sec. (Max.)			Positioning Repeatability
	FL	FM	FH	FL	FM	FH	
Horizontal—Standard Arm	48"	48"	48"	36"	30"	30"	Within ± .050"
Vertical—	48"	48"	48"	36"	30"	30"	Within ± .050"
Swing—	300°	300°	300°	90°	60°	60°	Within ± .050"
Traverse—	As required			24"	18"	18"	Within ± .050"

1. Horizontal and vertical strokes can be less or greater depending on the speed, load, and application, but require prior approval of VERSATRAN engineering.
2. Positioning repeatability can be held to closer tolerances by tooling design.
3. Velocities can be increased with AMF VERSATRAN engineering approval.
4. All primary motions of the Series F VERSATRAN units are modular in concept and can be furnished in a configuration to suit the application.
5. Maximum payloads at full rated velocities (payload = piece part plus tooling) without optional end of the arm motions are shown below. Heavier loads can be handled at reduced velocities but require prior approval of VERSATRAN engineering.

Model	Payload at Rated Velocities (Lbs.)	Max. Load with Engr. Approval (Lbs.)
FL	250	500
FM	350	700
FH	550	1100

SECONDARY END OF THE ARM MOTIONS (Optional)

Model	Stroke (Max.)			Velocity per Sec. (Max.)			Positioning Repeatability
	FL	FM	FH	FL	FM	FH	
Rotate—	300°	300°	300°	90°	60°	45°	Within ± .050"
Sweep—	270°	270°	270°	90°	60°	45°	Within ± .050"
Yaw—	180°	180°	180°	90°	60°	45°	Within ± .050"

Load ratings—Secondary motions at rated velocities. (Inch pounds)

Model	FL	FM	FH
Rotate—	5000	5000	10000
Sweep—	2500	2500	8000
Yaw—	2000	2000	5000

1. Secondary motions are hydraulically actuated, servo controlled, with analog feedback.
2. Secondary motions may be two position solenoid operated (speeds may be affected).
3. Maximum velocities can be increased with AMF VERSATRAN engineering approval.
4. All secondary motions of the Series F VERSATRAN units are modular in concept and can be furnished in a configuration to suit the application.

ADDITIONAL SPECIAL MOTIONS
1. Orienting and positioning work holding devices and tools.
2. Probe pick-up and set-down points with locating sensors.
3. Conveyor tracking devices.
4. Palletizing capabilities.

WORKHOLDING DEVICES
Standard and custom designs for single or multiple parts handling tooling with controls. Actuation may be mechanical, electrical, hydraulic, pneumatic, magnetic, or vacuum, with or without part sensing.

SPECIAL CONFIGURATIONS
F Series mechanical units can be adapted to overhead and other special configurations.

MODEL 600 CONTROL
The Model 600 Control unit is designed to control the VERSATRAN mechanical units. This control is designed to have a high memory capacity, be easy to teach (or change) and be extremely flexible and adaptable to a wide range of applications. It employs a microprocessor to obtain the basic system performance and flexibility, together with modular construction to permit expansion capability. The control also operates peripheral mechanical devices and interlocks. The control may be located away from work place environment with 20 foot or longer electrical cables.

PROGRAMMING
The F Series mechanical units are manually positioned at each point or position required in a program by means of push buttons located in the console or on a remote hand held teach unit. Pushing a "record" button at each position records that position in memory. Program functions such as output commands or interlocks, input interlocks, end of the arm commands, time delays, velocity, adjustable acceleration and decelerations and "next program" can then be entered as required between positions. As an example, the standard control with an average of 4 functions between positions provides 260 positions for a three axis machine or 207 positions for a seven axis machine. The control is expandable to 1638 positions for a three axis machine or 1304 positions for a seven axis machine with an average of 8 functions between positions. Additional positions are available if fewer than eight functions are required between positions. Modifications to a program are easily accomplished. The VERSATRAN Model 302 or Point-to-Point Control can be used in applications which require less memory. The Model 401 Control is available for contouring path controls.

HYDRAULIC POWER REQUIREMENTS
Petroleum Oil (Mobil DTE Light or equal) @1500 psi, 75 -110°F, regulated, filtered to 10 microns maximum particulates, in sufficient volume for application. Fireresistant hydraulic fluid suitable for servo-controlled systems may also be used, but requires extra-cost modifications of hydraulic system.

UTILITY REQUIREMENTS
Electric Power—440/220 volts, 3 Phase, 60Hz.
Water—(For VERSATRAN water-cooled hydraulic power pack) 2gpm, 30-150 psi, 90°F Max.
Air cooled hydraulic power units are available.

The Versatran robot's seventh axis (traverse) can be set up to pace a moving line by means of an external device to monitor line speed. If set up parallel to the moving line, this feature permits working on a moving object or transferring parts as if the object or line were stationary.

The Model 600 control uses a microprocessor with a semi-conductor memory for program storage. This is a volatile memory, that is, the memory is lost if electrical power to the control is interrupted. To handle minor power interruptions, an automatically recharged standby battery is included in the control. This will protect the memory during 10 to 20 minutes of loss of external power. Prolonged power outages will necessitate that the programs be reinserted in memory with the cassette recorder. This can normally be accomplished in less than 5 minutes.

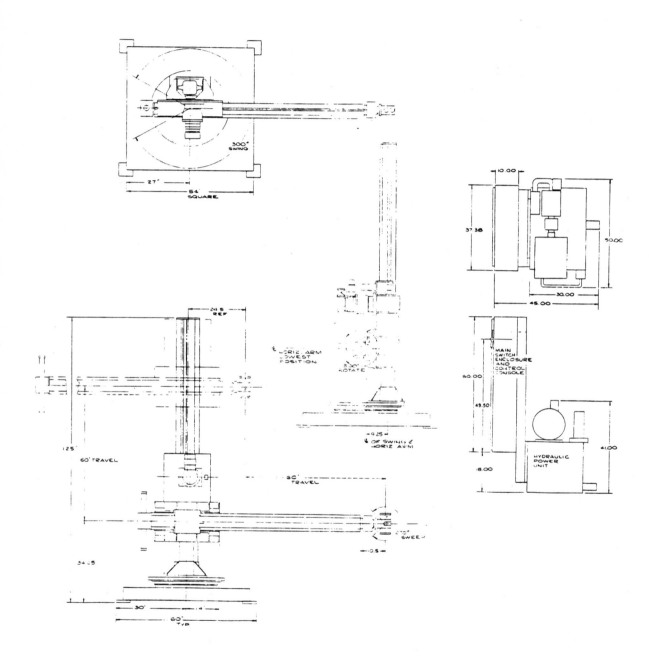

FIGURE 26

APPLICATIONS INFORMATION

When a potential robot application has been found (which appears profitable using the "rule of thumb" of one man per shift per robot) it is then necessary to develop the details of the robot installation. The first step is to determine what robot or robots are suitable, based on the application and the specifications. A layout of the proposed installation should then be made and fieldchecked for interferences. Particular attention should be paid to aisles, stock requirements, overhead clearances and access for maintenance. Back-up capability must also be considered.

Part transfer, orientation and fixturing requirements must next be determined. Facilities and tooling changes necessary to locate the part or assembly relative to the robot must be included in the project request. Working height of the robot and access to the job may require elevation of the robot above the floor or changes in part location from normal working height. Placement of the robot in a pit should be avoided if possible. A drip pan should be provided under the robot or the hydraulic power unit to contain oil leakages. Guard rails should be provided around the robot installation to prevent damage from material handling equipment and for personnel safety. A removable section in the guard rail large enough for lift truck access to the robot should be incorporated.

In spotwelding applications use of a self-equalizing gun, if feasible, will simplify installation of the gun on the robot. If not, an equalizing gun mount must be provided. Welding transformers may be installed overhead or on the floor, depending on the applications. In an operation involving large rotary motions of the robot, such as welding in more than one fixture, the welding transformer should be suspended over the center of rotation of the robot. This will reduce loads on the robot arm and the welding cable. No gun balancer is required; however, a small balancer may be desirable to keep the welding cable clear of the robot, fixture or part.

In all robot installations, it is necessary to provide signals to the robot to initiate its cycle and, usually, signals from the robot when it completes its task. Included in the interface signals to the robot is such information as part (or hook) in place, line speed, part clamped, part or model identification (for program selection), weld complete and operator clear of area. External signals from the robot include those required to initiate a weld sequence and end of robot program. These interfaces are normally handled through a series of relays, except for line speed information which requires a resolver/tachometer.

The following time data was developed by one robot user based upon actual operations with Unimate 4000 series robots. These times can be used for estimating cycle times for spotwelding with any robot, since maximum speeds are comparable for all present machines. Times for loading and unloading (or transfer), clamping and unclamping must also be added, in order to determine total cycle time.

Speed/Axis	Move Distance	Move Time (Minutes)
Max. Radial Velocity	Up to 12"	.015
	12" to 40"	.020
	40" to 52"	.030
Max. Vertical Velocity	Up to 20°	.015
	20° to 50°	.020
Max. Rotational Velocity	Up to 45°	.015
	45° to 90°	.020
	90° to 135°	.030
	135° to 175°	.040
	175° to 200°	.050
Max. Wrist Bend Velocity	Up to 100°	.020
	100° to 230°	.030
Max. Wrist Swivel Velocity	Up to 100°	.020
	100° to 200°	.030
	200° to 300°	.050
Max. Wrist Yaw Velocity	Up to 100°	.020
	100° to 200°	.030

Based on distances traversed in single motion. For each change of Velocity in motion add .0125 mins. If actions are performed simultaneously, use the greatest single value. Excessive loads will increase the above times but the exact amount is not available at this time.

SPOTWELDING	MINS. PER SPOT
.048 Gauge	.013
.060	.018
.075	.023
.090	.028
.110	.033

As mentioned earlier, part orientation and location must be consistent from job to job. In the case of transfer operations, the part pickup point may be different for different parts, but must be repeatable and the hooks into which the parts are deposited must be stabilized to prevent erratic movement.

In moving line applications, the entry of a part or hook into the work area must be detected and the speed of the line or transfer must be continually monitored. Those robots which track in only one axis must be oriented so that line movement is coincident with one axis of motion only.

For in-line, stop station applications, such as respot welding, speed-up rolls may be required before and after the work station. Although these sections of the line are normal facility items, the work station itself cannot be simply a lift table. Part orientation and location repeatability requires some tooling related to the part itself. This work station should be classified as a locating/fixturing device, rather than a conveyor or lift table. It may be convenient to incorporate the speed-up sections into this design, using transfer rails or similar devices rather than normal speed-up rolls. Installation of this station as a by-pass system, particularly in the case of multiple robot operations, should be considered as a means of protecting subsequent production and operations.

Utilities and service requirements for the various robots were specified earlier. Services to ancilliary devices such as spotweld guns or grippers must also be provided. As with any machine, normal disconnects and safety devices are required.

When specifying a robot for a particular application, bear in mind two things: A robot is a piece of capital equipment, with a design life of 40,000 operating hours or more, and; a robot has the flexibility to be used on a wide variety of tasks. In many (if not most) cases, the robot will outlast its initial application and another operation must be found for it. In order to enhance the potential of the robot, the initial specifications should include as much capability as can be purchased while still meeting the economic objectives of the cost saving. Examples of such capability include 6 axes of motion (instead of 5), extended memory, weld interface (if optional), and increased reach or load capacity. Such features, while not required, perhaps, on the initial application of the robot will make it possible to apply that robot to a wider range of operations at some later time.

DO'S AND DON'T'S

In developing and implementing robot applications, careful attention to detail will reduce the risk of problems in launching and maintaining production. The following list of "Do's and Don't's" should be used for frequent reference:

DO - Provide for comprehensive maintenance training of sufficient personnel to cover <u>all shifts</u>.

- Provide spare parts and test equipment for rapid trouble shooting and repair.

- Set up a system for return of defective parts which are handled on an exchange-repair basis and maintain up-to-date inventory lists.

- Provide backup for production; either standby facilities or floats.

- Protect the equipment from people and vice versa. Be sure that OSHA and other safety standards are observed.

- Consider alternatives to the usual floor mounting of robots -- overhead, side, etc., if such an approach will simplify the operation or reduce robot and/or facility requirements.

- Include "production" people in your planning. Get their ideas and help them feel that they are part of the action.

- Be honest in answering any and all questions coming from hourly people; they are genuinely interested and are vital to a successful application.

- Get your management to back you up -- total commitment is required for success.

- Ask for help from the "experts," the vendors, when in doubt.

- Try to plan your installation so that it will cause a minimum of disruption or down time and allow adequate time for programming, debugging and launching.

DON'T
- Try to do something to something which is not consistently located.

- Try to do something which requires the maximum capability of the machine -- you need something in reserve in case the job changes.

- Believe everything the vendor says his equipment can do.

- Rely on the vendor for service of the machine -- you aren't his only customer.

- Blame all your problems on the robot -- the fixture, etc., needs attention, too.

- Assume that the robot program will be good forever -- things change, sometimes gradually and sometimes suddenly.

- Neglect maintenance of tools and peripheral equipment -- the robot won't know and can't tell you when there is a problem, it will just continue to go through its program as if everything was allright.

FUTURE DEVELOPMENTS

One of the most important developments is a visual sensor and feedback system. Utilizing recently developed solid-state optical sensors and micro processors, this approach can provide a robot with the capability to "find" an object and adjust its programmed task to the location and orientation of the object. This system will simplify moving line applications and will make many more parts handling operations feasible.

Another area of investigation is fine wire welding. Although a limited capability already exists, additional sensors and feedback systems are required for adaptation to welding operations.

A third development which has significant potential is a tool changer system. This system will enable a robot to interchange spotweld guns or power tools under program control and will include disconnects for air, water and power. The tool changer will expand the capability of the robot by enabling it to perform a series or operations requiring more than one type of weld gun, power tool or part gripper.

Other areas of development are being pursued by the robot manufacturers. These include multiple-axis line tracking, computer controls, increased load handling and multi-armed devices.

CONCLUSION

Industrial robots are no longer just a product of the science fiction writer's imagination. They are being widely used in a variety of industries throughout the world. There are more than 50 robots presently in use in one company alone. (1976)

While robots are not the answer to every problem, they are a viable alternative to manual operations and to special purpose automation. They are versatile, flexible and reliable and are getting better and smarter all the time. They can provide a relatively easy means of improving productivity by reducing operating costs.

Successful applications, however, do entail some work. You must plan carefully, prepare thoroughly and anticipate problems. This chapter should make it easier, but, it cannot anticipate all situations or answer all questions. When in doubt about the feasibility of a potential application or the capability of a particular robot, it is best to consult the manufacturer.

It is hoped that this guide will stimulate your enthusiasm and provide some thought-starters while making the development of applications easier.

CHAPTER 8

Fastening Methods

Fastening the assembled components is the culmination of the assembly process. An enormous variety of methods is available. As is every other phase of the assembly project, the decision-making and cost-control begins with the product design as outlined in Chapter 1.

Fastening methods may be arbitrarily divided into several categories:

Permanent (or Throw-away) vs Temporary

Soldering, brazing, welding, adhesion, bonding, coining, staking, deforming, riveting, and press or shrink fits are examples of permanent fastening. Screwdriving, nut-running, treading, and retaining rings are typical "take-apart" methods.

Discrete vs Integral Fasteners

Chapter 1 discussed the comparative advantages of eliminating separate fasteners, where practical, in favor of integral threads, pressed tabs, riveted projections, etc. When required, discrete fasteners offer a wide range of methods with an equally wide range of cost and process capability.

Mechanical vs Fusion Methods

Product design consideration should be given to the advantages and disadvantages of the mechanical methods such as coining, staking, tab bending, etc., vs welding, soldering, brazing and bonding.

While volumes are written on each of the methods, it would be impractical to comprehensively cover the field of fasteners in this book. Instead, eleven papers are presented covering specific areas as follows:

- A. Adhesive Selection
- B. Water Base Adhesives
- C. Hot Melt Adhesives
- D. 6 Adhesive Cure Systems
- E. Anerobic Adhesives
- F. Injected Metal Fastening
- G. Soldering and Welding
- H. Laser Applications
- I. Electron Beam Welding
- J. Ultra-Sonic Applications
- K. Tensioning Threaded Fasteners

This admittedly limited fastener presentation may serve to emphasize the scope of the subject while offering some insight into the methods discussed.

ADHESIVE SELECTION

The most fundamental thing about choosing an adhesive is the simple fact that design or process engineers rarely make the choice alone. In fact, a given manufacturing company is rarely blessed with the talent and equipment needed to select an adhesive without assistance from a vendor or consultant. *Figure 1*, which lists a few of the adhesives used for various materials, illustrates the problem.

Adhesive	Material	Adhesive	Material	Adhesive	Material
ABS	polyester epoxy alpha cyanoacrylate nitrile-phenolic	Magnesium	polyesters epoxy polyamide polyvinyl-phenolic neoprene-phenolic nylon-epoxy	Polypropylene	polyester, isocyanate modified nitrile-phenolic butadiene-acrylonitrile
Aluminum and its alloys	epoxy epoxy-phenolic nylon-epoxies polyurethane rubber polyesters alpha-cyanoacrylate polyamides polyvinyl-phenolic	Nickel	epoxy neoprene polyhydroxyether	Polystyrene	vinyl chloride-vinyl acetate polyesters
		Paper	animal glue starch glue urea-, melamine-, resorcinol-, and phenol-formaldehyde epoxy polyesters cellulose esters vinyl chloride-vinyl acetate polyvinyl acetate polyamide flexible adhesives	Polyvinyl chloride, rigid	epoxy polyesters polyurethane
Brick	epoxy epoxy-phenolic polyesters			Rubber, butadiene-styrene	epoxy butadiene-acrylonitrile urethane rubber
Ceramics	epoxy cellulose esters vinyl chloride-vinyl acetate polyvinyl butyral			Rubber, natural	epoxy flexible adhesives
				Rubber, neoprene	epoxy flexible adhesives
Chromium	epoxy			Rubber, silicone	silicone
Concrete	polyesters epoxy	Phenolic and Melamine	epoxy alpha-cyanoacrylate flexible adhesives	Rubber, urethane	flexible adhesives silicone alpha-cyanoacrylate
Copper and its alloys	polyesters epoxy alpha-cyanoacrylate polyamide polyvinyl-phenolic polyhydroxyether	Polyamide	epoxy flexible adhesives phenol-, and resorcinol-formaldehyde polyester	Silver	epoxy neoprene polyhydroxyether
				Steel	epoxy polyesters polyvinyl butyral alpha-cyanoacrylate polyamides polyvinyl-phenolic nitrile-phenolic neoprene-phenolic nylon-epoxy
Fluoro-carbons	epoxy nitrile-phenolic silicone	Polycarbonate	polyesters alpha-cyanoacrylate polyurethane rubber		
Glass	epoxy epoxy-phenolic alpha-cyanoacrylate cellulose esters vinyl chloride-vinyl acetate polyvinyl butyral	Polyester, glass reinforced	polyester epoxy polyacrylates nitrile-phenolic		
		Polyethylene	polyester, isocyanate modified butadiene-acrylonitrile nitrile-phenolic	Stone	See "Brick"
				Tin	epoxy
Lead	epoxy vinyl chloride-vinyl acetate polyesters			Wood	animal glue polyvinyl acetate ethylene-vinyl acetate urea-, melamine-, resorcinol-, and phenol-formaldehyde
		Polyform-aldehyde	polyester, isocyanate modified butadiene-acrylonitrile nitrile-phenolic		
Leather	vinyl chloride-vinyl acetate polyvinyl butyral polyhydroxyether polyvinyl acetate flexible adhesives	Polymethyl-methacrylate	epoxy alpha-cyanoacrylate polyester nitrile-phenolic		

Figure 1. Adhesives Commonly Used for Various Materials. Most of the Adhesives are Available in a Number of Physical and Chemical Forms with Varying Process Requirements.

The second most fundamental aspect of choosing an adhesive is simply that adhesive application and cure systems are an integral part of the selection. Choosing to use a particular thermosetting adhesive implies a decision as to whether it is a one or two part product; whether it is catalyzed or cured; whether it is brushed, sprayed or applied as a film. Similarly, decisions regarding surface preparation, racking, fixturing, pressures, and cure temperature-time requirements are also an inherent part of the adhesive selection process. *Figure 2* presents some of the considerations important in adhesive process selection.

The situation described above exists because choosing an adhesive is complex at best and because the people who buy and use adhesives are not generally the ones who make and sell them. A self-evident truth now emerges: Clear communication between vendor and user is essential if the adhesive selection process is to succeed.

	Hot Melts	Pressure Sensitive	Brush	Spray	Roller	Dry Films	Pump
Skill Required	N	N	N	Y	N	N	N
Speed of Assembly	F	F	S	S	M	F	M
Messy Operation	Y	N	Y	Y	M	N	M
Clamping Required	N	N	V	V	V	Y	Y
Heating After Assembly	N	N	V	V	V	Y	V
Cure Time	R	R	V	V	V	S	S

N - No
Y - Yes
F - Fast
S - Slow
M - Medium
V - Variable, depends upon the adhesive
R - Rapid

Figure 2. Important Aspects of Various Adhesive Processes

The following comments are presented with the hope that they will help the reader adequately contribute to the adhesive selection process.

1. If you are working through a vendor give him as much information as possible about how your product is made, what materials are used and what performance is expected. Avoid generalities like "metal to plastic adhesive"; it is much more useful to say "1010 steel to high density polyethylene". The thermal history of your materials, prior to bonding, may be important in helping your supplier recommend the proper surface preparation.

2. Do not overdesign. Considerable money is wasted if joints are specified which are stronger or longer lasting than necessary.

3. Remember that heat curing adhesives can be used only if the temperature tolerance of the parts is above the adhesive cure temperature.

4. Interpret bond strength values with the knowledge that they pertain to specific adherends prepared in a specific manner. If your substrates, surface preparation, or cure conditions are different the bond strengths may be markedly affected. You must check this out in your plant with your parts.

5. Strength retention values obtained from technical literature are often for unstressed specimens. If joints are aged under load, particularly cyclic load at various temperatures, their strength retention behavior may be grossly affected. You should determine this behavior on joints made in your plant under conditions which simulate the service environment of your assemblies.

6. Remember that production line bonds usually show less consistent strength than bonds made in your process engineering lab. Process development people usually have the time to be more careful about surface preparation, mixing, pressure, cure time and bond line thickness. Do not choose an adhesive for production which is only marginally acceptable in a pilot operation.

7. Be aware that some adhesives contain chemicals which may affect some people physically. Good ventilation and minimum skin contact are often essential.

8. One-part adhesives are preferred over two-part formulations. Mixing, metering and dispensing operations allow operator and equipment errors to occur.

9. Tape and film adhesives are preferred over liquid and paste systems, because mixing, handling, outgassing and shrinkage problems are reduced. The result is higher, more uniform bond strengths.

10. Be aware that differences in thermal expansion properties of the adherends can produce severe stresses at the bond line when the joint undergoes temperature changes. Generally, the thermal expansion behavior of rigid adhesives should be about midway between that of the adherends if large temperature fluctuation is expected.

11. It is unrealistic to expect an adhesive bond to have high tensile and high peel strength. One or the other, or a compromise, is the rule. Problems will be minimized if the joint is designed with this tradeoff in mind.

12. Remember that the most common cause of poor joing performance is failure to properly clean the surface. The second most common cause is similar--failure to keep the surface clean.

13. An adhesive which is more expensive per pound may be cheaper to use because less material is needed for each bond or because curing requires less time, space or energy.

14. Time spent on training your workforce to properly prepare, assemble and cure bonds is well worth the cost and effort. This is especially true if you are converting from some other assembly method to adhesive bonding. The natural human tendency to resist change can make such a conversion extremely difficult. A good training program can minimize these problems.

15. Adhesives with the least critical surface cleanliness requirements are preferred. If possible avoid the use of dip or spray cleaning operations where solutions are reused. If economics dictate the use of these processes be aware that they are subject to gradual or sudden contamination.

WATER BASED ADHESIVES

Because of government legislation, the use of solvent based adhesives will decline in the future. Water based adhesives are one of the candidate replacements. These products are normally used where at least one of the surfaces to be bonded is permeable to water. For bonding of impermeable surfaces, if the required longer drying time can be tolerated or if forced drying is used, latex adhesives are available to replace some of the current solvent based adhesives. The types of water based adhesives, solution and latex, are discussed. Information is provided on the formulating, manufacturing, applying, drying and bonding of latex adhesives. Physical and performance properties of latex and solvent based adhesives are compared. Future needs for latex adhesives are outlined.

The DeBell & Richardson report states that the use of solvent based adhesives, in the developed countries of the world, will reach maximum consumption in the late 1970's and then rapidly decline, becoming almost obsolete by the year 2000.

E. I. du Pont's Delphi Study states that 360-million pounds (dry basis) of solvent based adhesives are believed to have been used in the U.S. in 1972. It has been predicted that by 1980, the usage level will drop to 80% of 1972 levels and by 1985, to 45% of 1972 levels.

Replacement products will be hot melts, 100% reactive adhesives, and water based adhesives.

TYPES OF WATER BASED ADHESIVES

There are two general types of water based adhesives; solutions and latices. Solutions are made from materials that are soluble in water alone or in alkaline water. Animal glue, starch, dextrin, blood albumin, egg albumin, methyl cellulose, and polyvinyl alcohol are examples of the former. Some examples of the latter are casein, rosin, shellac, copolymers of vinyl acetate or acrylates containing carboxyl groups, and carboxymethyl cellulose.

A latex is defined as a stable dispersion of a polymeric material in an essentially aqueous media.

An emulsion is defined as a stable dispersion of two or more immiscible liquids held in suspension by small percentages of substances called emulsifiers.

In the adhesive industry the terms latex and emulsion are sometimes used interchangeably.

There are three types of latices; natural, synthetic, and artificial.

Natural latex refers to the material obtained primarily from the rubber tree.

Synthetic latices are aqueous dispersions of polymers which are obtained by the process of emulsion polymerization. These would include polymers of chloroprene, butadiene-styrene, butadiene-acrylonitrile, vinyl acetate, acrylate, methacrylate, vinyl chloride, styrene and vinylidene chloride.

Artificial latices are made by dispersing solid polymers. These would include dispersions of reclaimed rubber, butyl rubber, rosin, rosis derivatives, asphalt, coal tar, and a large number of synthetic resins derived from coal tar and petroleum.

Since the majority of the solution type adhesives remain water soluble after drying, their use is restricted to end uses that do not require water resistance in the final bond. Most latex adhesives cannot be redispersed in water after they have once dried and therefore can be used in applications that require water resistance. It is expected that of the two types of adhesives, the latex types are the ones which will replace solvent based adhesives.

FORMULATION OF LATEX ADHESIVES

There are two general methods of designing or formulating a latex adhesive.

The first is to tailor make a polymer that will have the final desired physical and performance properties. This is done by blending monomers and using polymerization conditions that will result in the proper particle size, molecular weight, etc. Examples of this are some of the acrylate polymers used as pressure sensitive adhesives. Some of these polymers can be used as the final adhesive with only slight modification such as viscosity control.

The majority of latex adhesives are produced from polymers that were not designed for use in adhesives. They require extensive formulating in order to obtain the proper application and performance properties.

The first step in formulating a latex adhesive for a given application is selection of the polymer. Some factors in this choice are economics, physical and chemical properties, overall strength, static load resistance, weathering, ultra-violet and ozone resistance, light stability and specific adhesion.

Hundreds of resins are available for use in latex adhesives. Resins can help to attain many valuable properties. They can increase adhesion to various substrates, increase strength properties, increase the ability of dried or semi-dried films to adhere to themselves and increase tack.

A variety of fillers can be used; Clays, Calcium Carbonate, Talc, Carbon Black, etc. Fillers are used to reinforce, control viscosity, decrease cost, reduce extensibility, reduce tack and control flow properties.

Plasticizers are sometimes used to reduce stiffness, improve low temperatures flexibility, aid in film formation at low temperatures and increase tack.

Many polymers can be heat cured or vulcanized if curing agents are incorporated in the formulation. Curing of the adhesive can increase room temperature and elevated temperature strength properties, improve water, oil and solvent resistance, and improve resistance to weathering.

A variety of materials are available to modify the consistency of latex adhesives. Flow control is necessary if the adhesive is to be used in thick films on vertical surfaces. Examples would be ceramic wall tile adhesives or construction mastics.

Many other types of materials can be used in latex adhesives to control a variety of properties; anti-oxidants, anti-foams, freeze-thaw stabilizers, fungicides, and corrosion inhibitors.

MANUFACTURING OF LATEX ADHESIVES

Ingredients are added to latices as aqueous solutions, dispersions, or emulsions, depending on whether they are water soluble, water insoluble solids or water insoluble liquids.

Water insoluble powders are added to water which contains small amounts of dispersing agent and stabilizer. The resulting slurry is ground to produce a dispersion of small particle size using one of the following types of equipment:

(a) Those which breakdown aggregates of fine particles but do not affect any reduction of ultimate particle size. This type of equipment is called a Colloid Mill.

(b) Those which do effect a reduction of ultimate particle size as well as dispersing any agglomerates which may be present. Equipment of this type includes ball mills, pebble mills, ultrasonic mills, and attrition mills.

Emulsions are prepared by blending water, dispersing aids, stabilizers, thickeners, etc., with the liquid to be emulsified using high speed agitation.

APPLICATION OF LATEX ADHESIVES

Since latex adhesive viscosities vary from thin liquids to heavy pastes, they cannot be applied equally well by any single method. Application is governed by the type of adhesive, size, shape, and number of parts to be coated, and the required production rate.

The following are the common application methods for product assembly:

1. <u>Brush</u>

   Brushing is common for small area applications. For higher production rates, a flow brush is desirable. A flow brush is pressure fed and is usually recommended for continuous application to prevent drying on the bristles.

2. <u>Spray</u>

   Spraying, either air or airless, is one of the most economical methods of applying adhesive to large areas. Air spraying is the most common. Here the adhesive is atomized by air using pressures of 15-25 psi while relatively low pressure is used to pump adhesive to the spray tip. In airless spraying, adhesive is atomized by high adhesive velocity ejected under pressures ranging from 600-1200 psi. The high velocity adhesive atomizes when it strikes the air surrounding the spray tip.

High shear and certain metals can cause coagulation of some latex adhesives. Pumps that do not subject the adhesive to such a shearing action, such as a diaphram pump, are normally used. Stainless steel, plated metal, and plastic are the recommended materials of construction for pumps and spray guns.

3. Roll Coat

For low volume production or small jobs, paint rollers are convenient. For higher volume or multi-station production, pressure-fed hand rollers are available.

Roll coating is used on large, flat areas when large volume and fast production is required. Line speeds of 50-60 feet per minute are common. The adhesive is transferred from the surface of a roll to the substrate. Double roll coaters coat both sides of the substrate, simultaneously.

4. Curtain Coat

Curtain coating is faster than roll coating, but only one side of the substrate can be coated. A continuous "water fall" of adhesive coats the substrate as it passes underneath. Dry film thickness is controlled by the solid content of the adhesive, thickness of the curtain and speed of the substrate. This method is very useful with broad, flat surfaces.

5. Flow

A collapsible tube of adhesive is the simplest piece of flow equipment.

For intermittent or low volume applications, hand operated caulking guns are satisfactory. For high volume production, a continuous flow gun is used. A pump maintains constant pressure and adhesive flow.

6. Knife Coat

The simplest version of knife coating is a putty knife used with mastic adhesives on small jobs. Trowels are normally used for larger area applications such as walls and floors. Knife coating, with a mechanically driven unit, is used to apply high viscosity adhesives at high production rates. Adhesive is applied to the substrate in excess. The substrate then passes under the knife and all adhesive, except the required amount, is removed.

BONDING TECHNIQUES

The bonding techniques used for latex adhesives are about the same as for solvent adhesives. The bonding technique used depends upon the parts being joined, the adhesive itself, and the immediate bond strength requirements.

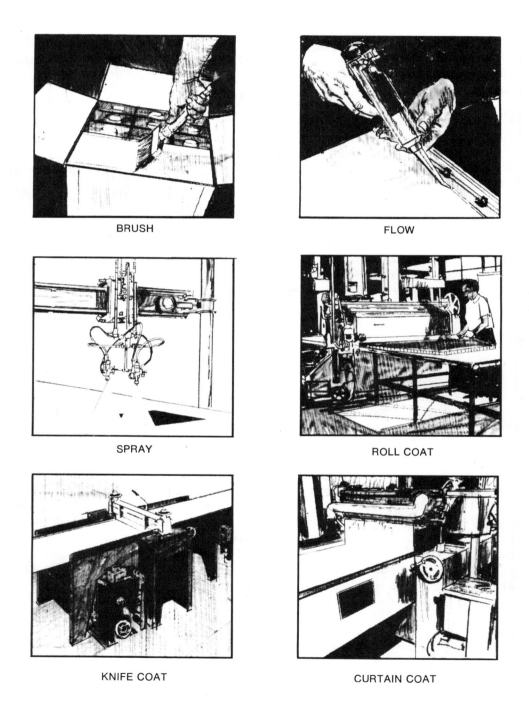

Figure 3. Application Methods

The following bonding techniques are used:

1. <u>Wet Bonding</u>

   When using the wet bonding technique at least one of the bonded materials should be porous. These materials would include wood, masonry, fabric, most flexible foams, leather, cardboard, fiberglass, felt, and cork.

   It is usually necessary to apply adhesive to only one surface. The surfaces to be bonded are assembled while the adhesive is still wet or "tacky". Immediate bond strength is generally low. In some cases it may be necessary to keep the parts being joined under pressure until enough strength has developed for subsequent handling.

2. <u>Open-Time Bonding</u>

   Open-time bonding is when the adhesive is applied to both surfaces and allowed to stand "open" until suitable tack or stickiness is achieved. This technique works best when at least one surface material is porous or semi-porous. It is probably the most widely used method of making small or medium-size bonds. The simplest test to determine the right tack is to touch the adhesive with the knuckles. If the adhesive feels sticky, but does not transfer to knuckles, it is ready.

3. <u>Contact Bonding</u>

   Both surfaces are coated and the adhesive is permitted to become dry to the touch. Within a given time, these surfaces can be pressed together and near ultimate bond strength is immediately achieved. Contact bonding is preferred when bonding two non-porous surfaces. Contact bond adhesives frequently are based on specially compounded neoprene latices. Drying times prior to bonding are around 30 minutes under ordinary conditions of temperature and humidity. This can be reduced to seconds by force drying with heat.

4. <u>Solvent Reactivation</u>

   The adhesive is applied to the surface of the part and let dry. To prepare for bonding, the adhesive can be reactivated by wiping with solvent or placing the part on a solvent impregnated pad. The surface of the adhesive tackifies and the parts to be bonded are pressed together. The tack range is short, so the technique is useful for relatively small size applications. It is not generally applicable on large surface areas since the short bonding range gives inconsistent results.

5. <u>Heat Reactivation</u>

   A thermoplastic adhesive is applied to one or both surfaces and let dry. Coated parts can be stored several weeks or longer before bonding. To prepare for bonding, the part is heated until the adhesive is soft and tacky. The bond is made under pressure while hot. On cooling, bond strength is as good as the adhesive will provide. This method is used with non-porous heat resistant materials when ultimate bond strength is needed quickly.

Heat reactivation can also be a continuous in-line operation. The adhesive is applied in liquid form to a film or sheet, force dried to remove the water, and then laminated to a second surface while still hot. Temperatures are usually in the range of 250° F. to 350° F.

FORCED DRYING OF LATEX ADHESIVES

The heat of vaporization of water is 540 cal./gram. This compares to about 100 cal./gram for the common solvents used in adhesives. This does not mean that latex adhesives cannot be dried as quickly and effectively as solvent based adhesives. All it takes is additional heat to achieve the same results.

One of the most efficient methods of drying latex adhesives is through the use of an infra-red heat source, in conjunction with air flow over the adhesive. As the water evaporates, air movement removes the high humidity directly above the adhesive surface and speeds the drying.

PROPERTIES OF LATEX ADHESIVES VERSUS SOLVENT BASED ADHESIVES

The overall strength properties of a latex adhesive based upon a given polymer will be similar to that of a solvent based adhesive containing a solution polymer of like composition.

The solids content of a latex adhesive is normally in the 40-50% range. This would compare to about 20-30% for solvent based adhesives.

Since latex adhesives contain little, if any, organic solvent they are usually cheaper than a comparable type of solvent system. The cost differential becomes even greater, in favor of the latex adhesive, if the comparison is made to a non-flammable solvent adhesive that contains expensive chlorinated solvents.

Latex adhesives have good brushability and usually require less pressure to pump or spray than solvent based adhesives. Prior to drying they can be cleaned up with water. They are non-flammable in the wet state and the vapors are not toxic.

The main disadvantage of latex adhesives is the longer drying time that is required before tack or strength develops.

A solvent adhesive designed for bonding of fiberglass insulation develops very fast tack and bonds can be made immediately after spraying. When using a latex adhesive, a drying time of 2-5 minutes might be required before a bond can be made.

A solvent based neoprene contact adhesive when used to bond high pressure decorative laminate to plywood requires a drying time of about 10 minutes before bonding. A latex neoprene contact adhesive would require a drying time of approximately 30 minutes before bonding under the same conditions.

TABLE OF SOLVENT PROPERTIES

		HEAT OF VAPORIZATION, CALORIES/GRAM	EVAPORATION RATE (ETHYL ETHER*=1)
1.	HEXANE	88	1.9
2.	TOLUENE	87	4.5
3.	XYLENE	82	9.2
4.	1,1,1-TRICHLOROETHANE	59	2.6
5.	ACETONE	125	1.9
6.	METHYL ETHYL KETONE	106	2.7
7.	ETHYL ALCOHOL	204	7.0
8.	ETHYL ACETATE	102	2.7
9.	WATER	540	45.0

*HIGHER NUMBERS INDICATE SLOWER DRYING.

Infra-red heat sources are usually specified by a certain wattage per unit. Total wattage is expressed as kilowatts per square foot or watts per square inch. Units are commercially available with capacities up to 100 watts per square inch. Using equipment of this type, neoprene latex contact adhesives can be dried in a few seconds, even on heat sensitive substrates such as polystyrene foam. This rate of dry enables line speeds for latex adhesives equal to those normally used for solvent adhesives.

Figure 4. Table of Solvent Properties

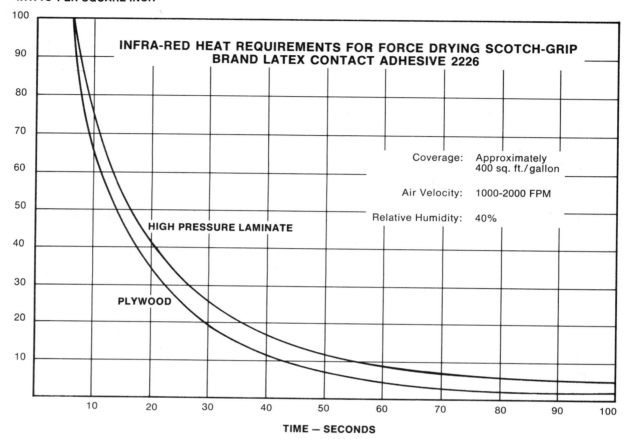

Figure 5. Infra-Red Heat Requirements for Force Drying Scotch-Grip Brand Latex Contact Adhesive 2226

In most cases, latex adhesives must be protected from freezing during shipment and storage. Some materials will recover after freezing, but most of them will coagulate and become unusable.

Solvent based adhesives tend to have better overall adhesion to substrates such as oily surfaces, metal, rubber, and certain plastics.

In general, the current state of the art of latex adhesives is considerably behind that of solvent based adhesives.

APPLICATIONS FOR VARIOUS TYPES OF WATER BASED ADHESIVES

Animal glue is used in large volume in the furniture and woodworking industries. It is the primary adhesive in gummed tapes. The coated abrasives industry consumes large quantities in the manufacture of sandpapers and sandcloths. Other uses include book and magazine binding and bonding paper or cardboard containers.

Casein adhesives are used mainly for glueing of wood and paper products.

Natural rubber latex is used in the manufacture of self-adhesive envelopes, floor tiles adhesives, and adhesives for leather.

Polyvinyl acetate adhesives are the "white glue" of industry. They are very versatile materials and are used extensively in the bonding of wood and paper products.

Neoprene is the common name for polychloroprene rubber. In comparison to other elastomers, neoprene latex adhesives are the most popular and offer an excellent balance of properties. Some of these properties are:

1. Excellent resistance to aging.

2. High strength.

3. Excellent resistance to continuous load stress.

4. A contact bond characteristic which allows two non-porous surfaces to be joined without trapping water.

5. Good elevated temperature strength.

The major applications of neoprene latex adhesives are packaging (foil laminating and bag duplexing), carpet installation, construction mastics, and general purpose industrial and consumer contact adhesives. The contact adhesives are used to bond foamed plastics, plastic laminate wood, plywood, wallboard, wood veneer, plastic, canvas and many other substrates.

Adhesives based on Styrene-Butadiene latices are among the lowest cost of the rubber based adhesives. SBR latex adhesives are not as strong as neoprene latex adhesives. They are usually used to bond light-weight materials. These would include synthetic and natural fabrics, felt, plastic foams, fiberglass insulation, canvas, etc.

Adhesives based on Butadiene-Acrylonitrile latex can be formulated to have good resistance to oil and gasoline. Applications include the bonding of polyvinyl chloride film, aluminum foil, leather, wood, cardboard, paper and metals.

Quite a variety of acrylate polymers in latex form are available for formulating latex adhesives. The end-uses are quite varied. They include pressure sensitive adhesives, laminating of aluminum foil, paper fabrics, foams, and P.V.C. film. Other applications include adhesives for flocking and installing floor coverings.

FUTURE USE OF LATEX ADHESIVES

Any predictions regarding the future use of latex adhesive have to be based on many guesses and opinions rather than facts.

The current latex adhesive markets, furniture, packaging, fabric lamination, floor tile, etc., should continue to grow.

Latex contact adhesives will probably replace solvent adhesives in the "Do-It-Yourself" markets.

The use of latex adhesives should grow in areas where large quantities of solvent based adhesives are being sprayed or coated, and where forced drying of the adhesive could be done. Some candidate areas might be: vinyl top and trim bonding in the automotive industry, sandwich panel manufacture for industrial housing and recreational vehicles, flexible foam bonding, bonding of various types of insulation, etc.

Latex adhesives that can handle many of these applications are currently available. Others can be developed using existing latex technology. Adhesive chemists will continue making further improvements in the performance of latex adhesives in such areas as faster tack, improved wetting, and better overall adhesion.

There is no doubt that latex adhesives will eventually take over a large share of the existing solvent based adhesive market.

HOT MELT ADHESIVES

As hot melt adhesives are only a portion of the current adhesive technology available, this paper will first compare adhesive bonding versus mechanical fastening. Next, the paper will contrast hot melt adhesives with adhesives in general. A special family of hot melt adhesives, polyamide resins, along with details on physical and adhesive bonding of selected resins, plus some specific applications will be discussed. A scenario will be presented on how a company might evaluate this technology for its own application.

ADHESIVE BONDING VERSUS MECHANICAL FASTENING

A comparison has been recently made of adhesive bonding and mechanical fastening. As with any comparison, there is not universal agreement on all these points. I have summarized the advantages of adhesive bonding in *Figure 6* and the advantages of mechanical fasteners in *Figure 7*.

HOT MELT ADHESIVES

Hot melt adhesives can be simply defined as 100 percent non-volatile thermoplastic materials which typically are solid at room temperature. The thermoplastic nature of melting when heated and solidifying upon being cooled is inherent in hot melt adhesives. Hot melt adhesives have advantages and limitations when compared to other types of adhesives. A partial listing follows: *(Figures 8 and 9)*.

- Adhesives less costly.
- Distribute stress over wide area.
- Attenuate mechanical vibration.
- Thin substrates joining are less distorted.
- Adhesives weigh less than metal fasteners.
- Don't mar appearance of finished product.
- Reduce galvanic corrosion between dissimilar metals.
- Provide good chemical barrier.

Figure 6. *Advantages of Hot Melt Adhesives.*

- Mechanical fasteners are stronger.
- Mechanical fasteners are useful in wider temperature range.
- Some mechanical fasteners can be removed and reused.
- Fasteners can serve a dual engineering function such as component mounting.
- Established technology requires no additional training or personnel.

Figure 7. *Advantages of Mechanical Fasteners.*

Up to this point, this paper has dealt with adhesives or hot melt adhesives without regard to the nature of the polymer. I would like to narrow the focus to a particular polymer class of hot melts that appears suitable for product assembly.

POLYAMIDE HOT MELT ADHESIVES

Certain polyamide resins have found use as specialty hot melt adhesives. These resins are used where their specific performance properties are required. The presence of the amide linkage in polymer chain gives interchain hydrogen bonding that results in high polymer strength at low molecular weights. In addition, these amide groups contribute adhesion to polar substrates. Polyamide resins formulated using dimer acids have been found useful as hot melt adhesives. The dimer acid contributes flexibility moisture resistance, grease resistance and adhesion to a variety of substrates.

General Mills Chemicals, Inc. manufactures polyamide resins based on dimer acid in three different molecular weight classes. These products are useful hot melt adhesives. The following Figure (*Figure 10*) indicates the typical property range of each of these molecular weight families.

- Speed of bond formation.
- Ability to bond impervious surfaces.
- Moisture resistant barrier properties.
- No solvent hazard--air pollution problem.

- <u>Speed of bond formation</u> -- As hot melt adhesives need only to cool to reach a set state, bonds can be obtained rapidly which will then withstand further processing. This can translate into a faster rate of production and improved productivity. Thermoset adhesives generally require a longer cure time to reach a set stage.

- <u>Ability to Bond Impervious Surfaces</u> -- As hot melts are 100% solids, impervious substrates can be bonded at reasonable production speeds. Other adhesive systems have to cure or remove solvents prior to bonding.

- <u>Moisture Resistant, Barrier Properties</u> -- Hot melts, in general, do not contain water sensitive additives, therefore, would be more resistant than emulsion adhesives. Barrier properties are dependent upon the chemical nature of the adhesive rather than form in which it's applied. Thermoset adhesives are generally more chemically resistant than are thermoplastic adhesives.

- <u>No Solvent Hazard or Air Pollution Problems</u> -- As 100% non-volatile materials, hot melts do not have volatiles to control. Increased government regulations will make this feature more attractive. The need to control solvents will retard the growth of solvent adhesives.

Figure 8. Advantages of Hot Melt Adhesives.

- Limited Service Temperature.
- Subject to creep.
- Critical application temperature.

- <u>Limited Service Temperature</u> -- Thermoplastics soften with heat. This limits practical service elevated temperature. The temperature limits are dependent upon the application and the particular adhesive used.

- <u>Subject to Creep</u> -- Hot melt adhesives are subject to creep under a continuous load because of their thermoplastic nature. This can limit the upper service load range.

- <u>Critical Application Temperature</u> -- Application of hot melt adhesives at too low a temperature will cause poor wetting that will result in a weak bond. Application of hot melts at too high a temperature may degrade the adhesive or result in a starved bond because the adhesive squeezed out due to too low a viscosity.

Figure 9. Limitations of Hot Melt Adhesives.

Property	Low	Intermediate	High
Molecular Weight			
Melt Viscosity-Poise			
(a) 160° C	5-Solid	120-Solid	N.A.
(b) 210° C	1-10	20-110	250-5000
(c) 260° C	N.A.	5-25	20-1000
Polymer Strength @ 75° F			
Break, psi	160-1600	450-3000	3400-6500
Elongation	5-100	25-1000	300-600
2% Modulus	5000-50,000	2000-55,000	15,000-100,000

Figure 10. *Range of Typical Properties.*

I would like now to present some data on a resin from our high molecular weight family which I will identify as Polyamide Adhesive A. This is tensile data obtained using ASTM D638 procedure with the cross head speed at 5 inches per minute. The specimens were aged 500 hours at each temperature.

	Ultimate Tensile		Elongation	Modulus	
	psi	Kg/Cm^2	%	psi	Kg/Cm^2
-65° F	10,400	731	12.5	14,200	999
0° F	6,700	471	15.9	14,000	985
75° F	4,400	309	710	14,000	985
140° F	3,200	225	820	22,600	1590
260° F	650	45.7	230	7,700	542

Figure 11. *Tensile Properties of Polyamide Adhesive A*

The next Figure (*Figure 12*) shows typical lap shear adhesion data on metal substrates for Polyamide Adhesive A obtained using ASTM D1002 procedure.

	Tensile Shear Strength	
	psi	Kg/Cm^2
Aluminum/Aluminum*	3130	220
Cold Rolled Steel/Cold Rolled Steel+	3055	215

* (0.064" thick) degreased and dichromate etched
+ (0.064" thick) degreased and bonderized

Figure 12. *Adhesion Data for Polyamide Adhesive A*

The next Figure *(Figure 13)* shows the effect of a Silane primer on various thin gauge metal substrates using a different high molecular weight Polyamide resin Adhesive B. The primer system does appear to improve the peel strength of these bonds.

	Lap Shear@		T-Peel	
	psi	Kg/Cm^2	pi	Kg/Cm
Aluminum/Aluminum*				
No primer	>126	>8.9	1.6	0.28
Primer	>126	>8.9	30.0	5.36
Copper/Copper+				
No primer	>140	>9.8	4.3	0.77
primer	>140	>9.8	8.7	1.55
Brass/Brass△				
No primer	>360	>25.3	0.8	0.14
Primer	>400	>28.1	21.8	3.90
Brass/Brass#				
No primer	1010	71	7.0	1.25
Primer	>1315	>92.5	36.9	6.60

* 3003 aluminum 5 mils thick
\+ copper 2.5 mils thick
△ brass 5 mils thick
\# brass 25 mils thick
@ Data reported > indicates substrate failure

Figure 13. Effect of Silane Primer on Various Thin Gauge Substrate Using Polyamide Adhesive B.

Even though the aluminum samples were cleaned and etched prior to bonding, the T peel specimens gave low results. The primer system used and markedly improved the T peels probably because of better wetting of the adhesive to the primer surface.

Primed, adhesively bonded specimens prepared in the laboratory were immersed in a 50% solution of ethylene glycol at 250° F for 24 hours. This test was used to simulate a proposed operating situation. Adhesive test results were obtained on these specimens at 250° F. *Figure 14* compares specimens not exposed to those specimens which were exposed.

Figure 14 indicates that the combination of Adhesive B and the Silane primer retained satisfactory adhesive properties under a severe laboratory test condition.

Another test was run on laboratory bonded specimens to compare how storage at various temperatures with testing at that temperature might affect adhesive results. *Figure 15* shows how the T peel values change on primed aluminum/aluminum samples bonded with Adhesive B.

Aluminum/Aluminum*	Not Exposed		24 Hr. Exposure	
Lap Shear	>80 psi	>5.6 Kg/Cm²	>80 psi	>5.6 Kg/Cm²
T-Peel	40 pi	7.2 Kg/Cm	35 pi	6.25 Kg/Cm
Copper/Copper+				
Lap Shear	>110 psi	>7.7 Kg/Cm²	>90 psi	>6/3 Kg/Cm²
T-Peel	20.5 pi	3.6 Kg/Cm	7.9 pi	1.4 Kg/Cm
Brass/Brass∆				
Lap Shear	>370 psi	>26.0 Kg/Cm²	345 psi	24.2 Kg/Cm²
T-Peel	23.5 pi	4.2 Kg/Cm	22.0 pi	3.9 Kg/Cm
Brass/Brass#				
Lap Shear	610 psi	42.9 Kg/Cm²	335 psi	23.5 Kg/Cm²
T-Peel	33.1 pi	5.9 Kg/Cm	19.3 pi	3.45 Kg/Cm

* 3003 aluminum 5 mils thick
\+ copper 2.5 mils thick
∆ brass 5 mils thick
\# brass 25 mils thick
@ Data reported as > indicates substrate failure

Figure 14. Exposure Data - Adhesive B Immersed in 50% Solution of Ethylene Glycol at 250° for 24 Hrs - Primed Substrate@

	T-Peel	
-40° F	17.4 pi	3.1 Kg/Cm
0° F	16.8	3.0
75° F	30.0	5.36
250° F	46.0	8.2

Figure 15. Effect of Temperature on T Peels Primed, Aluminum/Aluminum Using Adhesive B

The date presented in *Figures 13, 14 and 15* were part of a cooperative study to define a suitable adhesive. The results of this study were sufficiently encouraging, so that prototypes have been prepared which have performed satisfactorily in actual use testing.

SPECIFIC HOT MELT APPLICATIONS

I would like to discuss a few specific applications for polyamide hot melt adhesives. The purpose will be to illustrate some of the performance capabilities of these adhesives. While all of these examples have at least one metal substrate, the use of these adhesives are not so limited. In each of these applications a specific polyamide adhesive was found to meet the defined needs of the individual use. For some of these applications "tailor made" polyamide adhesive resins were developed.

Lap Seam Containers -- An adhesively bonded lap seam container has been developed by a leading commercial can maker. The primary use has been in the beer and beverage segment of the metal can industry. Billions of these containers have been produced. The process allows lower cost materials to be used for three piece can construction. These containers may also be used for packaging other food products.

The internal can pressure may reach as high as 90 psi during the pasteurization process. The polyamide hot melt adhesive has to perform under high speed manufacturing process with a bonding operation that is effected in only 20 milliseconds.

Laboratory testing of the lap seam bond under tensile shear conditions results in failure of these metal substrates before the adhesive bond will fail.

Oil Filter -- Polyamide hot melt adhesives are being used to bond the paper filter elements to the metal end caps in the manufacture of automotive type spin on filters. The adhesive bond must maintain its integrity and give a leakproof seal while in service between -40° F and 275° F. The bond must not be deleteriously affected by engine lube oil.

Stud Bonding -- A polyamide hot melt system has been developed which allows sandwich panel fasteners for light to moderately heavy loads (25 to 150 lbs pull off strength) to be installed at a savings in cost and weight.

In light to moderately heavy duty applications, these fasteners eliminate the use of heavy potting compounds since their performance does not rely on attachment to the core material and/or back cover sheet of the sandwich panel. The high strength of the bond permits reliable attachments of fasteners to a large variety of panels including solid fiberglass sheets which have thick epoxy gel coats.

The polyamide hot melt adhesives have the important advantages when compared to the usual potting compounds.

o Indefinite storage life
o No mixing or metering required
o No pot life limitations
o Eliminates the long curing cycle
o Bond reaches full strength in seconds
o No clean up required

A special installation tool has been developed which converts electrical energy to heat and applies the heat directly to the fastener flange. The heating cycle (550° F) is automatically initiated by the tool when pressure is applied to the fastener. The duration of heat applied is typically 15 seconds. Pressing the trigger will activate an air blast for 10 seconds. This lowers the temperature of the fastener to a normal level. A total cycle time of 27 seconds is claimed.

Rearview Mirrors -- Mirrors are bonded to metal or plastic housings using adhesives. Current production practices allow only a short time between application of the adhesive to the case and the assembly of the mirror to the case - in the order of 13-18 seconds. The assembled mirror is removed from the conveyor line and wrapped in a protective cardboard sleeve, also in 13-18 seconds. The adhesive must set rapidly so that the mirror does not slip out of position during the wrapping process.

The finished mirror must be able to withstand a 50 lb load at 180° F. It must also withstand shock testing by repeatedly slamming a car door after storage at -40° F for 12 hours.

Several types of adhesive systems will produce assemblies that will meet these finished product requirements. A polyamide hot melt system can meet these requirements, does offer potential savings, and result in less product loss in production.

<u>Ceramic - Aluminum Bond</u> -- An unusual application involves a combination hot melt adhesive - sealant for bonding a ceramic lens in an aluminum housing. This application is for a radar installation. Simulated long term aging tests were performed on the adhesive-sealant prior to choosing this particular system. The polyamide hot melt is pre-formed into a suitable gasket shape for ease of application on this difficult to bond assembly.

Obviously these are but a few examples that perhaps show the diversity of applications and performance requirements capable of being met by polyamide hot melt adhesives.

NEW APPLICATIONS

How does a company get started in finding out if hot melt adhesive bonding is suitable for their application? I don't profess to know the answer to that question, but in reviewing past successful adhesive applications several factors seem to be important.

NECESSARY FACTORS FOR SUCCESSFUL ADHESIVE BONDING APPLICATION

- Commitment
- Suitable Adhesive
- Cost effective assembly

<u>Commitment</u> -- Many companies are not in a position to evaluate hot melt adhesive technology for they are comfortable doing business in the same old way. It would seem that to survive in our present economy a means of evaluating better ways of making things is mandatory.

Without a commitment to evaluate new technology, a company will not know which method of manufacture is most cost effective. For example, special design changes would not be considered to make adhesive bonding more efficient. Study of new handling equipment will not be made to know which system is best for a specific application. Personnel training will not be done so that the new joining system will work efficiently and effectively. Hot melt adhesive bonding is not a universal solution for all your joining problems but it does merit careful evaluation.

Where does a company start, if it has made a commitment to evaluate this new technology? I would suggest the following scenario.

SPECIFIC APPLICATION

A specific application will generally define the factors necessary to evaluate in selecting the right adhesive. Which applications should be chosen? Some people counsel that a company should begin evaluating adhesive technology in a small application. This theory suggests that in any learning sequence false starts are likely. A single failure should not discourage a company. Others suggest choosing an application where potential savings can justify a major commitment. Both approaches have merit and have been used by various companies.

In the small application approach, assistance from adhesive suppliers probably would be limited to supplying general information and small commercial samples. In the large application approach, specific laboratory to laboratory cooperative programs might be expected. This approach can result in an adhesive specially designed to meet the unique requirements of this application. This approach may multiply the effective effort that a single company might be able to justify.

SELECTING THE RIGHT ADHESIVE

o Performance Requirements
o Substrates
o Joint Design
o Laboratory Evaluations

Performance Requirements -- Knowledge of the performance requirements of a specific application can assist in determining whether hot melt adhesive bonding may be a feasible candidate for this application. The performance parameters can be used to choose potentially suitable adhesive systems.

Performance requirements can also be useful in setting up laboratory screening tests to evaluate laboratory prototype of bonded specimens. Adhesive suppliers can be too optimistic about the suitability of a particular adhesive, performance of the bonded specimens will result in better selection.

Substrates -- Knowledge of the substrates to be joined can also aid in selecting suitable hot melt adhesive candidates. To develop a strong adhesive bond, the adhesive must wet the substrates properly. Adhesive suppliers can assist in selecting adhesives which can properly wet the specific substrates.

Not all substrates of the same generic name are the same. The best adhesive recommendations will result by supplying samples of the particular substrate to the potential adhesive supplier. Substrate preparation may be necessary to achieve the best adhesive bonds possible. Variable bonding surface will lead to an inconsistent bond that would not give a reliable assembly.

Joint Design -- Joints should be specifically designed for adhesive bonding. A satisfactory joint for a mechanical fastening may not be satisfactory for an adhesive bond. The joining should be designed so that stress is distributed uniformly over the entire bonded area and that the basic stress is primarily in shear or tensile with cleavage and the peel stress minimized.

Laboratory Evaluations -- The use of hot melt hand guns can be of considerable value in preparing laboratory prototypes. Laboratory specimens can also be made using hot melt adhesive in film form. While these specimens need not represent the proposed scale up, they can be used for preliminary screening adhesive suitability.

COMMERCIAL SCALE

Scale up -- Assuming laboratory evaluations appear promising, the next step would be to determine whether it can be done on a commercial scale. During this stage, the adhesive supplier can provide information on equipment suitable to apply the hot melt and the preferred operating range, plus details about application conditions the user should try to avoid. Hot melt equipment suppliers can offer assistance on features of their equipment and how this equipment might best be tied into the overall system. In many applications, the unique needs to integrate the application equipment into the overall manufacturing system are solved by the user himself or by working in coordination with hot melt equipment suppliers.

Some Operating Considerations -- For reliable, consistent results many factors must be controlled. For the adhesive to give proper bonds (1) it must be applied within the proper temperature range, (2) the proper amount of adhesive must be applied, (3) the substrates must be joined within the open time of the adhesive, and, (4) the joined substrates held until the adhesive is set.

The substrate must be handled properly (1) the surface must be clean and uniform (2) the substrate temperature must be in the proper range - some substrates may require preheating.

Changes in the environment might affect the quality of the bonds. In many production situations the work place is not air conditioned. Changes in the ambient temperature conditions may change the actual application temperature, the open time, or the set time of the hot melt. Raising the ambient temperature may increase the open time and the set time. Should stresses be put on the bond prior to the time the adhesive is set, a lower strength bond will result. A substantially cooler room temperature might lower the application temperature of the hot melt and decrease the open time. Increasing the velocity of air movement over the bonding operations has been found to change the operating parameters causing poor bonds (e.g. use of fan during summer time).

Accidental contamination of the substrates because of problems in the work space could lead to adhesive failures. Improper handling of the substrates can result in weak bonds.

Training -- The above indicates that there are special considerations to obtain reliable, consistent adhesive bonds. The operators must be properly trained as these important considerations are different for hot melt adhesive bonding than for mechanical fastening.

In any automated system, a means of evaluating the quality of material being produced is mandatory. If such control tests can distinguish more than "good - no good" it would be helpful. Such tests might guide the operator in adjusting the operating conditions to improve quality.

SUMMARY

A growing number of companies find that hot melt adhesive technology lowers product cost and improves product reliability. Polyamide resins are unique hot melt adhesives suitable for product assembly. Specific application areas were discussed where these resins are used. A scenario was presented to indicate how a company might evaluate this technology from concept to production using a specific application. A user can develop a cost effective system by selecting the proper adhesive and application method for a particular application.

SIX ADHESIVE CURE SYSTEMS

For chemically curing adhesives, there are six basic cure systems, each of which runs a whole gamut of advantages and disadvantages.

This section attempts to put each of these systems into perspective for the manufacturing engineer who has to make decisions on storing, handling, dispensing, curing, testing and otherwise preparing for adhesive bonding in the production environment. Solvent and hot melt and other non-curing drying materials are excluded.

Adhesive systems are advancing at a rate which puts the manufacturing engineer at a disadvantage if he does not have up to date information. He must make decisions on storing, handling, dispensing, curing and testing of adhesives which may not have been in existence a few years ago. Today, there are six basic curing systems for structural adhesives.

The cure systems are:

1. Anaerobic
2. Surface activated
3. Ionic or moisture activated: cyanacrylates, room temperature vulcanizing silicones
4. Ultra-violet light activated
5. Interaction of reaction chemicals: epoxy, acrylic, urethane
6. Heat curing: epoxy, acrylic, urethane

Each heading can include a great variety of materials with different uncured properties as well as cured. Each has its thick and thin and thixotropic and dilatant properties before cure. Each has its stiff, strong, brittle, flexible, weak variation after cure. In many cases, one adhesive will exhibit satisfactory cures by more than one system. For instance, some U.V. curing materials respond well to an activated cure and are indeed sometimes used this way. All chemical reactions respond faster as heat is applied so all may cure or change to one degree or another at elevated temperatures.

Because each system is formulated to cure under controlled conditions, its chemical make-up is extremely complex and even sometimes unstable under storage conditions. Let us perceive each as a mixture of active materials which ordinarily would combine chemically except for the presence of a potent stabilizing material. Stability can also be obtained by the absence of a key element. In the condition of activity, molecules will begin to combine forming denser and denserstructures or chains, while the physical properties change from liquids to gels to carmel to cheese to final hardness. All can occur in a matter of seconds or minutes but the last few percent of change may take days or weeks.

ANAEROBICS

This word comes from the Greek "an-aero-bios" which means, life without air. Although the term is common in biology for organisms which live in airless atmospheres such as septic tanks, it originated in the adhesives industry with Professor Vernon Krieble who founded the Loctite Corporation with adhesives based on anaerobic technology. It is a fitting name because the key stabilizing element in the formulations is oxygen. As long as oxygen in minute quantities is available, it ties up the free radical in the mixture which is trying to combine with the monomer to initiate the cure. If oxygen is excluded in a manner such as confinement between metal parts, the free radical can start polymerization. This is not a reversable process, therefore, when carried to completion, the monomer becomes a polymerized thermoset plastic. That is, it cannot be liquified again by the application of heat as in a thermoplastic. It is also insoluble in common solvents.

General Properties and Storage

Anaerobic structurals can be characterized as low energy curing materials with modest to high strength. They withstand temperatures to 400° F for considerable time. They tend to be brittle when cured and as liquids have a consistency and color of maple syrup to honey (100-10,000 cp).

Properties of Anaerobic Structurals

Type of Adhesive:	Hot Strength	Quick Setting	High Strength
Tensile Shear psi ASTM D-1002	2000 (13.8 Pa)	2600 (17.9 Pa)	3000 (21 Pa)
T-peel lb/in ASTM D1876	3 (.5 kgf/cm)	N.A.	7 (1.3 kgf/cm)
Impact ft-lb/in^2 ASTM D950	10 (2.1 J/cm^2)	10 (2.1 J/cm^2)	6 (1.3 J/cm^2)

Fatigue Strength 800-1000 psi (5.5 - 6.9 Pa)

Anaerobics are generally easy to apply with equipment designed to handle their peculiar characteristic of curing when confined. Ordinary valves, piping, and pumps will not work because they always confine the material away from oxygen and between active surfaces.

Because of their liquid nature, anaerobics wet parts easily. However, by the same virtue, large gaps cannot be filled and will not cure. Maximum gaps run .005" (.13 mm). They are therefore used on rigid close fitting parts. The anaerobic cure of structural adhesives is usually so slow (hours) that heat or activator is almost always used. This reduces fixturing time to seconds or minutes.

ADVANTAGES OF ANAEROBIC STRUCTURAL ADHESIVES

1. Single component, no mixing or pot life problem.

2. No energy is needed to cure.

3. The system is 100% reactive - no volatile or non-reactive solvents.

4. Anaerobic characteristic eliminates cure outside bondline. Clean up is easy.

5. Materials are non-toxic although most are eye irritants and will cause dermatitis in sensitive or traumatized body areas. They are non-flammable.

6. Seals as well as bonds.

7. Easily applied to large areas by screening, rolling and vacuum impregnation.

LIMITATIONS OF ANAEROBICS

1. Gaps of .005" maximum (.13 mm) without activators or surface cure.

2. Rigid resins for rigid substrates.

3. Slow cure on inactive surfaces without heat.

APPLICATIONS

The uniqueness of anaerobic materials allows for unique (to adhesives) application methods. The silk-screen process is very old and normally uses slow drying inks. With only slight alteration to the equipment and tools, anaerobic liquids and pastes can be applied with extreme precision, speed and intricacy. This technique has been used for printed circuits and air gage bonding. Extensive use of silk-screening is used in sealing and gasketing of flat joints.

Another process to which anaerobics are uniquely suited is vacuum impregnation. Commonly used for powder metal, laminated and die-cast parts, the material is pushed into mocroscopic pores (.005" or less (.13 mm)) after air has been excluded by a vacuum cycle. This is done in a large mesh basket in a vacuum vat with the parts immersed. Excess material is spun off leaving a relatively clean surface. Where drying or heat curing materials would plug up large holes and cure on the surface, the anaerobic material in contact with air can be rinsed off leaving a dry clean part.

One of the largest pure anaerobic applications is used to seal and retain cup or core plugs in automotive engines. The material is neatly rotosprayed on the bore of the core hole. A cup is pushed through the ring of material. The liquid material effectively wicks into the surface roughness and hardens in minutes to secure the seal and bond the plug.

Figure 16. Cup or Core Plug Bonding and Sealing. Rotospray Equipment is Shown in the Application Mode.

SURFACE ACTIVATED ADHESIVES

These materials are formulated to respond to steel, copper, brass, aluminum and chemically activated surfaces. This characteristic overcomes most of the disadvantages of "pure" anaerobics.

The advantages are the same as anaerobics with the following additions:

1. Fast fixturing - 3 min. to 15 sec. without heat.

2. Cure through .030" gaps (.76 mm).

3. Some cleaning and surface priming can be accomplished during chemical activation.

4. Chemical activation gives a certainty of cure and strength and eliminates some of the problems caused by steel heat treatment and other non-related adhesive surface treatments.

5. Cost of total system is low.

Limitations

1. Most activators contain solvents which must be flashed from the surface. Good ventilation and shop hygiene is required.

2. Humidity, heat and solvent resistance of final bond does not quite equal heat cures although resistance to water, alcohol, oil and Freon is good.

Applications

Rigid and porous parts are good applications of the activated anaerobic systems. Loudspeaker ferrite cores and backing plates are assembled on production lines without the use of heat or extensive fixtures. Porous honing stones are attached to channel holders and a plastic cooling fan impeller is attached to a phenolic insulator tube on universal motor armatures.

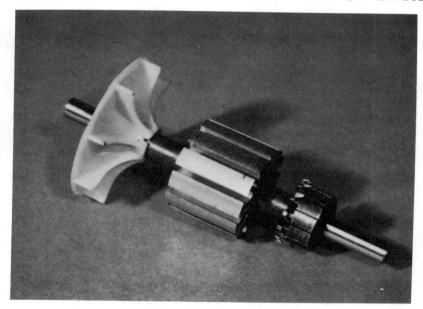

Figure 17. Plastic Fan Impeller to Phenolic Sleeve is Activator Cured in Place with Anaerobic Adhesive.

IONIC CURE - CYANOACRYLATES

Cyanoacrylates are the fastest growing of all the engineering adhesives. They are characterized by high speed, high strength and the ability to adhere most rubbers, plastics, metals and wood. They are cured by the presence of water vapor and/or a chemically basic surface. Just a few molecules of water on the surfaces to be joined will trigger the polymerization of a thin joint. Liquids vary from watery consistency to syrup (1-2500 cp) and are clear.

Advantages

1. Fast fixture - 1 min. to 5 sec.

2. High strength - 2500 psi (17.2 Pa) which exceeds the strength of most plastic and rubber materials commonly bonded with cyanoacrylates.

3. Single component.

Limitations

1. Temperature resistance of 180° F maximum (82° C).

2. Low resistance to solvents and moisture and weather.

3. Shelf life is 1 year if stored in moderate temperature and humidity conditions in the original unopened container. Once open, packages become contaminated with moisture and should be used within a few months.

4. Low impact strength.

5. Excellent shop hygiene must be used because the odor can be irritating (although non-toxic) and skin stick is very fast and frightening. The antidote for skin stick, spilled material and fumes is water, H_2O, and more water. Fingers can be unstuck with hot salty water or by <u>peeling</u> apart. Heat induced sweating also helps. Polyethylene gloves are helpful if holes can be reasonably controlled. Cotton or rubber gloves - never! Cotton wipers - no. CA's are extremely fast and exotherm in the presence of cotton.

6. Material cures too fast on clothes to clean out. After cure, scissors are the only workable spot removing agent.

With all of these limitations, one would wonder why the material is so popular. Most of the limitations are removed when good hygiene and mechanical applicators are used and all the advantages remain.

Typical applications are:

1. Structural bonding of hygiene rubber goods in high volume production.

2. Semi-automated decorative button bonding.

3. Fully automated lipstick case and cap bonding.

4. Repair of vinyl trim on automobiles.

Figure 18. Decorative Delrin (R) Fronts are Adhered to a Non-Sewn Back Snap by Fully Automated Assembly Machines. Many Decorator Designs are Assembled From One Basic Engineering Design. Fast Cyanoacrylate Adhesive is the Key Element.

(R) DuPont

ULTRA VIOLET LIGHT CURE

UV curing of adhesives is relatively new and has followed the development of UV sensitive inks and varnishes. The adhesives can be characterized as fast curing, clear liquids which have stabilizers which are destabilized by 365 nm wave length at an intensity of 10,000 μ watts/cm^2. This is roughly the same amount of energy in sunlight without the heat or visible portions. Cures are very rapid and strengths are high usually exceeding the strength of glass.

Advantages

1. Very fast one component.

2. Low safe energy to trigger reaction.

3. Strong glass and thermoset plastic bonds.

4. Can be formulated for potting and wire staking with cure to surface.

5. Excess is easy to clean up.

6. Resists environmental conditions very well.

7. Cures well regardless of gap.

8. Safe to use. No fumes. Mild skin and eye sensitivity only.

Limitations

1. One substrate must be transparent to UV light.

2. Some formulations will not cure a fillet since they are anaerobic. For fillet cure, high intensity mercury vapor lamps are a necessity.

Applications

The bases of miniature light bulbs are adhered to the glass bulb on automatic assembly machines. Adhesive was previously applied externally to the assembly machine. The ease of automating the application of UV materials made the cost of material insignificant with respect to the system savings.

Other applications are on automobile windows where hinges, latches and decorative trim are adhered directly to the glass.

UV materials can also be used for shallow potting or staking of wire ends. This gives physical support to soldered wires where they connect to motor armatures or solenoid coils.

Figure 19. Metal to Glass Bonding with Ultra-Violet Adhesive on Fully Automatic Assembly Machines.

REACTIVE CHEMICALS - 2 COMPONENT EPOXY AND URETHANE

Epoxies are the adhesives which made the great leap ahead in the 1930's. Their high performance compared to animal glues, glyptol and solvent cements pushed their use into structural areas. They run a wide range of uncured properties from heavy clear syrups to filled pastes. The mixture of the two reactive components is often dilatant which means it runs very slowly and it may not stay in place too well as it cures. Cured strength and hardness can be varied considerably by formulation. Material shelf life is good under normal temperatures (30-85° F) (-1 to 29° C) and dry conditions.

Advantages

1. Large volumes can be cured. The major limitation is holding it in place and not producing destructive temperatures from the exotherm.

2. Strength is moderate, 2-3000 psi (13.8-21 Pa) tensile shear.

3. Cost of material is modest.

4. Solvent resistance can be good and aging excellent.

5. Formulation can be made to fit almost any bonding application.

Limitations

1. Most formulations require precise metering and mixing of the resin and reactant. This is not easy with pastes, even with the best of equipment.

2. Post life after mixing varies but excess material is always wasted.

3. Clean up of joints is very difficult. It is most often done by machining or abrading after cure.

4. The materials are more than mildly toxic and irritating and must be handled remotely with gloves or equipment.

5. Unless heat is used and sometimes even with heat, cure time is slow - minutes to hours, depending on the compromise with pot life.

Applications cover everything from maintenance to high production such as furniture. Substrates include steel, aluminum, glass, ceramic, plasters and wood.

REACTIVE CHEMICALS - ACRYLICS

These materials are different from the epoxies in chemistry and application. They are more analogous to the free redical accelerated or activated surface curing material. The surfaces to be bonded are coated with a thin evaporative reactive chemical.

The adhesive is then applied over the reactive material and the parts immediately joined. The adhesive is often a thick yellow liquid which stays in fairly large gaps of .030" (.76 mm). Handling strength is reached in

3 to 15 minutes. Tensile shear values reach 2-3000 psi (13.8 - 21 Pa).

Advantages

1. Ease of application compared to epoxies. No metering or mixing required. No heat. Activator/primer can be applied days ahead of assembly.

2. Good adhesive qualities on less than chemically clean substrates.

3. High peel and tensile values.

4. Inexpensive material.

5. Environmental and solvent resistance is fairly good.

Limitations

1. Formulations are extremely toxic and odoriferous. They are skin sensitizers. They are flammable and must be used in well ventilated areas and skin contact rigorously avoided.

2. The resin thickens and dries on application causing stringing like rubber cement. The fumes are flammable.

3. Materials must be refrigerated for reasonable shelf life.

4. High shrinkage occurs during curing.

5. Materials are difficult to automate.

Applications

Applications include a preponderance of wood and plastic bonding where safer materials cannot provide as good performance.

Superior adhesion on hard to grab surfaces promoted their use on chrome plated wear strips on a silicon slicing machine.

Fast fixturing and low cost have been beneficial in furniture gluing of wood and plastic. It has also replaced slow curing epoxy in fiberglass reinforced components of a helicopter.

HEAT CURE - ANAEROBICS, EPOXIES, URETHANES

These one component materials can be very effectively cured by the application of heat. Heating not only cures the material but assists in the surface wetting action so that merely degreasing is an effective part preparation. The highest possible strengths are obtained with heat.

Applications of heat cured anaerobics are generally to stiff metal substrates where high performance is required. These include rotor shaft bonding, ceramic magnets to motor housings and roller bearing retainers. Glass bonding is also accomplished by the residual heat from the glass annealing process.

Advantages & Features

Anaerobics

Unlimited shelf life

Strength 3-4000 psi
(21-28 Pa) tensile shear)

Med. temp. cure 200-300° F

30 sec. at 300° F (149° C)

Easy dispensing liquid

Good environmental resistance

Overall system cost - low

One Component Epoxy or Urethane

Often refrigerated

Strength 3-500 psi
(21-34 Pa) tensile shear

High temp. cure 350-450° F

1-2 hours at 350° F (177° C)

Viscous liquids and pastes

Excellent environmental resistance

Limitations

Anaerobic

Usually hard and brittle

Epoxy or Urethane

Often must be refirgerated to obtain reasonable shelf life.

Cost of high temperature long term heating is high for production.

Heat cured epoxies have the highest performance available and are used for aircraft honeycomb and skin bonding as well as many fiber wound and composite parts.

Figure 20. Heat Cured Anaerobic Adhesive is Used to Retain a Phenolic Sleeve to Shaft and Armature and Commutator to Sleeve in this Double Insulated Universal Motor.

ANAEROBIC ADHESIVES

Anaerobic adhesives have simplified the process of assembling rigid cylindrical parts and flat structurally bonded parts through the unique stabilization and cure system discovered more than twenty years ago. While often providing a simpler and more reliable method of assembly, anaerobic structural adhesives have been limited by their brittleness. Now new technology has been discovered which allows greater flexibility in anaerobically cured polymers. This section traces some of the development of anaerobic adhesives and discusses data for improved anaerobic structural adhesives.

ANAEROBIC TECHNOLOGY DEVELOPMENT

An anaerobic adhesive can be defined as an adhesive which can be cured by a chemical reaction when excluded from air between metal parts. Originally anaerobic adhesives, first discovered more than twenty years ago, were formulated for locking small threaded parts. The low viscosity liquid cured slowly and only between ferrous and copper alloys. Only high strength permanent locking could be achieved with the first anaerobic adhesive formulation. The "adhesive" achieved its locking capability not by merely adhesively bonding threaded parts together, but by curing without significant shrinkage and creating a plastic interference fit.

Application requirements led to the development of a variety of anaerobic adhesives to serve the many needs of industry: higher viscosity formulations were developed to give locking capabilities on larger fasteners; plasticizer technology allowed products which would prevent vibration loosening but allow disassembly for repair; new catalyst technology led to products which would cure rapidly on virtually any metal surface.

These anaerobic machinery adhesives offer to users many benefits over conventional adhesives, solvent cements, or mechanical methods of threadlocking or assembly. These are:

1. A single-component, easy-to-use Adhesive System which fits easily into automated assembly systems.

2. The elimination of mixing and pot life problems which can occur with two-component adhesives.

3. Adhesives which will completely fill joints when cured, providing a positive seal and preventing corrosion.

4. Adhesives which will provide stronger, more reliable assemblies than mechanical methods such as press fitted parts of lockwashers.

5. Reduced toxicity over many conventional adhesives, no volatile solvents or strong skin irritants are present in anaerobics.

6. Rapid cures with fixture times as short as ten seconds when using surface activators.

The development of twenty years' research is summarized in *Figure 21*. Modifications to the original anaerobic product have led to more than one hundred products formulated not only for threadlocking, but for sealing flanged, threaded, or porous assemblies, and retaining smooth cylindrical parts.

ANAEROBIC ADHESIVE DEVELOPMENT

PROPERTY		1953	1965	1975
VISCOSITY		15 cps	4-20,000 cps	4-300,000+ cps
Full Cure Time	Inactive Metals	No Cure	No Cure	6-24 hrs.
	Steel	24-72 hrs.	12-72 hrs.	2-72 hrs.
Thixotropic Ratio		None	2-1	10-1
Tensile Strength		1,400 psi	2,000 psi	8,000 psi
Shear Strength		1,000-2,000 psi	75-3,000 psi	75-6,000 psi
Temperature Limits		-65°F to 300°F -55°C to 150°C	-65°F to 400°F -55°C to 200°C	-65°F to 450°F -55°C to 230°C
Impact Strength		1 ft./lb.	1-2 ft./lbs.	1-20 ft./lbs.

Figure 21. Property Development of Anaerobic Adhesives.

ANAEROBIC STRUCTURAL ADHESIVES

Attempts to bond flat parts using anaerobic machinery adhesives quickly pointed out some of the limitations of the anaerobic adhesives which are based on acrylic diesters. The adhesives did not develop sufficient tensile adhesion, and their brittleness resulted in poor impact and peel strengths. Therefore, applications were limited to rigid cylindrical parts. This self-supporting type of assembly was improved through the use of the chemical locking element. Rather than simply creating adhesion, the cured polymer created a plastic interference which served as a cushion against vibration.

To achieve a bonding capability for flat parts, a new chemical technology was developed with the combination of the anaerobic cure system with urethane prepolymers. This led to a line of adhesives with acceptable structural properties *(Figure 22)*. These adhesives offer greater flexibility and surface adhesion, higher tensile and tensile shear strengths, and improved impact and peel strengths over that of the polyether diacrylic systems.

| | HOT STRENGTH || QUICK SETTING || HIGH STRENGTH |||
| | 2 mil BONDLINE || 2 mil BONDLINE | 20 mil BONDLINE | 2 mil BONDLINE | 2 mil BONDLINE | 20 mil BONDLINE |
	HEAT CURE	ROOM TEMP. CURE	ROOM TEMP. CURE	ROOM TEMP. CURE	HEAT CURE	ROOM TEMP. CURE	ROOM TEMP. CURE
TENSILE SHEAR (psi) ASTM D-1002-65	1800	2000	3000	2600	3500	3000	2300
TENSILE (psi) ASTM D-2095-62T	2800	3000	5000	—	8000	6000	4500
IMPACT (ft lbs/in.^{2}) ASTM D-950-54	4	2	10	4	8	6	5
T-PEEL (lbs/in.) ASTM D-1876-61T	3	4	—	—	10	7	5
FATIGUE STRENGTH (psi) MMM-A-132	1000	800	1000	—	1100	850	—
ELONGATION (%) ASTM D-638-64T	.2	.2	1	1	2	2	2

Figure 22. Cured Properties of Anaerobic Structural Adhesives (First Generation). All Values Obtained on Steel Test Specimens.

Anaerobic structural adhesives have replaced conventional structural adhesives in automated applications where numbers of parts are being assembled rapidly. User benefits are:

1. The single-component adhesive eliminated mixing and application labor by allowing automation.

2. Waste is eliminated. No adhesive is discarded due to pot life.

3. Rapid fixturing (ten seconds to five minutes) on the assembly line eliminates clamps, fixtures, and storage area on the assembly line.

4. Room temperature cures with surface activators save energy and capital required for cure ovens.

5. Toxicity is lower than many other reactive adhesives.

In spite of high material costs, anaerobic structural adhesives have reduced manufacturers costs significantly by lowering labor and energy costs in bonding areas such as: stemware, electric motors, speakers, and others.

However, a series of limiting performance factors have been noted for the urethane acrylic structural adhesive formulations. Although elongation and adhesion had been improved over earlier anaerobics, the adhesives were often found unacceptable in areas where adhesion to bright plated surfaces was required, high impact or peel strength was needed, or high strength through bondline gaps of .020" or more was required. The requirement for a surface activator, a separate production step, would also occasionally eliminate the use of an anaerobic structural.

IMPROVED ANAEROBIC STRUCTURAL ADHESIVES

Recent research has developed new resins which are leading to a generation of anaerobic structural adhesives which overcomes many of the problems mentioned here. Now available as developmental samples, these adhesives represent the most efficient and effective structural adhesives available today.

The "Second Generation" anaerobic structural adhesives have retained the cure properties of the earlier structurals. The new resins show improvements primarily in cured properties. *Figure 23* illustrates comparative data for the range of properties available. Major improvements are apparent in toughness and elongation which results in significantly better impact and peel strengths. The new resins effect improved overall performance through gaps up to .060".

Figure 24 compares impact strengths at increasing bondline gaps for the present and emerging structural adhesives. Due to the improved flexibility and toughness of the new resins, the impact strength improves as gap increases up to .060". Due to brittleness, the practical limit for the "First Generation" structural adhesive is .020", although cures could be obtained through greater gaps.

Property	First Generation	Second Generation
Tensile Shear	1800-4000 psi	2000-4500 psi
Tensile	3000-8000 psi	3000-10,000 psi
Impact Strength	4-15 ft-lbs/in^2	8-25 ft-lbs/in^2
Fatigue Strength	800-1200 psi	800-1500 psi
T-Peel	3-15 pli	5-25 pli
Temperature Range	-65°F to 400°F -55°C to 200°C	-80°F to 450°F -60°C to 225°C
Gap Curing & Bonding	to .020"	to .060"
Elongation	2%	50+%

Figure 23. Cured Property Comparison, First and Emerging Second Generation Anaerobic Structural Adhesives

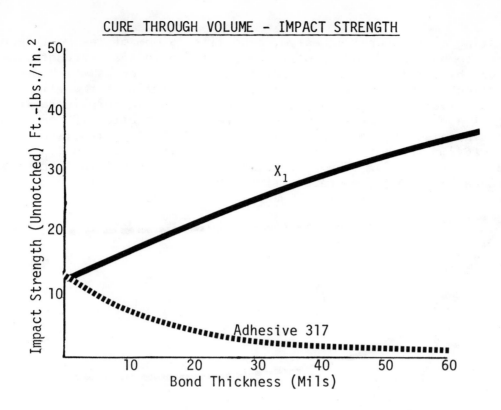

Figure 24. *Impact Strengths, Anaerobic Structural Adhesives Steel Impact Blocks per ASTM-D 950-54*

Peel forces have been a severe limitation of many structural adhesives and generally designers have designed bonded assemblies to eliminate peel. Peel strength has been significantly increased through gaps of .005" and greater for the new resins (Figure 25). While designers should continue to avoid this stress, the increased toughness and elongation of these resins will reduce problems inherent in any peel stresses which may occur.

Adhesive	Bondline Gap	Peel Initial/Break	Prevailing Peel Average
317	0	13	3
317	.005"	20	12
X_1	0	31	16
X_1	.005"	40	25

Figure 25. *T-Peel Strengths, Anaerobic Structural Adhesives - lbs/in Per ASTM D-1876-61T*

In testing a structural adhesive, bond life is one of the most important factors to be tested and is often unrelated to initial strength. The new adhesives have illustrated good performance in two areas which provide good indication of bond life. The temperature limitation for earlier anaerobic adhesives has been 400° F. New hot strength anaerobic resins are under test at 450° F and exhibit minimal strength loss after 750 hours aging. Flexible resins also show good performance at 450° F. Testing is continuing (See Figure 26A). The hot strength of these adhesives are also improved as illustrated in Figure 26B. The combination of improved polymer backbone and greater elongation under temperature stress has improved performance at all temperatures.

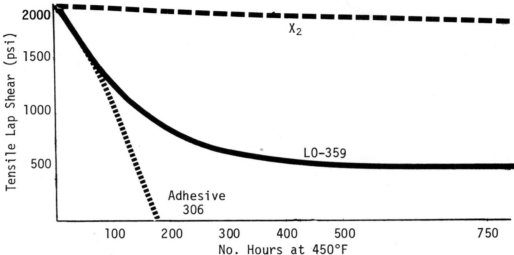

Figure 26A. *Heat Aging Properties of Anaerobic Structural Adhesives, Tensile Shear Strength on Steel, Tested at Room Temperature*

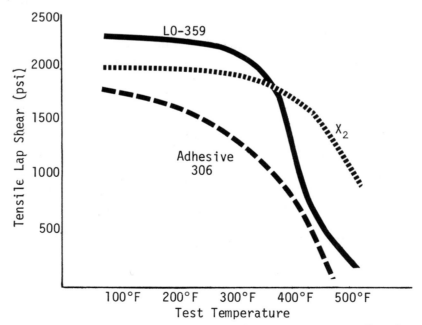

Figure 26B. *Hot Strength of Anaerobic Structural Adhesives, Tensile Shear Strength on Steel*

Salt Spray tests also illustrate improved performance over previous adhesives. These tests indicate improved weathering properties and improved moisture resistance and resistance to organic solvents is expected. Tests are continuing in these and other areas.

The "Second Generation" anaerobic structural adhesives promise to fulfill applications in which earlier anaerobics failed. The anaerobic cure properties have been brought to a line of tough, flexible adhesives approaching the optimal properties of two-component adhesives.

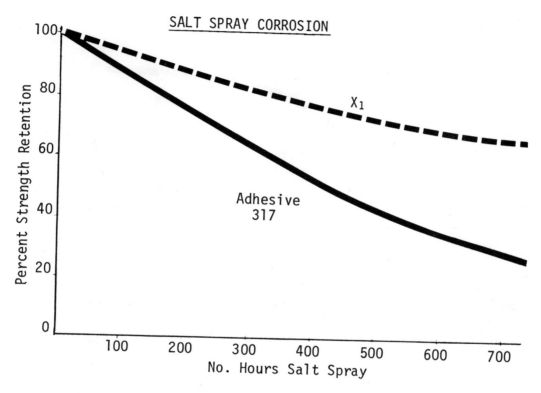

Figure 27. *Salt Spray Resistance, Anaerobic Structural Adhesives.*

OTHER DIRECTIONS IN ANAEROBIC RESEARCH

A new cure technology has been developed which eliminates all need of surface activation and develops rapid cures at ambient temperatures on glass. An adhesive which cures upon exposure to ultraviolet radiation is finding use in stemware and automotive industries. This rapid curing adhesive *(Figure 28)* eliminates costly glass welding operations in the stemware industry and fits easily into highly automated systems in miniature lamp plants and automotive glass plants. The adhesive exhibits cured properties comparable to other anaerobic structural adhesives.

A new anaerobic pressure sensitive adhesive is also now available. This adhesive eliminates problems with movement during cure and expensive clamping fixtures. This is the first true pressure sensitive structural adhesive. When applied to an activated surface, the adhesive fixtures immediately and cures anaerobically over a period of 72 hours. Strengths can reach up to 1,500 psi on steel.

As applications develop and anaerobic technology advances, anaerobic structural adhesives will improve to the point of true single component adhesive much like threadlocking adhesives. These will cure rapidly without the use of surface activator or heat. In addition to simplifying bonding processes, anaerobic technology is proving to be remarkable adaptable to many types of adhesive forms.

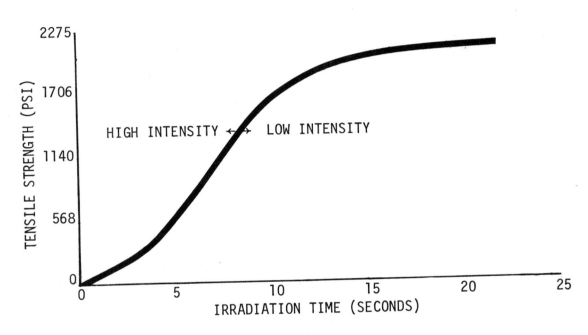

Figure 28. *Cure Speed and Strength of Ultraviolet Curing Anaerobic Adhesive. Adhesive Exposed to 6,000 μ Watt/cm² Light Centered at 3660A° Wave Length.*

INJECTED METAL FASTENING

Rapid, accurate assembly of two or more components of similar or dissimilar materials by the injection of molten zinc or lead alloy. Components to be assembled are located and held in a fixture containing a mold cavity. Solidification locks injected metal into undercuts, ridges, grooves, knurls or keys in the components. Hole location and tolerances can be eased yet accuracy and repeatability of assembly are consistent. Components can be eliminated. Cams, ratchets, pinions, keys, etc. can be formed by the injected metal when assembling parts. No secondary operations are needed. Rejections are virtually eliminated, thus costly inspection is greatly reduced. Up to 80% savings have been reported by users.

What is Injected Metal Assembly? It is precisely what the name implies-it is the assembly of several components by the addition of a hub or joint through the injection of molten metal, usually a zinc alloy, to produce a ready-to-use assembly.

Figure 29. *Drive Gear Assembled on Injected Metal Assembly Equipment. Injected Metal Hub Showed no Sign of Failure During Torque Test. Shaft Sheared at 26 Foot-Pounds.*

'Injected Metal Assembly' is the application of pressure die casting technology to the assembly operation. It is not new, since practical applications were made by a leading clock manufacturer nearly 100 years ago.

Virtually all watt-hour meter manufacturers in North and South America, Europe and Japan make use of Injected Metal Assembly equipment. The first machine to produce watt-hour meter disc and shaft assemblies by injected metal assembly was put into service nearly 30 years ago in Canada. This same machine is still in full production.

DESCRIPTION OF PROCESS

Injected Metal Assembly brings the repeatability of die casting to the assembly operation. The components to be assembled are loaded in a locating fixture which incorporates a mold or cavity and are held in the required relationship. The tool or fixture is closed and molten metal is injected into the cavity, flowing in and around some portion of all the parts being assembled. *See Figure 30.* After cooling the assembly is removed or ejected from the tool.

As rapid cooling and solidification occurs, a minute 'spherical' shrinkage takes place which locks the injected metal into undercuts, ridges, grooves knurls or keys in the parts being joined. This shrinkage is approximately 0.007 inches per inch (0.07 mm per centimeter) or slightly less than one percent. The shrinkage is directed toward the theoretical centre of the injected hub and is at 90° to the load lines of force developed by the assembly. With press fit and most other assembly processes, the strength of the joint is dependent on component precision, eg. - interference for press fit or the proper clearance for silver brazing, etc. With IMA, the injected metal flows into the void between the components completely filling it and the cavity. This permits inaccuracies in hole diameters and alignments, while compensating for them.

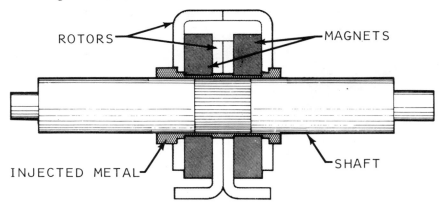

Figure 30. Rotor Assembly of Seven Components. The 'D' Shaped Shaft has Been Positioned by the Tooling in Relation to Steel Rotor. The Zinc Alloy Injection Makes a Permanent Assembly.

For higher loading, one of the zinc alloys is used. Given design freedom, a properly designed zinc alloy IMA joint can be stronger than the components themselves; load tests consistently show that the components fail before the joint. The resultant assemblies are permanent since the joints are of solid zinc or lead alloy. Lead is commonly used where parts operate under light load or where good bearing surface are paramount. The Injected Metal Assembly process leaves no sprues, gates, runners or flash, thus costly trimming or cleaning operations are eliminated.

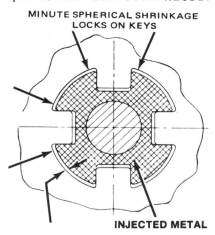

Figure 31. This Cross Section of Injected Metal Hub Illustrates the Principle of Spherical Shrinkage.

Since the molten metal cools very rapidly (in milliseconds) a wide range of materials such as paper, plastic, rubber, glass and ceramics can be assembled as well as all types of metal parts. Savings of up to 80 percent have been realized by some users through simplified design or through the elimination of a component by producing it as an integral part of the injected metal joint. Cams, pinions, ratchets and keys are often formed by the injected metal. There can be large savings through the elimination of secondary operations, freedom from need of skilled operators and minimization of inspection due to the consistency of the assemblies. *(Figures 32 and 33)*.

Figure 32. Injected Metal Pinion and Brass Shaft Replaced a More Expensive, Extruded and Machined Pinion Rod

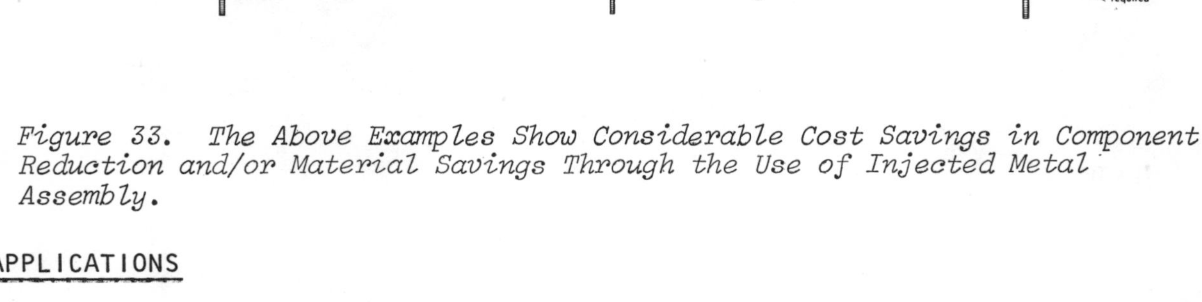

INJECTED METAL ASSEMBLIES COMPARED TO CONVENTIONAL STAKING

Figure 33. The Above Examples Show Considerable Cost Savings in Component Reduction and/or Material Savings Through the Use of Injected Metal Assembly.

APPLICATIONS

The potential applications of this assembly method appear to be limited only by the imagination of the design and manufacturing engineers. The following are but a few of the successful uses of Injected Metal Assembly:

Fan Hub

A good example of cost savings is a fan hub. In the conventional method of assembly, a screw machine hub is made and then pressed into the centre hole of the blank. This requires:

- Close control of the bore to the O.D. of the hub:
- Close control of the O.D. of the hub; and
- Close control of the centre hole of the fan blade both in size, position and concentricity.

With IMA, both the manufacture of the hub and close fit tolerances are eliminated. The fan blank is located by its outside diameter with the tool determining the required concentricity. The hub is then produced by injecting molten metal into the fixture, resulting in a concentric and parallel centre bore. The process applies to many types of fans and impellers.

Rotor Cup

This cup is a rotor of a small synchronou electric motor used in a timing device. Very close concentricity is achieved between the shaft and the cup by locating the shaft by its journal diameter and the cup by its inside diameter. The centre hole of the cup is oversize thus an open positional tolerance and size tolerance is permissible. Result - improved concentricity, reduced component cost and consistent accuracy from the Injected Metal Assembly process.

Scrap and rework are eliminated. In the case of the rotor cup, the holding fixture becomes a parts inspection fixture as it locates and fits the parts in the relationship required in the final assembly. Accuracy of assembly is consistent for the life of the tooling. *See Figure 34.*

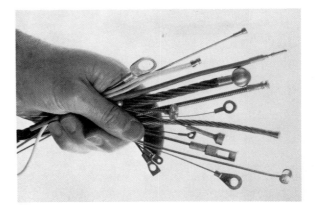

Figure 34. Shown are Some of the Applications Described.

Figure 35. An Assortment of Cable Terminations, Each Formed by an Injection of Zinc or Lead Alloy.

Telephone Component

A good example of positional accuracy is the plate and case assembly found in most dial telephones. The I.D. of the cup must be accurately located in relation to three mounting holes in the plate. The plate is produced with a hole sufficiently large to ensure clearance of the cup, thus eliminating the need for size and positional accuracy of the hole.

The two components are placed into the assembly fixture located by the same surfaces or points that locate them in the final assembly. The gap is then filled with injected metal in the assembly process. *See Figure 35.*

Hub and Clutch Plate

The steel hub and plate shown in *Figure 34* had to be assembled to withstand a torque of 108 foot-pounds (146 Newton-metres). In torque tests using a steel shaft, the injected metal hub showed no sign of failure when the shaft sheared at 160 foot-pounds (217 Newton-meters).

Rivetted Plate

In this assembly, rivets have been injected in place completely filling the holes in the two parts. Although the holes may be eccentric to each other and of different diameters, accurate assembly is possible because the parts are located by the functional surfaces rather than by the rivet. The gap must be held to a close tolerance and with the rivetted method had to be adjusted after assembly. This is now held automatically as part of the Injected Metal Assembly process. The savings are obvious.

Cable Ends

Virtually any shape of terminal can be injected on to most types of wire, bare or jacketed, single stranded or braided. Tests have shown that zinc terminals can be used for electrical purposes due to the excellent conductivity resulting from the complete contact that is achieved between the injected terminal and the strands in the conductor. In high load applications, the cable is upset or 'birdcaged'. Tensile tests have shown that in every case, the cable itself will break before the injected metal terminal releases. *See Figure 35.*

Abrasive Points

In the assembly of abrasive points the components are held in the required relationship while molten zinc alloy is injected into the void between the abrasive material and the mandrel. The zinc alloy penetrates the grit and solidifies, joining the wheel to the mandrel. The assembly can be handled immediately. *See Figure 36.*

This penetration into the porous abrasive and shrinkage on to the knurled mandrel results in a strong mechanical lock. High production rates from 400 to 600 assemblies per hour, low cost zinc alloy and the elimination of curing time contributes to a reduction in assembly costs. In contrast, cementing methods require curing ovens and long cure times whereas IMA permits short fast runs with concentricities of 0.003 in to 0.004 in. TIR (0,075 to 0,1 mm).

Figure 36. Examples of Abrasive Points Assembled by Injecting Molten Zinc Into the Void Between the Abrasive Material and the Mandrel. The Zinc Penetrates the Grit and Shrinks on to the Mandrel.

Gear Train

A gear and shaft assembly from a watt-hour meter gear train is one of the best examples of savings from the use of Injected Metal Assembly. By using IMA rather than the conventional method to secure the gear to the shaft, savings were realized through:

- elimination of parts;

- elimination of locking shoulders;

- reduction of raw materials for shafts (pinion stock formerly used);

- forming pinions as an integral part of the joint;

- lower screw machine costs;

- one standard shaft or spindle rather than 4 or 5 different shafts;

- tolerances relaxed or eliminated;

- reduced inspection costs;

- fewer rejects.

Pins in Plate

The design of a small pocket dictating machine required that three steel pins be assembled into a sheet metal chassis. The original design called for the pins to be staked into three precise staking bushings which were staked into the plate. The method required very tight diameter tolerances and precise location of the three holes. A complicating factor was that one of the staking bushings was located close to a functional slot which had to be free of interference. This required a 'flat' on one side of the bushing and special care in orientation during assembly, or alternatively, a secondary machining operation during assembly.

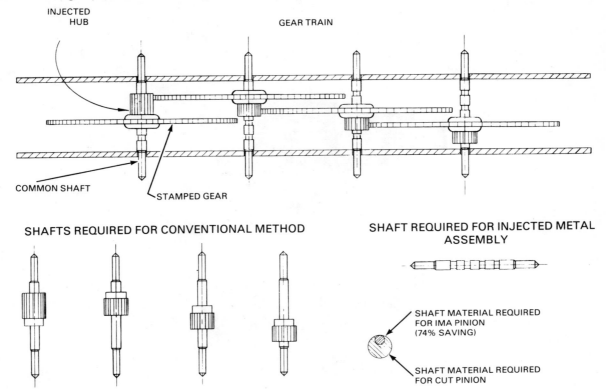

Figure 37. *Gear Train of a Watt/Hour Meter. Reduction of Machining and Fewer Parts Resulted in Considerable Cost Savings.*

A change to the Injected Metal Assembly process resulted in major cost savings as well as other advantages. The three bushings are replaced by zinc bosses injected directly in place while assembling the pins. The bosses are firmly locked in place by four keys in each of the three mounting holes. The boss at the slotted hole has a cast-in-flat to avoid interference with the slot. Since the pin-to-pin dimension is now controlled by the IMA tooling, hole position and diameter tolerances were relaxed.

The three pins, which are manually inserted into bushings in the tooling, and the plate are located in the exact positions required in the final assembly prior to injecting the zinc alloy.

The cost savings on this application can be summarized:

- elimination of three staking bushings and machining of a flat;

- elimination of need to orient the staking bushing with the flat;

- relaxation of the dimensional tolerances of the three holes and their location;

- elimination of rejections due to damaged parts;

- reduction of inspection time to nearly zero because of the built-in control features of Injected Metal Assembly tooling.

Although the foregoing examples are all basic metals, materials such as glass, magnets, nylon, and even paper have been assembled by the Injected Metal Assembly process.

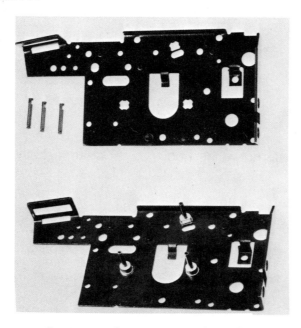

Figure 38. A Before and After Photo of a Chassis of a Small Pocket Dictating Machine. Three Pins are Manually Inserted into Bushings in Tooling. The Pins and Plate are Located in Their Required Positions as Zinc is Injected to Form Zinc Bosses. Considerable Savings over a Conventional Method were Realized.

APPLICATION FEASIBILITY

Injected Metal Assembly is not the answer to all problems but has proved to be beneficial particularly:

- if one or more components can be replaced by a shape which is formed as an integral part of the joining;

- if the functional surfaces of components can be used to produce the required assembly tolerances;

- if an injected metal hub can join several parts and therefore eliminate several assembly operations;

- if staking, brazing, soldering, or electron beam welding is being used to join components.

- if several dissimilar parts are to be joined, eg., plastic to ceramic to metal;

- if delicate parts sensitive to deflection by conventional methods are to be assembled.

If one or more of these criteria apply, it is almost certain that considerable assembly cost savings can be made.

INJECTION PROCESS

At this juncture, a few terms used in the Injected Metal Assembly process should be defined:

Injection Unit: The device which is immersed in the molten alloy and pumps or injects the metal through a nozzle into the cavity.

Sprue: That portion of injected metal that fills the conical passage (sprue hole) that connects the nozzle with the mold and is removed from the mold during the cycle.

Shear: The rotary action in the fixture which shears the sprue.

Advance Mechanism: A mechanism which moves the operating head on to the injection nozzle and retracts it when injection has taken place.

Operating Head: The device which locates and actuates the tool and/or mold. It closes when parts are loaded and opens to eject the completed assembly.

The actual metal injection process is basically quite simple. Once the parts to be assembled are located within the tool, the operating head is advanced by a mechanism on to an injection nozzle. Injection occurs through actuation of an injection unit by an external cylinder which forces the molten metal through a nozzle into the tool containing the cavity and the parts being assembled. The injection time of an assembly cycle is rarely more than a few hundredths of a second.

MACHINES

The mechanical operations of the machine utilize pneumatic or hydraulic motive power for individual functions. The timing of each function is precisely controlled by electronic time delay units. Injected Metal Assembly machines can vary from a completely manual operation through semi-automatic to full automatic with parts feeders. The only functions requiring an attendant in fully automatic machines are replenishment of the injection metal in the machine's melting pot, the loading of parts into feeders and the removal of boxes of completed assemblies. However, fully automated systems are much less flexible. In general, they are created for one assembly only, and a changeover to produce a totally different assembly could be too expensive both in capital cost and lost production. Another consideration is that the parts feeding equipment can handle only limited variations. It is therefore of advantage to use the manual or semi-automatic system to obtain full versatility.

In manual or semi-automatic machines, operating heads can be changed in about ten minutes while an appropriate electronic control panel is plugged into place for programming. A tool changeover alone can be made to produce an entirely different assembly. Although tool changeover takes slightly longer, the capital investment is less.

It is usual for the machine to be loaded and unloaded manually when joining delicate and extremely accurate assemblies. Partial automation in which 'difficult to feed' parts are manually loaded and 'easy to feed' parts are automatically loaded is an ideal answer.

Depending on the application, the total machine portion of the cycle takes from one half to one second, with the remainder of the process time taken up by loading and unloading the assemblies. Production rates may vary from a few hundred to over a thousand an hour, depending on parts complexity, the number of components per assembly, orientation of functional surfaces and degree of automation utilized.

A major feature of Injected Metal Assembly is the high degree of reliability and consistency. Once initial sample have been made and accepted, results are repeatable over the life of the tool as the injected metal always assumes the shape of the mold.

STRENGTH OF ASSEMBLIES

The injected metal joints are solid metal, either zinc or lead alloy. The difference between the application of the two metals is strength, the cost being within a few cents per pound for each metal. Zinc being the stronger, it is used in the majority of joint applications. Molten zinc has excellent fluidity making it easy to inject into thin cross-sections and intricate shapes. This property permits a freedom of design combined with a strength of up to 54,000 psi (37,000 kilopascals) tensile strength which, in most cases, gives a joint which is stronger than the original parts. See *Figure 39*.

Test data on the strength of injected metal joints is difficult to compile due to the variety of material combinations to be joined as well as the myriad of joint sizes and shapes. The data shown in *Figure 39* were compiled from destructive tests in which the assemblies were stressed to load the joints. In all cases, the steel shafts failed. This confirms that zinc alloy joints can be designed to be stronger than the components being assembled.

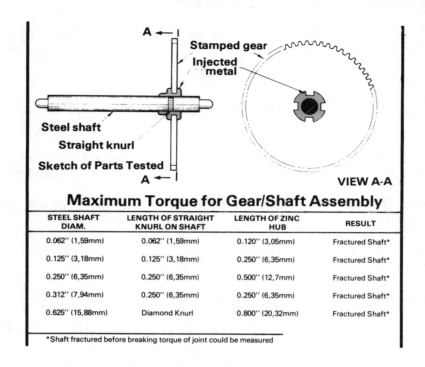

FIGURE 39. *A torque table showing results of tests on various sizes of gear and shaft assemblies.*

ADVANTAGES

The advantages of the Injected Metal Assembly process can be summarized as follows:

- A great variety of dissimilar materials may be joined;

- The process is quite rapid contributing to reasonably good production rates;

- Integral injected metal parts can be substituted for machined shapes;

- Dimensional repeatability of assembly considered by users to be 'absolute';

- The quality of joint is extremely consistent;

- The tool is often used as a component checking fixture;

- In most cases the components in an assembly are located from their actual working surfaces to achieve desired locations rather than by a 'fit-contact' means;

- Tolerances can usually be opened or reduced;

- Unskilled operators can be trained rapidly to achieve optimum performance;

- The process is clean, non-toxic and odor free;

- The IMA alloy is low in cost compared to many other joining materials;

- Parts to be joined need no special cleaning or chemical preparation before assembly since the joint is a mechanical clamp;

- Assemblies can generally be handled immediately after the assembly operation;

- Assemblies are ready-for-use with no secondary operations required;

- Possible installation of IMA process into production line.

LIMITATIONS

An injected metal assembly is usually permanent once the injected metal has solidified. The resulting assembly cannot be disassembled without destroying the injected portion or part of the components.

Although solidification time of an injected metal portion of an assembly is very short and permits the use of many heat sensitive materials in the assembly, a means must be found to either extract the heat through the tooling or use the assembly components as heat sinks.

As molten metal is injected under pressure, a close fit between the tool and the components is of paramount importance. Component size can affect this requirement. Failure to maintain a close fit can result in metal flashing between the tool and components thus causing unsatisfactory assemblies.

As Injected Metal Assembly is a hot chamber process and since lead and zinc alloys are used, it is very important that the end application of the injected metal assembly is one in which the materials are compatible with the environment or the conditions under which they will be utilized.

It was discussed earlier in this paper that the Injected Metal Assembly process requires a keying or locking configuration on the components to be assembled.

PROJECTED APPLICATIONS

The potential application of the Injected Metal Assembly process are limited only by our imaginations. Quite a number of interesting developments are under investigation. One that seems to have great potential is the injection of shading coils into solenoid cores, relay cores, etc. to replace the conventional formed copper ring. This gives an obvious cost reduction through the elimination of the copper ring, lower cost and greater availability of zinc versus the copper now used, and the elimination of the usual brazing operation after assembly. An added benefit is the capability of injecting the lamination rivets at the same time. *See Figure 40.*

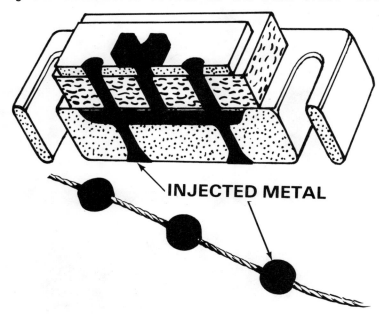

FIGURE 40. *Dissimilar materials shown sandwiched by injecting zinc alloy rivets. Several rivets can be injected at the same time. Holes are not required to be aligned or even of the same size.*

Another interesting application that shows great promise is the joining of shunt leads to carbon brushes.

Other applications under development include:

- Terminations on extremely fine and delicate wires in the 0.010 in (0,25mm) diameter range;

- Injection of lead alloy bushings into large grinding wheels. This has a potential for greatly increased productivity and accuracy over current production methods;

- Producing a modified type chain by casting lugs or other shapes on to continuous lengths of cable. *(See Fig. 40, lower)*;

- Injecting metal pistons on to piston rods for small pneumatic cylinders.

CONCLUSION

In review, Injected Metal Assembly is continually expanding in scope of application, volumes of metal injected, sizes of components assembled and variety of materials used.

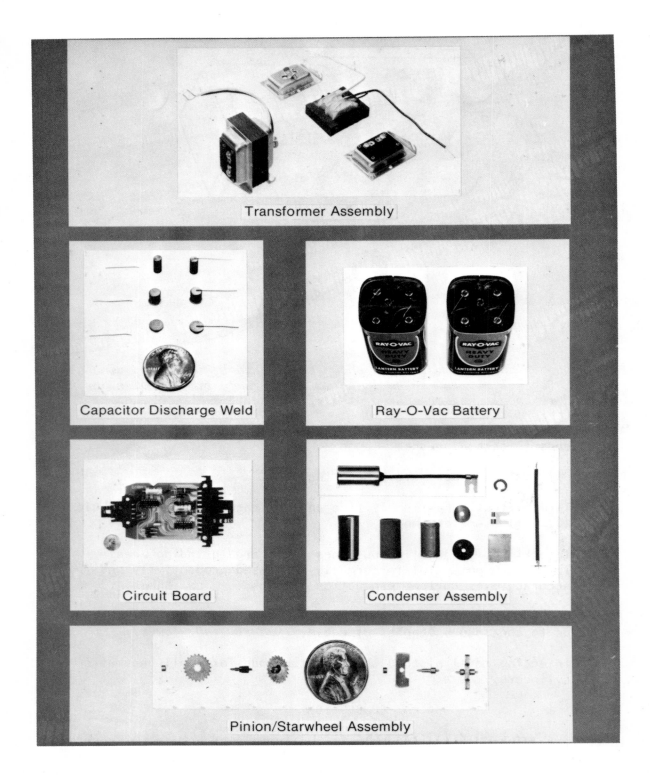

Figure 41.

SOLDERING & WELDING

There are many assembly operations that utilize joining techniques of one kind or another. *(See Figure 41)*

This paper will discuss the criteria for automating some of these manual manufacturing operations, and is primarily limited to those techniques which involve a process.

The most important point to automating a manual joining operation is proper application and utilization of the process controls.

Application of a joining operation in one or more stations of an assembly machine without proper application and utilization of the process controls inherent to a particular joining technique would be disastrous. It is of prime importance that the machinery builder is knowledgeable in the joining technique to facilitate implementation of the process controls. It is not only desirable to have a mechanically well designed joining station, but a station that produces reliable joints.

Many times the builder relies on the user for joining process information. Care must be exercised to ascertain that the information imparted about the manual joining technique is applicable to high speed automatic assembly operation.

SOLDERING

For example, on a hand soldering operation, involving a solder preform, it may be acceptable to incorporate the flux within the solder preform, whereas in an automated operation the production rate may cause a problem since the flux, melting at a lower temperature, could expell itself under high pressure causing solder blow holes, or dispelling the flux onto the fixturing causing part feeding problems, or even raising the flux to such a high temperature that the flux becomes a varnish before the solder melts. *(See Figure 42)*

It may be more feasible to apply the flux in one station and the solder in a subsequent station. The operation could be more reliable. We have provided some process control by reducing the functions required in any one station. *(See Figure 43)*

A good understanding of the particular joining methods is required to implement the process controls. It is necessary to understand the basic parameters of the joining process, how these parameters are interrelated, and how they can be varied for desired effect.

It is also necessary to analyze the joining process with respect to high speed operation. Does it lend itself to automation? Is it a controllable joining technique? What effect does high cyclic speed have on the process?

Figure 42.

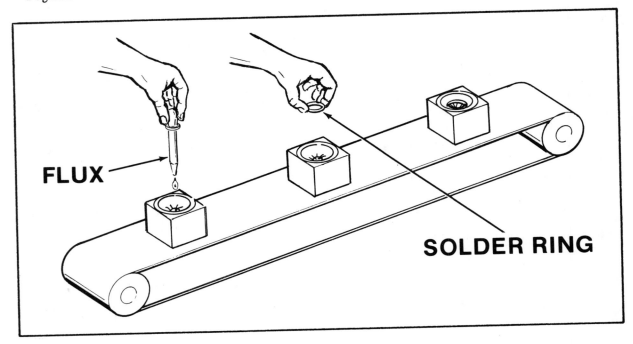

Figure 43.

Can the joining materials be handled properly? What effect does the joining process have on fixture design? Can the joining technique be reliably handled in a single station, or are multiple stations required to minimize station function and to maintain process control? Should a different joining technique be used in the automated operation, or, more importantly, should variations of the manual joining process be adapted for automatic operation?

The soldering process is one manual joining technique which can be readily automated.

Soldering is simply a joining technique by melting another, more easily melted, metal between the work pieces to form an alloy bond. Soldering is classified as soft soldering (low temperature) and hard soldering (high temperature).

The soldering process requires clean materials. Additionallly, the surface oxides must be removed to expose the bare metal. Fluxes are used to remove these oxides and also to provide a means of conducting the heat to surface areas not in direct contact with the heat itself.

Fluxes also play an important part in reducing the surface tension of the solder so that it flows readily, "wets" to the work pieces, and through capillary action, penetrates small cracks and recesses.

The soldering process requires some basic considerations:

> The flux is basically a cleaning agent. At room temperature, however, its cleaning ability is prolonged and limited. The flux must be raised to its proper activating temperature. Overheating will boil off the flux solvents too fast, preventing a full cleaning operation and may even cause the flux to bake as an enamel.

> There are many varieties of solder fluxes. The proper flux depends on such factors as the metals being soldered, the type of solder used, the extent of part oxidation, military and/or other user specifications, and the method of flux application.

> The important point to the machinery builder is the realization that the flux being used in the manual operation may not be the best for the automatic operation. The flux manufacturer should be consulted.

> There are also many varieties of solder. The selection again depends on such factors as strength required, base materials, allowable temperatures and the flux that can be used. Again, the solder manufacturer should be consulted.

> The machinery builder should consider using a eutectic solder. This solder alloy does not have a plastic range. It changes to a liquid and back again to a solid at one specific temperature.

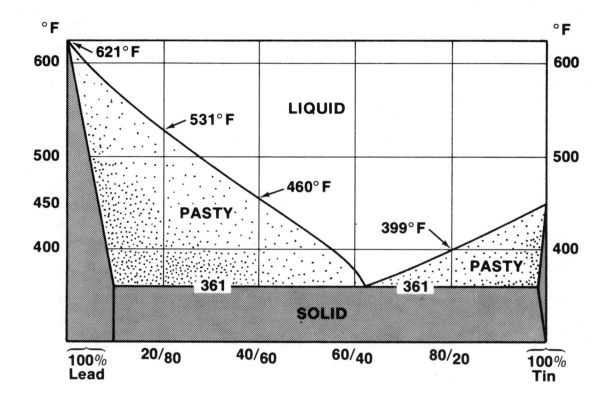

Figure 44.

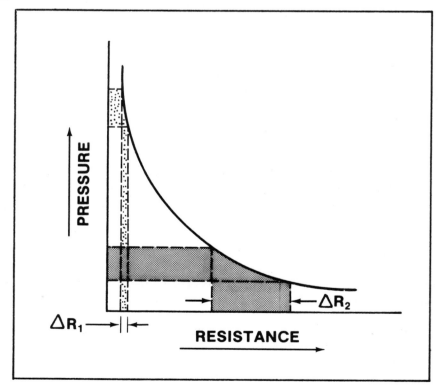

Figure 45.

The advantage, of course, is that machine vibrations will not produce a cold solder joint, or a solder joint that is brittle, with course grain structure, gray in appearance and subject to corrosion, poor strength, and giving rise to a high resistant connection. *(See Figure 44)*

The soldering process requires that the parts be heated at the correct temperature long enough for the flux to clean the parts and the molten solder to flow and "wet" the base metals. The finished joint should be silvery and smooth with rounded or tapered edges.

Generally, it's desirable to primarily heat the solder, causing the flux to act like a medium to conduct the heat to the various surfaces, cleaning as it progresses and, thereby, allowing the solder to flow behind it "wetting" to the base metals. The heat should be sustained sufficiently to eliminate all flux solvents and to allow capillary action to fillet the solder around edges, cracks, and recesses. This approach allows minimum heating of the piece parts.

Once the type of flux and solder are selected, the machinery builder becomes concerned with handling these materials, selecting and adapting the heating source, and controlling the process.

Flux can be applied in many ways including contact, spraying, dipping, and as an integral part of the solder. The builder must be concerned with the method of application as it applies to fixturing problems, cleaning problems, atmospheric conditions, handling and feeding problems, and process control criteria.

The use of solder preforms when automating manual operations is quite prevalent because of the controlled size, shape, and amount of solder. These preforms can include a flux as part of the preform. The necessary precaution has been previously referred to and is especially noteworthy where hermetically sealed joints are required. None-the-less, it eliminates a flux handling problem and should be considered.

Spraying flux has many drawbacks. Besides poor control over the volume applied, there is danger in applying the flux to the fixtures causing parts feeding problems in other stations as well as venting problems since many of the fluxes will atomize into the air and can be toxic.

Dipping also has severe limitations in quality control problems, contaminating fixtures and mechanisms, and, in general, process control.

Applying flux by contact, on the other hand, has some advantages. It provides a control over the quantity applied and need not cause fixture, mechanism, and house cleaning problems.

Contact can be made by a relatively hard sponge material acting like a squeegee or by a roller mechanism. Excess flux can be removed by brush in contact with the work. In the contact system, a flux reservoir is used and

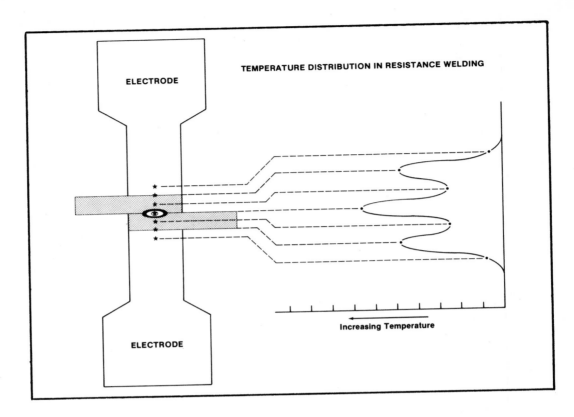

Figure 46.

HOW ELECTRODE POLARITY AFFECTS WELD STRENGTH
Non-polarity-sensitive material combination

	Nickel wire (0.025 in. dia.) cross-wire welded to nickel ribbon (0.010 x 0.031 in.)			
Electrode Force in Pounds	Weld Energy in Watt-Seconds	Average Pull Strength in Pounds (Torsional Shear Test)		Percentage of Normal Pull Strength with Polarity Reversed
		Normal Polarity Nickel Ribbon (-)	Reverse Polarity Nickel Ribbon (+)	
8	10	18.4	18.0	98
8	12	17.6	17.3	98
8	14	16.9	16.6	98
8	16	16.4	15.9	97

Polarity sensitive material combination

	Tinned-copper wire (0.026 in. dia.) cross-wire welded to nickel ribbon (0.010 x 0.031 in.)			
Electrode Force	Weld Energy	Average Pull Strength		Percentage of Normal Pull Strength with Polarity Reversed
		Normal Polarity Nickel Ribbon (-)	Reverse Polarity Nickel Ribbon (+)	
7	32	12.9	6.3	49
7	34	12.4	8.4	68
7	36	13.3	7.5	56
7	38	12.3	2.6	21

Figure 47.

the user should be cautioned to check the specific gravity of the flux on a scheduled time basis.

Besides solder preforms, solder can be handled in paste form, wire form, dipping and others. As noted, preforms provide the advantage of quantity, size, and process control. All solder forms can be obtained with or without inherent flux. A consideration must be given to the binders used in paste type solder which may be for venting and/or subsequent part cleaning due to the corrosive action of the binders.

Using wire solder with proper feed control for both quantity and rate provides a good system for process control. A mechanism must be included to backoff the wire solder prior to solidification.

There are, of course, just as many ways to supply the heat as there are fluxes and solders.

To the machinery builder, induction heating has the advantage because it's fast, provides localized heating, requires very little maintenance, and can be integrated very readily with the machine cyclic and interlocking systems. It's usually used in conjunction with solder preforms.

These machines are small radio stations with controlled radiation in the form of an antenna called a work coil. Work coil design is important and the manufacturer should be consulted. This equipment is available in several frequency ranges. Four hundred kilocycle equipment is generally used for ferrous material of relatively small mass, and two to five megacycle equipment is employed for non-ferrous materials. Low frequency equipment is used for large mass heat treating equipment.

The important point here is the realization that you're working with the transfer of heating energy by high frequency magnetic principles using an antenna or work coil. Care must be exercised in eliminating magnetic materials within the lopp created by the work coil or its extensions to the equipment to prevent such materials or mechanisms from absorbing energy from the generator leaving little for the work. Consideration must be given to fixture design and possible use of non-magnetic stainless steels or non-ferrous materials. The coil leads should be kept as short as possible and as close together as possible without shorting to prevent losses. It's a good heat supplying system providing exceptional process control when properly used. The equipment manufacturer should be consulted.

There's nothing wrong with using hand soldering irons, resistance soldering equipment, or wire solder, if properly tooled and if consideration is given to process control.

In these cases, the irons or tools should be station mounted with actuation provided by reciprocating tooling plates or other mechanical and/or cam operated devices. Proper consideration should be given to preheating, fixed adjustable solder feed and feed rate, and possibly post heat. The wire solder must be backed-off prior to solidification.

A good system for automation is reflow soldering. In this system the parts are previously tinned in a prior operation thereby eliminating the problems of handling and selecting solders.

With this technique, flux is still applied to clean the oxides, transfer the heat, and act as a capillary agent for the solder.

WELDING

Resistance welding is another joining technique that is often automated. It has the basic advantage of being fast and does not require handling filler and/or other materials. It's a very controllable process.

There are basically two types of resistance welding - A.C. and D.C., or capacitor discharge. Each has its application. A.C. is usually used on similar materials and larger masses. Capacitor masses and high conductivity, non-ferrous metals.

Both systems are basically similar in that the heat generated is a function of the current passed between the two metals and the electrical resistance of the metals. There are several resistance welding forms such as spot welding, projection welding, series welding, and other. Each variation has its application, advantages, and disadvantages.

Resistance welding, as the name implies, depends on the electrical resistance at the interface of the two metals. Proper design criteria for automation requires use of the highest welding pressure possible consistent with electrode deformation to prevent resistance variations and problems with process control. *(See Figure 45)*

As a weld is made, the metals expand and contract as heat is imparted to the metals and conducted away by the metals. The inertia of the pressure delivery system is of major concern to the machinery builder because the pressure at the weld zone must follow these contractions and expansions in order to prevent expulsion, and to minimize brittle welds cause by grain growth resulting from lack of pressure.

In fact, the so-called weld nugget is, in reality, an area of high grain growth caused by the lower pressure at the interface of the metals. It normally is the point where the weld separates when pull tested since it is the most brittle area.

In capacitor discharge, welding nuggets are rarely attained since the heat is applied many times faster than in A.C. welding and due to conduction, does not remain in the weld zone long enough to cause grain growth.

The trick of resistance welding is heat distribution control. In order to make a proper weld, it is necessary that sufficient heat be applied to both metal parts simultaneously so that both reach the plastic point at the same time, thereby allowing a weld or metallurgical joint as the forging pressure is applied.

This does not pose a problem with similar masses and materials, but it is a consideration when welding dissimilar materials and masses. *(See Figure 46)*

Heat balance can be attained by selection of electrode materials to control the heat sink effect and by using electrode contact area to control current density.

A.C. welding equipment has the advantage that it can make many welds per minute and is limited primarily by the mechanical handling time involved. It is simply a transformer to convert normal in-plant power to high current power needed to generate heat.

A.C. welding equipment has the disadvantage in that it is subject to in-plant power variations which are multiplied by the transformer and which can cause process control problems.

D.C. or capacitor discharge welding equipment is not subject to in-plant power variation since with this type of welding the source of welding energy is a charged capacitor which acts like a battery or fixed energy source.

The disadvantage of the D.C. system is the repetitive rate of the equipment or the number of welds per unit of time it can make. Each weld or capacitor discharge period must be followed by a charge period to refill the capacitor energy. Normal repetitive rates are up to 60 welds per minute but equipment is available to as high as 400 welds per minute if power requirements are low enough.

That's the other restriction to D.C. welding equipment. Its capacity is limited and, therefore, the mass that can be welded using D.C. is limited.

A consideration when using D.C. equipment is power supply polarity. Some metals are polarity sensitive. Proper polarity may increase the weld strength by 100% or more. Of course, in A.C. welding, the polarity is continually changing and the strength effect of polarity sensitive materials cannot be utilized. *(See Figure 47)*

As noted, there are several forms of resistance welding such as projection welding, spot welding, series welding, and etc. Projection welding is very adaptable to automation and is mandatory when multiple welds are required by part design. It has the advantage that the current carrying electrodes are in the form of dies, usually with replaceable wear inserts. Projection welding dies are fairly maintenance free compared to spot welding electrodes which may occasionally require tip dressing.

Projection welding also has the advantage that the current flow and, more importantly, the current density is controlled by the projection design rather than by the electrode shape and is, therefore, more nearly a constant resulting in increased process control.

Designing for spot welding should dictate that the electrode face be flat and cylindrically shaped to maintain current density as the tip is dressed.

Series welding and step welding are forms of spot welding used when only one side of the work is accessible. It's advantageous to the machinery builder when parts are located in fixtures not readily adapted to conventional spot welding.

Some important criteria exists for the machinery manufacturer when automating manual resistance welding techniques. Criteria that may not be apparent to the user in a manual operation or to the builder by observation of a manual operation.

A large magnetic field exists in the electrical loop created by the welding transformer, leads, and electrodes whenever current is flowing. Understanding the effects of this magnetic field is imperative to proper design for automation.

Basically, whenever ferrous materials are included in this loop they electrically act as a short circuit on the transformer thereby draining energy from the work pieces. Additionally, the ferrous materials will get hot due to this action and may possibly cause severe tooling problems.

It is essential that the secondary circuit, or welding loop, be physically void of all magnetic materials and that the leads are properly dressed and as close as possible without shorting. The leads should be short to minimize power loss.

A.C. welders always require water cooled transformers, current carrying leads and electrodes or dies. It is unnecessary to water cool capacitor discharge equipment because the heat duration is very short.

Welding current, being very high in quantity, causes a power loss even in such high conductivity materials as the copper alloys. This power loss is generated in the form of heat which causes the resistance of the current carrying members to increase, adding to the power loss and subtracting from the available power for welding. It's apparent that process control has been lost.

Electrode material is another important consideration to the machinery builder. There are a variety of different electrode materials available for welding different metals controlling heat sink effect, minimizing pick-up, and with various hardness.

In an automatic operation, the electrodes stay hotter than in a manual operation. Care must be exercised to select the right materials to prevent pick-up and electrode deformation or excessive wear.

Material platings require further consideration especially in automatic operations because of the higher heat and higher production rate which can cause excessive pick-up and even a lack of welding.

Sufficient heat and heat duration must be applied to break through the plating and make a weld in the bare material. Special welding controls are available for these types of materials.

Aluminum welding also has problems. Aluminum has an oxide insulating coating which appears almost immediately after cleaning and which builds very rapidly with heat. It is difficult to weld with standard A.C. equipment due to this oxide although three phase A.C. welders and D.C. welders, which have an output wave shape of much shorter rise time, weld aluminum quite readily.

The machinery builder should also be aware of transformer power ratings or KVS (thousands of volt-amps).

These ratings tell how much power can be taken from the transformer without overheating at a 50% duty cycle. They have nothing to do with its actual output which is controlled by the impedance in the secondary welding circuit.

The cost of switching to a next larger size transformer than used in the manual operation is very small, but the advantage is very high.

FIGURE 48. Bell Transformer Assembly Machine

A semi-automatic "in-line" assembly machine for door bell transformers. Parts are hand-loaded on conveyor by 4 operators. Parts are then aligned, riveted, electrical connections soldered, completed units tested and rejects separated at unloading station - all automatically. The unit has a production rate of 900 cycles per hour.

Uses flux core solder wire fed from spool and guided to soldering point through a metal guide tube. Adjustable controls are provided for solder feed rate (in./sec) and solder feed time duration. A back-off mechanism is used to break solder away before solidification.

This unit uses conventional hand soldering irons mounted on a vertical, cam-operated, up/down mechanism. Adjustable preheat (iron contact) is applied prior to soldering and adjustable post heat (iron dwell) is maintained after feed completion.

FIGURE 49. *Automatic Ignition Condenser Assembly Machine*

This 36 station in-line type assembly machine utilizes a 6-bay "Bodine" basic machine chassis to provide the index type motion to carry 52 parts fixtures past the 36 working stations. Six (6) parts are automatically loaded, two (2) parts are manually placed by two (2) operators. The working stations include an induction soldering station, a roll crimping station, a parts turnover station, an ultrasonic wire tinning station, a terminal blanking and crimping station, a reflow soldering station, a condenser charge and check station and a "good" - "bad" select station.

All working motions are cam operated and are protected against overloading. Automatic probing stations verify each work station.

The production rate of this machine is 2200 cycles per hour.

The disc shaped contact with a preformed solder ring. Flux is applied at a separate station by a contact pad. Soldering of the contact with the preform ring is accomplished by induction heating.

The outer housing is roll crimped to complete condenser body assembly. The wire lead is first ultrasonically tined with solder and then the terminal is formed and crimped around the tined lead. A resistance welder is used to reflow the solder for a permanent terminal connection.

LASER APPLICATIONS

Industrial applications of lasers are reviewed in the following areas: (1) materials processing (welding, cutting, drilling, machining), (2) non-destructive testing (spectrochemical analysis, and strain measurement), and (3) precision measurement, alignment and inspection. The major advantages and limitations of the use of lasers in each area are analyzed. Specific examples of successful applications are discussed. Many references to sources of additional information in the published literature are cited.

The laser is beginning to find broad applications in a number of areas of manufacturing technology. The objective of this paper is to discuss a limited number of these applications briefly in order to illustrate the types of manufacturing problems that can be successfully resolved through the use of lasers. It is impossible to provide an exhaustive treatment of industrial laser applications in a paper of this length. Consequently, no claim of completeness is made here. A number of excellent review articles and books have been written on this subject (e.g., refs. 1-3), and the interested reader should consult those as well as the many papers and books referenced throughout the body of this paper for additional information. The applications discussed in this paper are chosen to be fairly representative of the overall field of industrial laser applications. Through the mechanism of discussing these specific applications we hope to help identify the major advantages and limitations of using lasers in each of the major application areas of materials processing, non-destructive testing, and precision measurement, alignment, and inspection.

Section II treats the use of lasers in materials processing. Welding, cutting, drilling, and machining applications are discussed. Examples of achieved welding and cutting rates are provided for various materials and laser systems. Non-destructive testing is treated in Sections III and IV with emphasis on spectrochemical analysis (Section III) and interferometric and holographic strain measurement (Section IV). Precision measurement, alignment, and inspection are treated in Section V.

II. MATERIALS PROCESSING (WELDING, CUTTING, DRILLING, MACHINING)

High power laser systems are being developed, tested, and evaluated for performing classical materials processing (including the manufacturing and industrial tasks of welding, cutting, drilling, and machining) and for performing new, unique materials processing (such as drilling and welding through a glass enclosure, punching holes in baby-bottle nipples, tagging salmon, branding and dehorning cattle). Manufacturing and industrial tasks that are normally accomplished with standard, conventional high-heat-flux sources (namely, reacting gas jets, plasma arcs, electrical discharges, and electron beams) may be performed better and more economically utilizing high-power laser systems. One major advantage of material processing with laser systems results from the focused laser beam's ability to generate very localized, extremely high-power densities without mechanical contact with the material surface; thus, no material contamination by the working tool and no working tool wear. The laser energy can be delivered to the work materials in a continuous or in a pulsed (seconds, milliseconds, microseconds, nanoseconds, or picoseconds pulse-length) beam to permit heating, melting, and/or vaporization of the materials. However, today for some manufacturing materials processing, the commercially available laser systems may have limited utility due to their physical size, space requirements, efficiency, costs, maintenance requirements, and/or performance (such as the depth of welds may be limited due to the laser energy being reflected).

Welding: The percentage of welded joints continues to increase while the percentage of screwed and riveted joints decreases in the U.S. In the aerospace industry, new manufacturing technologies require new jointing techniques. Today a high-power laser system for welding (such as a CO_2 laser) permits precision welds with minimum material deformation and minimum material heated zone. It also permits welding of materials that are difficult to weld (except with electron-beam welding). Welding with laser systems has been reported for welding steel, titanium, plastic, quartz, and glass. At Battelle Laboratories, a 1.2KW, CO_2 laser system was used to weld two 0.6mm thick stainless steel sheets at a rate of several centimeters per second, and to weld 0.5 mm wall thickness, pre-rolled stainless steel sheets to form seam-welded tubes, without any ridge on the inner surface, at a rate of 5 centimeters per second. Because of a higher optical reflectivity, the 0.6 mm thick titanium-aluminum alloy sheet was welded with a 1 KW, CO_2 laser beam at a slower rate, a rate of several millimeters per second. Clean, durable welds were obtained, however the weld was rather broad, being the order of 2mm when viewed from the front. Normally, metals must be welded in an inert gas atmosphere (e.g. in a stream of argon gas) to prevent oxidation. Laser beam welding permits precision welding of metals in inert gas atmospheres rather than in vacuum - an important advantage laser beam welding has over the electron-beam welding. CO_2 laser beams are very useful for welding plastics; for example, a 500 W laser beam can weld 0.2mm polyethylene film at a rate of 0.5 meters per second - a welding rate better than that of other techniques. Clean, durable, and economical welds are obtained with polyethylene films utilizing laser beams. Also, the CO_2 laser beam can be used to weld quartz; for example, to weld optical windows onto quartz tubes--conventional (gas flame, torch) methods would normally damage the optical windows; however, the laser permits energy concentration, and the weld is achieved without damage to the optical window. Employing a high-power, 20 KW, CO_2 laser system, deep-penetration welding of 304 stainless steel has been reported. For example, a bead-on plate weld with a penetration of approximately 3/4 in. with approximately a 6:1 depth-to-width ratio has been achieved at a speed of 50 in./min. With an 8 KW CO_2 laser system, a bead-on plate weld in 3/8 in. thick 304 stainless steel with nearly full penetration was achieved at a speed of 30 in./min. Presently, the Caterpillar Tractor Corporation in Peoria has a high-power CO_2 laser system for welding, and the Ford Motor Company is evaluating a CO_2 laser system for welding. Today, laser welding compares "favorably" with that of the electron-beam in materials up to 1/2 in. thick; and the laser system is adaptable to automation. The CO_2 laser system at the Caterpillar Tractor Company's Manufacturing and Materials Development Department (denoted HPL-10, manufactured by AVCO-Everett Research Laboratory, Inc.) is also being used for cutting, heat treating, and machining; for example, to cut 1-3/4 in. thick steel at 15 in./min. At the United Aircraft Research Laboratories in East Hartford, tests were performed with a 5 KW CO_2 laser in burn-through welds, edge welds, and bead-on plate welds.

By trading penetration depth for speed, a 4.7 KW CO_2 laser achieved penetration depths of 0.06 in. at speeds of 300 in./min. Theoretical models for laser welding with deep penetration have been developed. A Nd: YAG laser system (1 to 2 joule, 10 pulse-per-second repetition rate) has been used for microelectronic package sealing. Major advantages of laser welding for microelectronic welding include: very localized, high energy welds; high reliability; rapid; non-contact welding of many dissimilar metals.

Cutting: High-power CO_2 laser systems are also suitable for cutting many materials including metals such as steel and titanium; combustible materials such as plastics, textiles, wood, and paper; and hard, brittle materials such as aluminum-oxide and silicon-carbide. The cutting width is determined by the size of the laser beam spot size. In cutting, material is removed at least partially by evaporation. In laser beam cutting the laser output optics and focusing lenses must be protected from the material being cut—this is usually accomplished by using a gas stream (e.g., argon with combustibles and oxygen with metals) to shield the optics. The cutting rate is increased when cutting metals by flowing oxygen over the cut to produce an exothermic reaction. Using a 1.2 KW, CO_2 laser system, a 5 mm thick steel sheet can be cut at a rate of several centimeters per second with a 500 W CO_2 laser beam. The major advantage of laser beam cutting is that the cut width is smaller than oxyacetylene or plasma cutting. The use of a laser can reduce the cost of cutting titanium by approximately 75% in saved materials and lower cutting/finishing costs. Employing a 15 KW CO_2 laser system, without gas (oxygen) assist, a 1/2 in. thick aluminum plate can be cut at 90 in/min, 1/4 in. thick carbon steel at 90 in/min, and boron epoxy composite (0.32 in. thick) at 65 in/min. With a 20 KW CO_2 laser system, 3/16 in. thick 304 stainless steel can be cut at 50 in/min, 1/2 inch thick fiberglass epoxy composite at 180 in/min, and 3/8 in. thick glass at 60 in/min. With an 8 KW CO_2 laser system, 1 in. thick plywood can be cut at 60 in/min, 1 in. thick plexiglass at 60 in/min, and 1-1/2 in. thick concrete at 2 in/min. The continuous cutting of quartz usually does not present any difficulties; however, when cutting glass, it is useful to cool the cut edges with a gas jet and use high cutting speeds. Cutting of hard, brittle ceramic materials is best performed with a repetitively pulsed laser—a line of holes is formed that are spaced at small intervals and then the material is separated along this line of holes. Laser cuts are normally of higher quality than those obtained by diamond scratching followed by breaking. Paper cutting with a CO_2 laser at cutting rates of 0.2 to 5 meters per second has been reported in Ref. 4 utilizing a 100 W laser. Oxy-laser metal cutting studies have been performed by GTE Sylvania, Inc. in Mountain View with their Model 971 CO_2 laser system, a 1 KW system.

Drilling: Pulsed CO_2, ruby Nd:glass, and ND:YAG laser systems are being used to drill holes in metals such as stainless steel, aluminum, copper, nickel, and brass and in non-metals such as diamonds, alumina ceramic, and rubber. Lasers can drill holes as small as 10 microns across in the hardest materials with precision and safety, for example, drilling diamonds for wire-drawing dies; making holes in rubber nipples and gaskets; drilling surgical needles; and for drilling zirconium, nuclear fuel rods under stringent safety controls. Laser drilled hole quality depends on the type of laser system employer—high-energy pulsed lasers (0.1 to 100 joules per pulse) can blase holes in materials and leave a "volcanic" lip structure around the hole that may have to be removed later. However, utilizing a high-repetition-rate pulsed laser (typically, millijoules of energy per pulse), holes can be trepanned in materials without the lip buildup around the hole. In trepanning, holes are cut into a material by scanning the laser beam in a circle at the material's surface. A model to describe the temperature profile and thermal stress propagation for laser drilled holes in alumina ceramic substrate material has been developed. Experimental data obtained with a pulsed ruby and a pulsed CO_2 laser system compare well with the theoretical analysis. The effects of laser beam energy and pulse duration in laser drilling are important, and optimum laser drilling parameters are desirable.

Machining: Material removal by high-power laser systems (e.g., pulsed ruby, Nd:glass, Nd:YAG, CO_2, argon) is usually by laser-induced vaporization and explosion of solid material. In Ref. 18, an analytical model is developed for material removal from the front surface of a solid by a high-intensity laser beam. Under certain laser and material parameters, the subsurface temperatures exceed the surface temperatures and explosive removal of material can occur, which is a very rapid and efficient removal process. The analytical results for laser beam machining are in good qualitative agreement with experimental results reported by other researchers. The dynamics of material removal for a transparent medium by high-power lasers was investigated analytically and experimentally by researchers at the Institute of Applied Physics, University of Berne, Switzerland. Experiments showed that transparent materials can be vaporized by intense laser beams (in spite of their normally low absorption). In Ref. 20, the researchers state that lasers for micromachining of some materials require very specific laser beam, time-dependent intensity patterns. The direct observation of laser-induced explosion of solid materials (namely, copper and alumina ceramic materials) with a high-speed framing camera (25,000 to 830,000 frames per second) were described in Ref. 21. Qualitative agreement of experimental data/observations with their theory for laser-induced explosive removal of material from solids was reported. Utilizing laser systems (primarily Nd:YAG) for materials maching, thick- and thin-film electrical resistors can be "trimmed" or adjusted, precisely and rapidly for electronic circuit fabrication. At Bell Telephone Labs., conductor patterns for integrated circuits were created with relatively low-power laser systems. With a Nd:YAG laser system operated in the TEM_{oo}, Nd:YAG laser system mode, a 14 micron resolution test pattern was produced and reported in Ref. 22. At the Western Electric Company in Princeton, a 20 W TEM_{oo}, Nd:YAG laser system operated in a pulsed mode was used for pattern machining of thin-films with a positional accuracy of ± 6.4 microns for all lines. In Ref. 24 computer drawn modulated zone plates, or in-line, "holograms," for material processing were investigated. Experimental results showed that the modulated zone plates, the holograms, are useful for many potential industrial applications. At the Bell Telephone Laboratories in Murray Hill, laser machining of complete, small electronic circuit patterns (integrated circuits) directly onto ceramic substrates has been reported. Utilizing a pulsed ruby laser system, micro-machining (micro-drilling and micro-welding) was performed by researchers at Ecole Polytechnic in Paris. Photographs of 10 micron diameter holes in ruby plate, 25 microns holes in brass plate, and 50 micron diameter wire welds were presented in Ref. 26. A reliable wire-stripper was developed by the Orlando Division of Martin Marietta Aerospace that uses a 50 W CO_2 laser system for material cutting/removal

The system eliminates the problem of cut and nicked wire, and meets the rigid military standards for missile system wiring that could not be met by mechanical wire-strippers. Utilizing a CO_2 "TEA" laser system with 3 nanosecond pulse length, the order of 300 millijoules of energy per pulse, and 250 pulses per second, spinning gyro-rotors of gyroscopes have been balanced by laser removal of materials. With this laser system, 3 mg of material was removed with each pulse.

<u>Laser Systems for Materials Processing</u>: Presently, the CO_2 laser system provides high power output beams at efficiencies in the range of 15-20%, which is one-to three-orders of magnitude better than most other laser systems used for industrial applications. CO_2 laser systems with continuous outputs in the megawatt range appear feasible using "TEA," chemical and/or gas-dynamic lasers. Commercially available laser systems for material processing include ruby, Nd:glass, Nd:YAG, CO_2, and argon. A partial list of laser system manufacturers for materials processing is given in Ref. 29, namely: Advanced Kinetics, Inc., American Laser Corp., American Optical Corp., Apollo Lasers, Inc., AVCO Everett Research Lab., Coherent Radiation, Technology, Inc., GEN-TEC., Inc., GTE Sylvania, Hadron, Lasermation, Inc., Lumonics Research, Ltd., Photon Sources,Inc., Q.E.D. Corp., Quantronix Corp., Raytheon Co., Spacerays, Inc., and Systemation, Inc.

In the near future, laser systems may supplement or even replace conventional materials processing techniques. The laser beam now offers certain advantages over the electron beam; such as, no need to operate the laser beam in a vacuum or with special devices, no need to provide x-ray protection, one can process materials through transparent enclosures, and simple lenses and/or mirrors can easily direct a laser beam to the workpiece (electro-static or magnetic deflectors or focusing systems are not required as with electron beams). In many industrial and manufacturing processes, the laser has already proven to be reliable, time-saving, and relatively inexpensive. For example, after studying more than forty techniques for cutting shapes in thin flexible material (cloth) researchers at the Hughes Aircraft Company recommended a focused CO_2 laser beam as the most suitable cutting tool for accurate, economical, computer-controlled material cutting. Also, some researchers believe that laser systems may be utilized for shock strengthening metals--the National Science Foundation has awarded a grant to Battelle Columbus Laboratoies to study shock hardening of metals with laser beams. Another basic study on high-power interaction with materials, with direct application to materials processing, involved reflectivity measurements of laser beams from various materials at the General Motors Technical Center in Warren, Michigan. Measured data for metal samples of 2024-T4 aluminum, OFHC copper, AISI 1045 steel, and AISI 1095 steel were reported in Ref. 32.

III. NON-DESTRUCTIVE TESTING: SPECTROCHEMICAL ANALYSIS

Laser microprobe spectrochemical analysis was an early application of laser systems for non-destructive testing of materials. Because of excellent sensitivity, the amount of sample material required utilizing the laser microprobe with an excitation spark or arc can be as small as 10 grams. This very small amount of sample material required for analysis has qualified laser microprobe spectrochemical analysis as a non-destructive testing method. Localized, qualitative and quantitative tests and analyses are possible with very simple sample preparation using the laser microprobe. Normally, pulsed ruby or Nd:glass laser systems (with Q-switching) are used to produce a "plume" of material from the test sample, which is then excited further to permit emission spectroscopy, by striking an ac spark or dc arc between two carbon electrodes placed over the focused laser beam spot on the sample surface.

Utilizing a mass spectrometer with the laser microprobe, time-of-flight analysis of solid samples in the size range of 10^{-8} to 10^{-10} grams can be performed. Also, absorption spectroscopy has been attempted with limited success on the laser produced "plume;" however, metals such as silver and copper (with long neutral atom lifetimes) and high vapor pressure metals such as zinc and magnesium have been detected in trace amounts of the order of 40 parts per million.

Laser Microprobe: The laser microprobe spectrochemical analysis system usually consists of a pulsed laser (e.g. a Q-switched ruby laser with peak power on the order of 4 megawatts and a pulse length of 35 nanoseconds) a microscope optical system, an electrode system with separate power source to provide ac spark and dc arc excitation, and a control console. Advantages of the laser microprobe over conventional ac spark or dc arc excitation for spectrochemical analysis include: 1) localized analysis--the laser beam can be directed and focused to vaporize a test sample from the desired region with a spot size the order of 50 microns in diameter; and 2) non-conducting samples can be analyzed--in the laser, optical power rather than electrical power is used to vaporize the test sample and electrical conductivity is not required. The amount of test sample vaporized is a function of the test sample material and the laser beam's energy level and pulse shape. Typically, between 0.1 and 1 microgram of test sample is vaporized from a region 35 to 150 microns in diameter. After a fraction of the partially ionized vapor reaches the electrode gap of the spark or arc power supply, the vapor is further excited by an arc spark or a dc arc. In Ref. 33, a laser microprobe system designed by Brech and Cross and manufactured by Jarrell-Ash Company is described in detail. Modifications of this laser microprobe system to improve safety and performance are also presented.

Spectrograph and Spectra: A grating spectrograph with photographic plate recording capability can be used to analyze the emission spectra. Because of the small (about 2mm diameter) and weak emission source, the optical system and the spectrograph must be designed to collect as much as possible of the emitted light from the source. Details of a typical spectrograph and optical instrumentation are presented in Ref. 33. Spectra with and without auxiliary ac spark or dc arc exitation are presented. The spectra without the auxiliary excitation contain the lines of the major constituents. Resonance-broadened and self-absorption are observed with zinc (3282A) and copper (3247A, 3273A). Addition of an auxiliary ac spark or dc arc provided sharper, less self-absorbed lines with increased intensity. Thus, the laser microprobe with ac spark or dc arc excitation is usually preferred for spectrochemical analysis.

Applications and Quantitative Aspects: Laser microprobe spectrochemical analysis has been used at the National Bureau of Standards, Washington, D.C.

for the following, non-destructive, materials testing: 1) identification of the materials in individual components of small semiconductor devices, including the Fe, Ni, and Co lead; Au lead plating, Si semiconductor disk; and dielectric junction, overcoat--all materials located within a 1 mm X 0.5 space; 2) a 75 micron diameter wire with gold and gallium; 3) grains of silver alloy that were roughly cylindrical with diameters of 40 to 50 microns; 4) a ceramic disk with platinum-rhodium alloy less than 100 microns in diameter; and 5) iron, aluminum, copper, zonc metallic constituents imbedded inside a fluorinated polymer bottle. <u>Quantitative</u> spectrochemical analysis utilizing a laser microprobe was investigated in Ref. 34. Random errors were largely due to variations in the laser microprobe's energy and from photometric errors. For precise analysis, the laser microprobe vaporization and the ac spark or dc arc excitation processes must be controlled and reproducible between the test samples and the standards. Since the ac spark or dc arc excitation processes were well controlled in Ref. 34, the effects of the laser microprobe vaporization processes could be determined. Single-spike, laser pulse operation was found to be preferred for most purposes. Correlations were established between the laser beam's energy, the physical size of the laser produced crater, or "pit", in the test sample, and the observed emission spectral intensity. Data were reported for many materials, including low alloy steels, high alloy steels, high temperature alloys, zinc, and pellets of powdered talc. The coefficients of variation for the derived concentrations were 15 to 40%. The effects of background and laser beam energy corrections of spectral line signals on the precision of measurements of the elemental concentration obtained by laser microprobe spectrochemical analysis were investigated with a pulsed ruby laser (Q-switched, 20 nanosecond pulse, 12 to 17 millijoules per pulse) by researchers at the Stanford University Medical School. With vaporized test samples of 50 to 60 microns diameter of iron, zinc, and calcium, the statistical treatment of the data in Ref. 35 indicated that optimal precision was obtained from measurements of <u>uncorrected</u> spectral line signals. The relationship of the emitted spectral line intensity to the weight of the test sample vaporized with a laser microprobe system was investigated by researchers at the Western Electric Company in Princeton. Using a Jarrell-Ash Mark II laser microprobe system with a 5 microsecond, multi-spiked, 10 millijoule, Nd:glass laser beam, test samples of iron, tin, zinc, aluminum, lead, copper, bismuth, and cadmium were examined. Statistical correlations show that the measured spectral line intensity is a function of the weight of the test sample material vaporized by the laser microprobe. Employing optical microscopy, the laser-produced crater size and shape were determined and then the test sample material vaporized can be calculated. The accuracy of the laser microprobe spectrochemical analysis technique was determined to be equivalent to most semiquantitative techniques, and many practical applications are possible.

Recently, laser excited, mass spectroscopy studies by researchers at the French National Glass Institute in Charleroi and at Catholic University in Louvain were reported. An 80 W CO_2 laser system was employed to analyze gases, such as hydrogen, helium, sulfur dioxide, and sodium oxide, desolved in solids. A CO_2 laser system was selected because its relatively long wavelength does not induce consistent photo-chemical reactions.

IV. NON-DESTRUCTIVE TESTING: STRAIN MEASUREMENT

Optical interferometry and holography are extremely useful for material strain measurements and non-destructive testing. Today, holographic interferometry, which combines laser optical interferometry and holography, is one of the principal areas of research, engineering, and applications activity in optical holography. Basically, holographic non-destructive testing consists of comparing the image of a test object reconstructed from a hologram with the test object itself or with a second image of the test object after the test object has been distorted, or changed. To be easily observable with holographic non-destructive testing, these distortions only need be of the order of microinches, i.e., the order of the dimensions of the optical wavelength for the laser utilized in recording the holograms. The test object may be distorted due to mechanical strain and/or vibration caused by applying testing forces; such as, mechanical stress, gas pressure loading, thermal heating and cooling, and/or mechanical vibration. Now, holographic non-destructive testing is being utilized for mechanical strain determination, deflection measurement, flaw detection (hidden voids or cracks), vibration and mode analysis, and contour-mapping. There are three major types of holographic interferometry; namely, real-time, double-exposure (or double-pulse), and time-averagine. During the last few years, holographic interferometry has developed to the point that now holographic non-destructive testing (HNDT) is actually performing tests which were formerly exclusively accomplished by other non-destructive tests; namely, tests by ultra-sound, x-rays crack-penetrant techniques, etc. A few of the holographic non-destructive tests that are currently being performed by several of the large manufacturing companies and by a number of testing laboratories include the following: analysis of stress on jet-turbine blades, vibration analysis of turbine blades and rotors; honeycomb bond flaw detection, metal-to-metal bond and rubber-to-metal bond studies, detecting debonds between brazed beryllium-alloy brake disks and sintered-iron friction pads, pneumatic-tire testing; sonar transducer evaluation; material strain and fracture studies; deflection measurements, optical component testing; and contour-mapping.

Optical Laser Interferometry and Holography: A typical optical laser interferometry or holographic system for non-destructive testing, strain measurement consists of the following items; a laser source, which may be a low-power continuous source for static strain tests or vibration mode analysis with time-averagine or a medium- to high-power pulsed source for dynamic strain tests and "stop-action" analysis; optical mirrors, beamsplitters, beam expanders, filters, laser-mode selectors, attenuators, mounting rails; control panel; and recording photographic plate/film holder/very simple "camera". Complete laser holographic systems are commercially available from several vendors, including to name only a few: Apollo Lasers (model 22 HD, double-pulse holographic system), Gaertner Scientific, GCO, Jodon Engineering Associates, Korad (Model K1200 QDH industrial double-pulsed), Metrologic Instruments, Newport Research, and TRW Instruments.

Applications: Recently reported holographic non-destructive testing applications for strain measurement include: dynamic strain measurements utilizing a pulsed ruby laser (Q-switched, double-pulsed, 30 millijoule per pulse) of three-dimensional, diffusely reflecting objects such as bending metal plate with a force applied at an edge; a rotating fan blade; the transient deformation of a motorcyclist's crash helmet after being struck by a hammer

--with the crash helmet being worn by one of the authors; and strain due to driving a nail into a block of wood--with the nail being hand-held. Double-pulsed optical holography is finding more and more uses as a "practical industrial tool: and as a "problem solver in the lab", e.g., precise measurements of the distortion of the surface of a brazed plate heat exchanger under 250 psi pressure with and without braze anomalies, by staff members at Garrett Air Research Company in Torrance, California for industrial quality control; visualization of the interactionof a gas shock wave with a re-entry vehicle model by personnel at the Air Force Flight Dynamics Laboratory, Wright-Patterson Air Force Base, Ohio; and the collision of two jets of neon flowing in a pressure vessel containing nitrogen at 20 N/cm by researchers at the Los Alamos Scientific Laboratory. On October 15-18, 1973, the Instrument Society of America offered four clinics on holographic applications, including technical presentations on the following: an introduction to holography, holographic techniques, and practical laboratory details for holographic application; non-destructive testing with continuous laser systems and holographic instrumentation with both quantitative and qualitative data analysis; optical strain measurement; and pulsed holography of moving objects and flow visualization. In Ref. 42, a number of industrial holographic measurements by researchers with the Royal Institute of Technology, Stockholm, Sweden are described, including: bending and twisting of a 2 meter long steel bar due to a 50 gram load; deformation of a large throttle valve in a pressure system; deformation of rock drills with carbide tips soldered and glued into place in the drills; strain in a plastic outlet tube due to approximately a 10°C temperature change; vibration patterns in a dynamic model of an airplane fin being flutter tested in a wind tunnel; and vibration analysis of a hand drilling machine in operation, showing the hand, the drill and the workpiece.

Holographic interferometry, non-destructive testing for strain measurement makes optical interferometry "amazingly simple." With holographic interferometry, precision optical components (optical surfaces flat to a fraction of an optical wavelength and free of optical defects) are _not_ required; optical alignment is _not_ very critical; the surface or test volume does _not_ have to be optically perfect but can be irregular or diffusive; and _in situ_, high-quality, non-destructive test data can be achieved for mechanical strains of the order of microinches or more, for vibrational modes and patterns, for subsurface bond separation and defects, and for contour-mapping of three-dimensional objects. Utilizing commerciallly available pulsed lasers, stop-action photography (including optical interferometry/holography) is possible on a sub-microsecond time scale, and on a picosecond time scale with recently developed mode-locked laser systems. Optical holography now permits quantitative and/or qualitative measurements on difficult, and some previously impossible manufacturing processes.

V. PRECISION MEASUREMENT, ALIGNMENT, AND INSPECTION

In addition to the types of measurements associated with non-destructive testing that were described in the previous section, lasers also find useful application in direct measurement of critical dimensions, alignments, surface conditions, shapes, material concentrations, and patterns. In these applications the use of laser systems often provides much greater accuracy than can be achieved by other methods or allows enhanced efficiency in the measurement operation.

Length Measurement: Lasers can be used in several manners for measuring length or distance. First, a light pulse can be generated and its time of flight over the distance in question measured. This measurement can then be used to compute the range traversed. Techniques based on this principle are capable of measuring long distances with high accuracies. Accuracies of one part in 10 are possible for ranges in excess of 100 miles. Normally round-trip time of flight measurements are made using light reflected from an object or reflector.

A second technique uses a continuous laser beam that is modulated with a high frequency radio signal. The phase of the modulated transmitted wave is compared to that of the received wave which has been reflected from the object whose range is to be determined. The phase difference measured is used to compute the range. In theory at least, measurement accuracies of 1/1000 of an inch at ranges up to 500 ft. are possible.

The third method of laser range or length measurement is the most accurate of all. It makes use of the highly coherent nature of laser light to compare the optical phase of a light wave transmitted to and reflected back from an object with that of a second laser beam propagated over a known range. The resulting interferometric measurement allows accuracies of a fraction of an optical wavelength to be achieved (less than 10 u inch). The advantages of this type of measurement include high accuracy, even over lengths where no tangible physical standards are available. Disadvantages include sensitivity to vibration, to changes in the refractive index of the air, and to expansion or contraction of the device to be measured. However, all these disadvantages can be overcome, so that this technique is practical for industrial use. Industrial applications of the technique include the calibration of positioning systems and secondary length standards, checking the accuracy of precision machine tools, and controlling linear displacement in machine tools, and measuring machines.

Size and Thickness Measurement: When coherent light is observed after passing through an aperture or around an opaque object, the light distribution in the Fraunhofer or Fresnel diffraction field is characteristic of the size, shape, and orientation of the aperture or object in question. This effect has been used to measure the size, shape, and location of long, small-diameter holes drilled in precision items. Holes with diameters ranging from 0.007 to 0.5 inches have been measured successfully. Holes with diameters of the order of 0.040 in and lengths of 1 to 10 diameters have been measured with accuracies of ± 20 uin on diameter, $\pm 0.5\%$ departure from roundness, $\pm 0.05°$ on angularity, ± 0.0005 in on location, $+0.001$ inch on impingement, and $\pm 2\%$ on taper. Burrs can also be easily detected.

The same basic effect has been used to measure the size of slit-like apertures and the proximity of an edge to a surface. Typical resolution is better than 1 um or less with a range of 0-1 mm. Wire thickness can also be measured using this approach.

Laser techniques can also be used to measure veneer thickness and profiles of moving or extruded materials. A scheme developed for plywood manufacturing uses laser range measurements to the moving material surface to achieve

5-mil resolution.

Alignment: The highly directional beams of gas lasers make them very useful alignment tools. A typical gas laser beam diverges at a rate of roughly one part in 2000. Slow divergence rates can be achieved if collimating optics are used. Since the beam travels in a straight line unless deflected by external optical elements, it is useful for a number of alignment problems. Laser alignment has been successfully used in surveying, in trenching operations, as an aid in drilling holes through walls or floors for pipe or tubing installation, and for aligning assembly jigs for aircraft construction. It has been applied to the alignment of the km long waveguide at the Stanford University Accelerator Center in Menlo Park, California, within an accuracy of 0.5 mm. Clearly, very accurate alignment can be achieved with laser techniques. In addition, such measurements are normally also relatively easy to employ in comparison to other techniques of similar accuracy.

Surface Inspection: A number of laser techniques also exist for measuring surface roughness. Such schemes as oblique incidence interferometry, in which a laser beam is scanned at an oblique angle of incidence across a rough surface and then observed for interference fringes, allow the flatness of rough surfaces to be inspected and measured. As an example of the sensitivity that can be achieved, a 50AA surface can be measured to approximately 0.0001 inch. Polished surfaces can be measured with standard optical and holographic interferometric techniques to accuracies in the 10 microinch range.

Spatial filtering techniques can also be used for surface inspection. An example is a scheme to detect defects (pits, projections, haze) on the surface of semiconductor wafers. A collimated laser beam is reflected off the surface to be measured at small angle (12.50) to the normal. The reflected beam is focused with a lens. If no imperfections exist on the reflecting surface, the collimated beam is brought to a point focus. The light thus focused is blocked by an opaque dot placed in the focal plane of the lens. However, if any defects exist on the surface under inspection, they scatter light which is not brought to a focus at this same focal spot and thus is not blocked. If this unblocked light is reimaged with another lens, it shows an image of the defects on the surface under inspection.

Shape Measurement: Laser and optical techniques can also be used to measure the shape of complex objects and forms quickly and accurately. Some of the holographic techniques discussed earlier with regard to non-destructive testing such as contour-mapping of three-dimensional objects can also be used to measure complex shapes. Contour line range separations as low as 7.7 mm have been demonstrated over scenes several feet across.

A related technique is Moire topography in which the object to be studied is placed behind an equispaced plane grating. The object is then illuminated and viewed through this grating. Viewing the projected shadow of the grating on the object through the grating results in a Moire pattern depicting contour lines. Contour lines separated by 2 mm have been demonstrated over objects 2-3 feet across. Although this is not strictly a laser measurement technique (other light sources can be used), it does show promise of great utility in industrial measurement.

The length measurement techniques described earlier such as the time of flight measurements of laser pulses can also be used to measure the shape of complex objects such as large antenna structures.

Bulk Absorption and Scattering Measurements: Lasers also have measurement applications based on the scattering and absorption of light as it is transmitted through materials. When a light beam propagates through a region containing a number of particles, it is scattered by the particles. The scattering can be described according to one of two theories, Rayleigh scattering for particles smaller than a wavelength of light and Mie scattering for particles larger than a wavelength of light. Various factors such as the angular dependence of the scattering, its intensity, its variation with wavelength, and the polarization of the scattered radiation can be used to study the scattering medium. It is possible to obtain information about the number of particles as well as their sizes. Possible industrial applications include the measurement of particle contamination in liquids and gases. Sandia has developed a laser particle counter for monitoring the number and size of airborne particles in clean rooms. Detection of particles as small as 0.3 um appears feasible, and large sampling rates up to 50 cu ft/min make real-time monitoring possible.

It is also possible to use scattering to measure particle or object velocities. Light scattered from moving particles or objects is Doppler shifted in frequency according to the particle or object velocities. Measuring the frequency shifts thus provides a measurement of the velocities. Instruments have been built on this principle to measure fluid velocities by scattering light from particles suspended in the fluids. This phenomenon can also be used to measure vibration of large objects. The measurement of velocities as low as 0.004 cm/sec. has been reported.

Laser Raman spectroscopy is another potential industrial laser application. Raman spectroscopy is based on the fact that momochromatic light scattered from any material contains the incident light frequency plus various side frequencies. The separations of the side frequencies from the incident frequency correspond to the vibrational frequencies of the molecules and thus are characteristic of the material. This can be used to analyze materials. If techniques can be developed to rapidly analyze the Raman spectra, it may be possible to utilize laser Raman spectroscopic systems in feedback loops to control chemical processes. It has also been proposed that remote laser Raman spectroscopy be used to provide a three-dimensional mapping of concentrations of atmospheric pollutants.

Absorption spectroscopy is also being investigated as a means of detecting atmospheric pollutants. Different chemicals have characteristic absorption bands which can be used to identify their presence. The absorption of laser beams propagated through the atmosphere at various wavelengths can thus be used to identify the presence of particular atmospheric constituents.

Pattern Inspection: The final example of laser measurement applications to be considered here is a technique for inspecting integrated circuit photomasks for random errors. The photomask consists of an array of a given pattern repeated at a constant interval. The problem is to detect errors or defects in the individual patterns. This cannot be accomplished practically by normal microscopic inspection due to the complexity of the patterns.

A technique has been devised whereby the photomask is illuminated with a collimated laser beam. The transmitted beam is brought to a focus with a lens. The light distribution on the back focal plane of the lens (consisting of a complex dot-like pattern) represents a two-dimensional Fourier transform of the spatially modulated light distribution exiting the photomask. It consists of an array of spots determined by the periodic spacing of the patterns on the photomask, with each spot modulated in intensity by the transform of the individual pattern. If a spatial filter consisting of an array of opaque dots on a transparent background is placed on this plane (one dot for each spot location in the Fourier transform) and if the light transmitted through is retransformed with a second lens, then the resulting image of the photomask will have all information removed which is repeated on the photomask with the same spacing as the individual patterns. Thus the only image seen will be that of the non-periodic features on the photomask, in other words the errors. In this way contamination, step and repeat errors, and missing and added features in the individual patterns can all be displayed. It is reported that errors smaller than 0.1 mil can be detected. This is a good example of the very complex measurement and inspection techniques possible using laser systems.

VI. CONCLUSION

The industrial uses of lasers have been discussed briefly in a number of applications areas. The individual topics treated have been chosen in an attempt to illustrate the broad variety of laser applications currently practiced in industry. Needless to say, many worthwhile applications have not been treated due to space and time limitations. Again interested readers are urged to consult the referenced literature for more complete treatments of industrial laser applications. Finally, it is probably worthwhile to emphasize that research in laser applications is very active, and significant new development can be anticipated in many areas. As these applications develop and as laser systems become more versatile, more reliable, more powerful, and cheaper, we can expect significant increases in industrial laser use.

REFERENCES

1. F.P. Gagliano, R.M. Lumley, and L.S. Watkins, Proceedings of the IEEE 57, pp. 114-147, February, 1969.

2. S.S. Charschan, Lasers in Industry, Van Nostrand Reinhold Company, New York, 1972.

3. M. Ross, Laser Applications, Vol. 1, Academic Press, New York, 1971.

4. C. Ruffler and K. Gurs, Optics and Laser Technology 4, pp. 265-269, December, 1972.

5. Edward V. Locke, Ethan D. Hoag, and Richard A. Hella, IEEE Journal of Quantum Electronics QE-8, pp. 132-135, February, 1972.

6. "Deep-penetration Laser Welders Inspire Brave Words . . .," Laser Focus 9, cover photo, pp. 3,26,28,30, August, 1973.

7. V. Bodecker and G. Sepold, IEEE Journal of Quantum Electronics QE-9, p. 650, June, 1973.

8. Conrad M. Banas, IEEE Journal of Quantum Electronics QE-9, pp. 650-651, June, 1973.

9. E.L. Baardsen and D.J. Schmatz, IEEE Journal of Quantum Electronics QE-9, p. 651, June, 1973.

10. E. Locke and R. Hella, IEEE Journal of Quantum Electronics QE-9, pp. 651-652, June, 1973.

11. Ethan Hoag, Henry Pease, John Staal, and Jacob Zar, IEEE Journal of Quantum Electronics QE-9, p. 652, June, 1973.

12. James Arruda and William Prifti, Electro-Optical Systems Design 5, pp. 34-35, May, 1973.

13. "A Laser for Metal Working and Fabrication," Optical Spectra 7, pp. 40-41, May, 1973.

14. Marce Eleccion, IEEE Spectrum 9, pp. 62-72, April, 1972.

15. J.D. Foster, R.F. Kirk and F.E. Moreno, IEEE Journal of Quantum Electronics QE-9, p. 652, June, 1973.

16. "Precision Laser Machining: Drilling by Trepanning," Electro-Optical Systems Design 5, pp. 31-33, May, 1973.

17. Un-Chul Paek and Francis P. Gagliano, IEEE Journal of Quantum Electronics QE-8, pp. 112-119, February, 1972.

18. Franklin W. Dabby and Un-Chul Paek, IEEE Journal of Quantum Electronics QE-8, pp. 106-111, February, 1972.

19. Ernst Kocher, Lorenz Tschudi, Jurg Steffen, and G. Herziger, IEEE Journal of Quantum Electronics QE-8, pp. 120-125, February, 1972.

20. G. Herziger, R. Stemme, and H. Weber, IEEE Journal of Quantum Electronics QE-9, pp. 649-650, June, 1973.

21. F.P. Gagliano, and U.C. Paek, IEEE Journal of Quantum Electronics QE-9, p. 649, June, 1973.

22. Walter W. Weick, IEEE Journal of Quantum Electronics QE-8, pp. 126-131, February, 1972.

23. J. Raamot and V.J. Zaleckas, IEEE Journal of Quantum Electronics QE-9, p. 648, June, 1973.

24. A. Engle, J. Steffen, and G. Herziger, IEEE Journal of Quantum Electronics QE-9, pp. 647-648, June, 1973.

25. "Laser Machines Circuits in One Step," Lasersphere **2**, pp. 1-2, May 15, 1972.

26. G. Boyer, J.R. Huriet, and B. Lamourex, Optics and Laser Technology **2**, pp. 196-199, November, 1970.

27. "Reliable Laser Wire Stripper," Optical Spectra **7**, pp. 37-38, May, 1973.

28. A Jacques Beaulieu, Electro-Optical Systems Design **5**, pp. 36-37, May, 1973.

29. "Laser Applications and Where to Get the Lasers From," Electro-Optical Systems Design **5**, pp. 38-39, May, 1973.

30. Loyd D. Malmstrom, IEEE Journal of Quantum Electronics **QE-9**, p. 674, June, 1973.

31. "Does Laser Shocking Strengthen Metals?" Lasersphere **3**, pp. 6-7, April 15, 1973.

32. Barbara A. Sanders and Victor G. Gregson, Jr., "Reflectivity Measurements Implement CO_2 TEA Laser Applications," Lasersphere, pp. 9-14, October/November, 1973.

33. Stanley D. Rasberry, Bourdon F. Scribner, and Marvin Margoshes, Applied Optics **6**, pp. 81-86, January, 1967.

34. Stanley D. Rasberry, Bourdon F. Scribner and Marvin Margoshes, Applied Optics **6**, pp. 87-93, January, 1967.

35. A.J. Saffir, K.W. Marich, J.B. Orenberg, and W.J. Treytl, Applied Spectroscopy **26**, pp. 469-471, July/August, 1972.

36. Kim L. Morton, James D. Nohe, and Bruce S. Madsen, Applied Spectroscopy **27**, pp. 109-117, March/April, 1973.

37. "Light Elements within Glass Detected with Laser-Excited Mass Spectrometry," Laser Focus **9**, pp. 10, 12, 19, February, 1973.

38. Otto M. Friedrich, Jr. and Arwin A. Dougal, Instrumentation in the Aerospace Industry **17**, Instrument Society of America, pp. 141-153, May 10-12, 1971.

39. J.W.C. Gates, R.G.N. Hall, and I.N. Ross, Optics and Laser Technology **4**, pp. 72-75, April, 1972.

40. Lance W. Riley, Optical Spectra **7**, pp. 27-30, December, 1973.

41. "ISA Offers Four Clinics on Holographic Application," Holosphere **2**, p. 3, September, 1973.

42. Nils H. Abramson and Hans Bjelkhagen, Applied Optics 12, pp. 2792-96, December, 1973.

43. D.A. Worth, Industrial Research 11, pp. 42-44, August, 1969.

44. L.S. Watkins, Proceedings of the IEEE 57, pp. 1634-1639, 1969.

45. J.S. Zelenka and J.R. Varner, Applied Optics 8, pp. 1431-1434, July, 1969.

46. J.S. Zelenka and J.R. Varner, Applied Optics 7, pp. 2107-2110, Oct., 1968.

47. L.O. Heflinger and R.F. Wuerker, Applied Physics Letters 15, pp. 28-30, 1 July 1969.

48. H.Z. Cummings, N. Knable, and Y. Yeh, Phys. Rev. Letters 12, p. 150, 1964.

49. S. Porto, Industrial Research 11, pp. 66-68, May, 1969.

50. "News in Focus", Laser Focus 6, pp. 19-22, March, 1970.

51. H. Takasaki, Applied Optics 9, pp. 1457-1472, June, 1970.

52. G. Watt and P. Langenbeck, "Testing with Lasers," SME Southwestern Engineering Conference and Tool Exposition, Dallas, Texas, Nov. 1969, paper IQ70-411.

53. G.L. Cline, 1969 IEEE Conf. on Laser Engineering and Application, Washington, D.C., May, 1969, paper #9.5.

54. T.R. Pryor, O.L. Hageniers, and W.P.T. North, 1971 IEEE/OSA Conf. on Laser Engineering and Applications, Washington, D.C. June, 1971, paper #11.5.

55. J.R. Kerr, 1969 IEEE Conf. on Laser Engineering and Application, Washington, D.C., May, 1969, paper #9.4.

56. H.A. Elion, <u>Laser Systems and Applications</u>, Pergamon Press, New York, 1967.

ELECTRON BEAM WELDING

Electron beam welding has moved from an exotic laboratory tool to a practical method for assembly and joining of components. During the recent past, the process has found widespread acceptance in the automotive industry, has solidified its use in the aerospace industry and has branched out further into the nuclear industry and oil industry.

Electron beams are used to weld parts in vacuum at various pressure levels, in air and in inert atmospheres. A good deal of basic research into the dynamics of the electron-beam welding process has been conducted and new equipment has been developed as a result.

A simple overview of these developments is presented. Several key applications in automotive production are discussed. Typical applications in high-speed welding and very heavy welding are shown. In addition, other uses of the electron beam are reviewed such as heat treating, vapor coating, vacuum remelting.

Electron-beam welding, how it works:

Electron-beam welding has found its way out of the high-technology, special process stage into the everyday production welding of the automotive, aerospace, nuclear and energy industries. Due to this increased usage, the principles of the process have become more widely understood. For this reason, it will suffice to give a very short outline of the nature of the process and the basic types of systems presently available.

Electron-beam welding utilizes a high-density beam of high-speed electrons which collide with the material at the weld seam. Considerable research has been undertaken, primarily in the USA, Japan, and Europe in the attempt to determine what takes place at the beam inpingement point. High speed conventional photography and real time X-Ray movies show that the beam evaporates material and forms a deep vapor hole with a film of molten material around its periphery. If this vapor hole is moved along the seam, the molten material will flow together after its passing and fuse, thereby creating a weld. Formation of the vapor hole occurs if a seam is present or if a solid sheet of material exists. For this reason, the large majority of welds made by electron beams utilize a flat, square, butt joint without need for shaped weld preparation. Filler wire is normally not employed but can be used when necessary. It is also possible to make "T" welds, penetrating through a solid material into a member positioned on the opposite side. The total heat input needed to make a weld is substantially lower than needed for a similar weld joint by other conventional welding methods. Therefore, the weld zone and heat affected zones are narrow. For this reason, distortion is low, and frequently pre-finished parts are welded together without requirements for final machining after welding.

The ability to pre-finish components, the high welding speeds attainable by electron-beam welding, the ability to weld without consumeables such as filler wire, or cover gas, and the ease of automation of theprocess has proven to be the basic features which make this process economically desirable.

Basic types of electron-beam welders:

Vacuum considerations:

Traveling electrons will collide with gas molecules, if a gas is present, and will be scattered, thereby making it difficult to concentrate the beam into a small area. For this reason, the electron-beam gun area must be evacuated in order to allow undisturbed beam generation. All electron beam systems therefore have a high-vacuum pumping set associated with their beam guns. Vacuum levels in this area are usually in the 10^{-4} or 10^{-5} mbar pressure range. Once the beam is generated, it can be brought out into higher pressure regions. The distance it can be allowed to travel without excessive scattering is inversely proportional to the ambient pressure. Four basic welding systems are in use at this time if pressure at the workpiece is the classification of the process *(See also Figure 50)*.

High vacuum e.b. welding:

The workpiece is at the same high vacuum as the electron-beam gun. Usually there is a vacuum chamber which includes a work manipulation system. The gun and chamber can be pumped by the same pumps. Advantages are: Very clean welds free of oxidation without gas contamination. The gun-to-workpiece distance can be very long, 6 to 8 feet in some instances. Disadvantages are: Relatively slow pumpdown times (15 to 30 minutes). Typically, this is a batch-load operation which is usually difficult to adapt to high quantity production.

"Partial", "soft", or medium vacuum e.b. welding:

The workpiece is at a partial pressure of approximately .13 to .04 mbar, while the gun is at high vacuum. The vacuum chamber is usually custom built for the part to be welded and its volume is minimized. With proper pumping capacity, pumpdown times of 5 to 10 seconds can be realized for chamber sizes of one cubic foot or less. Beam throw distance is somewhat reduced when compared with high vacuum operation. However, the distance is usually still long enough for most applications (typically 12" or less). Advantages: Short pumpdown times coupled with proper work handling systems allow high production rates, typical machines of this type in the automotive industry are rated at 200 parts per hour. Disadvantages: Pumpdown times are not constant, they vary with humidity changes of the ambient air and/or improper maintenance. Since pumpdown still is a significant part of the cycle, production rates vary unless the area is humidity controlled. Vacuum system maintenance is still important and usually must be learned by the respective maintenance crews.

Non-vacuum welding:

The workpiece is at atmospheric pressure. The beam is brought out to atmosphere through a series of vacuum stages with incoming leakage air pumped between nozzles. Production machines of this type usually have the highest production rates due to the absence of pumpdown times. For the same reasons, the production rate is steady and independent of ambient humidity. Advantages: Usually provides the highest production rates and least vacuum maintenance requirements. In addition, weld quality is less sensitive to cleanliness in the joint. Disadvantages: Weld bead wider than partial vacuum. It should be noted that this can also be a practical production advantage for certain applications. Distance from beam exit hole to work surface is 2.5 cm or less in most cases.

Quickvacuum welding:

A new development, yet to be incorporated in a production system, combines a nonvacuum e-b gun with a relatively poor vacuum of 33 to 66 mbar in the welding chamber. This process allows narrower welds than non-vacuum, or greater gun-to-work distances. The beam is considerably narrower than a non-vacuum beam. The vacuum requirement is less severe, and is unaffected by normal humidity variations. *Figure 51* shows beam dispersion at various pressures. Advantages: The same high production rates obtained with non-vacuum apply except for the addition of approximately 2 second to the production cycle time for the pumpdown added. Greater weld penetration, high weld speeds, reduced vacuum maintenance are other important advantages. Disadvantages: This process still requires a vacuum on the part, and does limit work distances considerably.

Voltage considerations:

The most common accelerating voltage levels used in electron beam welding guns are "high voltage", from 100 Kilovolt to 200 Kilovolt, and "low voltage", at approximately 60 Kilovolt. Low voltage systems operate only in the partial vacuum and high vacuum mode. High voltage welders can work very well in high vacuum and partial vacuum machines and they are the only systems utilized in the nonvacuum and quick vacuum mode to date.

There has been considerable controversy about advantages of low-voltage vs. high voltage. In brief, the differences stem from basic physical limitations of the gun design parameters: A key parameter in electron beam welding is power-density of the focused beam spot on the work surface. High power density means deeper penetration at equal weld speed or higher weld speed at equal penetration. Since beam power is the product of accelerating voltage and beam current, one can have equal beam power, for example, with 60 KV and 200 miliamperes (12KW) or 120 KV and 100 milliamperes (12KW). Here the low voltage gun needs twice as much current than the high voltage gun to produce the same power. Beam current is the number of electrons flowing, therefore low voltage needs more electrons, hence larger emitters in the gun. It is more difficult to get a small focus spot, the other part of the "power density" formula, if the emitter is larger. Therefore, more beam current is required, over and above the increase needed to equal power, to get the same power density in a larger focused spot. This is the basic reason why more power is required for low voltage guns, when compared with high voltage, producing equal weld penetration. On the other hand, high velocity electrons emit harder X-Rays than low velocity. For this reason, more X-Ray shielding is necessary for high voltage than low voltage guns.

For these reasons, low voltage machines have found use in the high production partial vacuum automotive applications where weld requirements are not as critical, penetrations not very heavy and non-vacuum welds are too wide. In addition, low voltage guns have been utilized in large chambers operation at high vacuum with an internally moving gun.

In many high vacuum aerospace and deep penetration applications, the high voltage beam is superior because of its excellent penetration ability, better depth of focus, and long gun-to-work distance capabilities.

For practical non-vacuum and quick vacuum operation, high voltage is a must. Typical accelerating voltages are from 150KV to 200KV. Low voltage beams would scatter at atmospheric pressure to an extent which will make them useless for welding.

Some typical applications of electron-beam welders:

    High vacuum - partial vacuum chamber machines:

Figure 52 shows a standard vacuum-chamber machine for operation in the high vacuum mode. This type of machine is widely used in the aerospace industry and in electron-beam welding job shops. The chamber is equipped with an X-Y table for work handling, and rotary attachments are available for circular welds and girth welds. Operational mode is batch-loading. Usually at least 2 work fixtures are used, one being loaded and unloaded, the other being in the chamber for pumpdown. Typical pumpdown time is 10 minutes, a factor limiting production rates and causing this equipment to be used for smaller production lots. The main advantage of this equipment is its flexibility in handling a great variety of parts which fit within the vacuum chamber. To simplify setup, a coaxial viewing system is integrated into the gun column, allowing the operator to see the part the way the beam sees it, prior to and during welding.

For larger workpieces, larger chambers are available, such as shown in *Figure 53*. Please note the rotary attachment mounted on the X-Y table and the use of a run-out platform to allow overhead crane access to heavy work pieces. The illustrated machine can operate in both high vacuum or partial vacuum mode.

Standard chamber machines in high-vacuum and partial vacuum modes are available with power ratings between 1KW and 45 KW. With a beam power of 25 KW, it is possible to make single pass, high quality welds up to 16 cm penetration in steel, *(Figure 54)*.

Partial vacuum production machines:

Figure 55 shows a typical high production partial vacuum welder. The work is handled with a dial index table, the illustrated machine has three stations. This design allows automatic loading and unloading as well as other associated assembly functions to be integrated into the cycles. The welding is done in a small chamber at the welding station. The pallet with the workpiece and fixture is lifted into the bottom of the chamber, the chamber is evacuated, typically in 5 to 10 seconds, and the welding cycle automatically initiated. At completion of welding, the chamber is vented and the pallet is lowered back onto the index table. The next part is then indexed into position and the cycle repeats.

Typical production rates are 150 to 200 parts per hour gross rate. A typical part is shown in *Figure 56*. With the high production rates of this equipment, it usually is dedicated and tooled for one specific part or a small group of parts.

Non-vacuum production machines:

The large majority of high production non-vacuum welders fall into two basic categories: The local radiation shield type and the X-ray-proof-room type. *Figure 57* shows a typical version of a local shield machine. The basic construction is similar to the index-table partial vacuum equipment. Work pieces are usually rotary parts such as torque converter housings, torque converter components or transmission parts. Again, the work is lifted off the index table into a local radiation enclosure. This time no evacuation is needed, saving 5 to 10 seconds per cycle. After the weld cycle, parts are lowered back onto the index table and the next part is indexed into position. Production rates depend on the length of the weld cycle but welding speeds are high, typically 12.5 to 17 cm/sec. (300 to 400 inches/min.)

Typical production rates range from approximately 3200 welds per hour on planet-carrier pin welds (short welds) to 240 parts per hour on a typical torque converter housing (long weld). Most machines of this type, as well as similar partial vacuum equipment, employ programmable controllers or mini computers for the machine logic function as well as fault indication. Such equipment aids greatly in quality control and facilities simplified troubleshooting procedures.

The X-Ray-proof room type of production non-vacuum welder is used where the weld contour is complex and the parts are relatively large. In this case, usually a computer controlled two or three axis table as well as computer control of weld speed and beam current is required. To avoid complex X-Ray shielding, the table and the welding gun are enclosed in an X-Ray proof room and the parts are transported into and out of the room by suitable conveyors or other transfer equipment. Typical applications for this type of machine are the rim welding of catalytic converters and the more recent welding of die-cast aluminum intake manifolds. *Figure 58* shows the outside of such a machine and *Figure 59* shows a die-cast two piece aluminum manifold with a very complex weld path. Production rates typically run approximately 150 parts per hour where most of the cycle time is actual welding time at speeds up to 19 cm/sec. (450 inches/min.).

Quickvacuum welding:

This type of welding utilizes a vacuum of 11 to 67 mbar (10 to 50 Torr) pressure in combination with a non-vacuum type gun to extend the stand-off distance from gun to workpiece or extend the penetration depth over a non-vacuum weld without the penalties of a longer and non-constant pumpdown time. *Figure 51* showed as comparison an electron beam at atmospheric pressure and at various reduced pressures. It can be clearly established that the beam is much less scattered at lower pressures and has a significantly better reach. Pumpdown times are very fast and unaffected by humidity of the environment, typically 2 seconds for a volume of 1 cubic foot. The equipment operating in this mode is very similar to the index type machines of the partial-vacuum variety, except that it employs non-vacuum type guns. *Figure 60* shows typical welds in 2.5 cm (1") thick low carbon steel made at about 67 mbar, at various power levels. It clearly shows a narrower weld zone than similar non-vacuum welds.

Other machines for welding:

The typical machines shown in the high vacuum and partial vacuum series all were of the high voltage type. *Figure 61* shows a 60 KV machine, available in 3KW, 6KW and 10KW power range, for use on shallower penetrations and where very light and delicate welds are needed. The 3KW and 6KW machines employ refined electron optics, highly sophisticated and regulated power supplies and beam current controls which make use of high-frequency generator is available for the 7.5 and 15KW high voltage machines, mostly used for supercritical welding and beam-puddle control in high vacuum and partial vacuum welders.

Other uses for electron beam in industry:

As previously mentioned, beam power density in welding is the primary process parameter for weld penetration or welding speed since a vapor-hole must be established and drawn through the metal. Increased beam currents for higher power bring with them a decrease in focal spot size, since emitters become larger for supply of sufficient quantities of electrons. If the final use of the electron-beam is for heating purposes, not welding, a much larger focal spot can be tolerated. Larger emitters can now be employed and e.b. guns of up to 300KW beam power have been built and used. Their main application is in vapor coating and vacuum remelting.

Figure 62 shows a typical evaporation gun with attached crucible for vacuum-coating of a variety of substrates. These guns are typical of those used in vacuum systems to coat substrates such as glass, thin-film plastics, automotive parts such as reflectors for headlights, decorative plastic parts which need a metallic appearance and other items such as the protective coating on high temperature turbine blades for jet engines.

Higher power guns are used to re-melt metal ingots in a vacuum environment to produce high quality and high purity steels. *Figure 53* shows typical uses for guns in this field. One of the largest remelting furnaces in the world, uses a total of 1,200 KW of electron-beam power for production of specialty metals.

A very recent development in the use of electron beams is its use to locally and selectively harden the surface of steels as well as use the beam as a heat source for surface alloying operations. The groundwork for this use was laid by applications of lasers. Some production systems, using lower power lasers of approximately 1KW and below, are presently in use in the automotive industry. De-focused electron beams from welding guns, and especially the higher power heating guns with naturally larger focal spots, are excellent candidates for this work, especially where power levels above present-day laser capabilities are needed. Typically, case-hardeness of 60-62 R/C with a case depth of 1.3 mm (.050") has been achieved in 4340 steel with a power density of 1.2 KW per square cm. (7.5KW per square inch) and a 3 second cycle. Equipment for this purpose will be very similar to partial vacuum welding machines.

The heating and melting guns operate at lower voltages (25 to 35 KV), and at higher powers (up to 300KW each). Another advantage is the very long filament lifetime, for example, a 100KW gun will typically operate 100 hours on one kathode. Finally, the cost of the power sources for these systems is important to compare: EB welding systems typical cost $3 to $6 per watt, the new high power CO_2 lasers are reported to cost $40 to $50 per watt, we estimate that the e-b heating and melting guns when applied for heat treating will cost between $1 and $2/watt of installed beam power.

Summary:

Electron beams are continuously finding new and wider applications as useful sources of power. The increasing use of these beams for welding is mainly caused by the high productivity and rapid welding speeds of the process. In addition, the electron beam is very efficient and it delivers 80%

to 95% of the power furnished to the gun to the weld spot on the workpiece. Additional cost savings are realized by the fact that in most cases consumeables such as welding wire and inert gas are not needed. Due to the fact that only limited, but intense, heat is delivered precisely to the area where it is needed, distortion of the welded parts is minimized and frequently parts can be machine finished prior to welding, a practice which usually saves additional money in machining costs.

Electron-beam equipment has come a long way, out of the laboratory into the real world of production and present day machines are rugged and reliable.

List of illustrations:

- *Fig. #50:* *Operational schematics, high vacuum, partial vacuum and vacuum guns.*
- *Fig. #51:* *Beam dispersion as function of ambient pressure.*
- *Fig. #52:* *High vacuum chamber machine.*
- *Fig. #53:* *Large chamber high vacuum machine.*
- *Fig. #54:* *Deep penetration electron-beam weld.*
- *Fig. #55:* *Partial vacuum index type bottom load machine.*
- *Fig. #56:* *Automotive part, tang and pin welds.*
- *Fig. #57:* *Non-vacuum local shielding production machine.*
- *Fig. #58:* *X-ray-proof room type production non-vacuum welder.*
- *Fig. #59:* *Die cast aluminum manifold welded on machine of fig. #58.*
- *Fig. #60:* *1" penetration "quickvac" welds.*
- *Fig. #61:* *60KV, 3 KW precision welder.*
- *Fig. #62:* *Evaporator gun with attached crucible.*
- *Fig. #63:* *Use of E.B. in melting.*

Further aided by modern electronics and computer control, these machines also become highly automated and produce repetitive quality welds with reduced skill-level requirements of the operator. These characteristics are likely to lead to over increasing use of this process in high-production environments throughout the world.

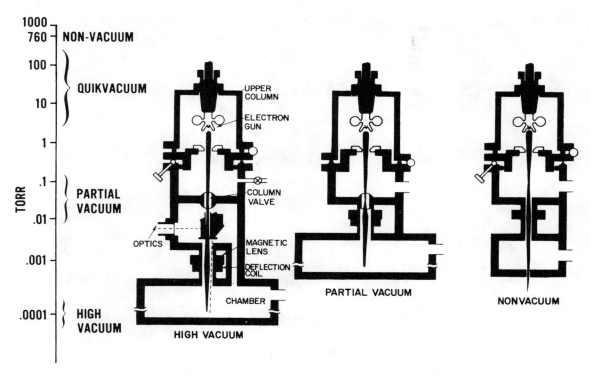

SCHEMATIC OF THE BASIC TYPES OF ELECTRON BEAM SYSTEMS, ALSO SHOWING THE RANGE OF OPERATING PRESSURES AT THE WORK PIECE.

FIGURE 50

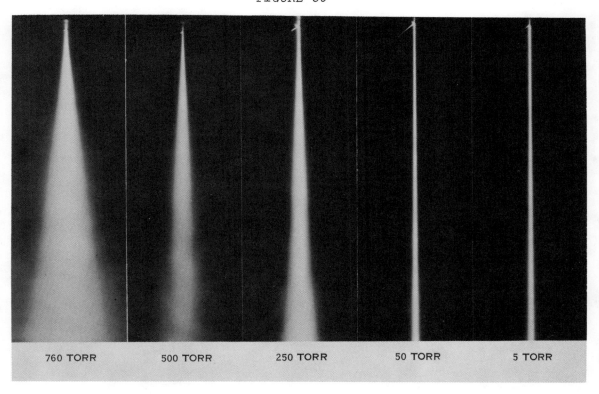

FIGURE 51

FIGURE 52

FIGURE 53

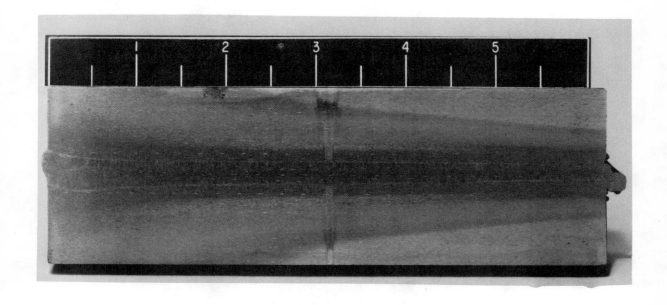

FIGURE 54

FIGURE 55

FIGURE 56

FIGURE 57

FIGURE 58

FIGURE 59

FIGURE 60

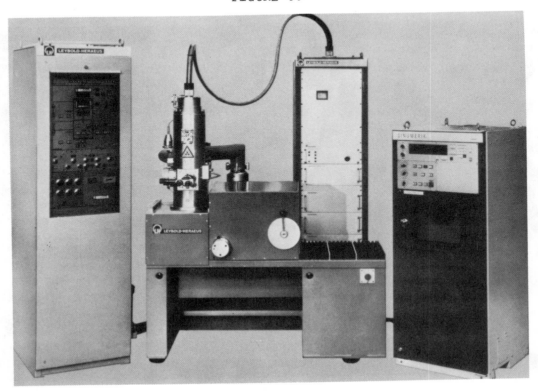

FIGURE 61

FIGURE 62

Practicable Processes

Fig. 2a Drip melting of horizontally fed bars in a continuous casting crucible.
The bars will be pushed over the edge of the crucible by the feeding system and melted by electron beam power. This method allows continuous casting of ingots up to 50 mm diameter and 320 mm length.

Fig. 2b Bath melting of granulated material, sponge etc. by electron beam power.
With a conveyor (accessory) it is possible to feed granules or sponge into the melting pool and to cast continuously ingots of 50 mm diameter and 320 mm length.

Fig. 2c Melting of button samples and bars.
With the straight combination crucible button samples and bars for material tests etc. can be melted. For this process the crucible can be moved along its axis under the electron beam.

Fig. 2d Floating zone melting.
Electron beam density and energy are adjusted such that only a short liquid zone will exist across the rod. By moving the beam generating system the liquid zone travels along the rod achieving remarkable purification. By repeated slow melting, rods of maximum purity and single crystals can be produced.

Fig. 2e Arc melting with the non-consumable electrode.
With a non-consumable electrode, made from tungsten or graphite, the electric arc melts the material in the crucible under inert gas in a pressure range of 50–100 Torr. This allows good comparison between arc melting and electron beam melting.

Fig. 2f Electron beam welding.
With some accessories, welding tests can be carried out.
Max. power of the welding gun: 6 kW,
acceleration voltage: 25 kV,
power density: max. 3×10^8 W/cm².

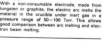
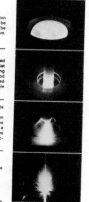

FIGURE 63

ULTRASONIC APPLICATIONS

Ultrasonic bonding is accomplished through specially designed equipment. This equipment produces an output signal in the form of vibrations, at the rate of 18,000 to 20,000 cycles per second. These vibrations are produced through three basic components; the generator, the transducer and the tool (or horn).

The generator or power supply converts 115 volts, 50 or 60 Hz to 20 Hz at various wattages. This vibration is transmitted to the interface of the two materials being welded. The friction caused by these two surfaces rubbing together at 20,000 cycles per second, causes the molecules of both materials to melt and inter-twine to form a bond. This bond is usually 80 to 90% of strength of the base material itself.

EQUIPMENT:

As described earlier, there are three (3) basic components used in bonding ultrasonics., the transducer, the generator and the horn. There may be a booster between the transducer and horn. The horn or booster usually attaches to the transducer by means of a threaded coupling. The horn or tool should be tightened firmly against the transducer, so that energy will not be dissipated through poor coupling.

The generator drives the transducer, by supplying electrical energy to the transducer which is converted to mechanical energy. It is connected to the transducer through electrical coupling, therefore; it may be located away from the transducer. The generator produces various wattages which determines the size of the parts capable of being welded.

The transducer, generator and horn combination, can be supplied in various forms. The conventional method is as a press. This is very similar to a common two hand bench press. The transducer and horn assembly may be mounted so that they are operating continuously, and the parts that are to be welded are passed under them. Also, they can be mounted so the parts to be welded are mechanically brought to them, or they are mechanically brought to the parts. In all cases, the horn or tool must be brought in contact with the parts to achieve a bond.

There are two basic types of transducers, the magnetostrictive and the electrostrictive types. The magnetostrictive transducer is usually made from nickel laminations brazed to a connector body made of monel. These laminations are placed in a magnetic field which induces motion which results in the ultrasonic vibrations.

A magnetostrictive system is usually 50% to 65% efficient.

The electrostrictive transducer uses two or more piezoelectric ceramic wafers. These wafers sandwich a copper washer, and each wafer has a polarity. When excited, the crystals flex producing ultrasonic vibrations. Barium titanate or Titanate zirconate material is usually used for the crystal material. This system is usually 80% to 90% efficient.

WELDING TECHNIQUES:

The method in which the materials are prepared, and the technique of welding the parts will vary according to the materials used, the end use requirements, and sometimes operational flow.

There are three (3) basic techniques used for joining plastic materials. The <u>first</u> technique is called "Plunge Welding". In this technique, the parts to be assembled are placed under the tool or horn. The tool either comes down onto the part, or the part is brought up to make contact with the horn.

In <u>Plunge Welding</u>, there are usually several functions that take place and control the quality of the weld. First, the pressure at which the part is to be clamped between the tool and anvil is determined. <u>Too much</u> pressure may restrict the parts from moving at ultrasonic rates. <u>Too little</u> pressure may not be adequate to hold the parts together to develop proper interaction between the faces of the parts. The rate at which the tool descends upon the parts, plays a minor role. For those parts that cannot be fixtured easily, a fast jerky motion may cause the parts to shift upon impact. A slow descending speed will cause longer production rates.

Another function, is the gap or the stop of the horn travel. Many parts can be welded without using stops to control the horn's travel. Most parts require that the travel be limited to maintain consistency from part to part.

The previous functions mentioned, were mechanical and pneumatic, others are electrical. There are usually three basic <u>timed</u> functions., dwell, weld, and delay or hold. <u>Dwell</u>, is the delay time between the time the horn reaches the part and the time the sonics come on. This insures that the two parts are adequately clamped and aligned prior to sonics, also, this varies directly with the parts being assembled.

The sonic time, of course, is dependent upon the time required to adequately bond the components. The timer for this cycle must be accurate and repeatable, or the quality of the weld will vary.

The <u>dwell or hold</u> function, insures that the parts are properly cooled and bonded before they are moved. With some parts and materials, this is very critical.

<u>The second</u> method of joining parts ultrasonically is called "Scan Welding". This is the process in which the tool scans over the parts. This is usually used when materials are to be joined by the yard on a continuous basis. The material is either fed under the horn, or the horn is passed over the material. The important factors to consider in this process is the tool and anvil gap, the down pressure and the speed of the material moving under, or the speed of the tool moving over the material. There is usually no time cycles involved in this type of welding, the sonics cycle is <u>always on</u>.

The <u>third</u> method, is called the "Sonic Removal" process. This procedure is used when the cross section of the pieces being welded would be critically marred using the plunge welding technique. In this process, the two faces of the parts to be bonded are brought against opposite sides of the vibrating tool. The tool is removed after the prescribed cycle has been reached and the two parts are brought together very rapidly. The parts are allowed to cool while being held together.

JOINT DESIGN:

To insure that the plastic parts are assembled to the optimum strength, they should be designed for ultrasonics. It is best to decide at the inception stage to use ultrasonics so that the joint design can be incorporated as part of the injection molds. Bonding is enhanced when the tool is close to the weld joint. Near field welding, when the tool is positioned ¼ inch or less from the weld joint is preferred when possible. Far field welding, where the tool is more than a ¼ inch from the joint, is more difficult to be accomplished. Of course, the success of far field welding depends upon the material being used, the type of joint, and the ultimate end use of the product.

ENERGY DIRECTOR:

The energy director, the most common type of joint used, is also the easiest to add to the injection molds. It consists of a small triangular projection which focuses the energy to a concentrated area. The director provides a high concentration of energy which results in almost immediate melting and a uniform flow in the joint area.

(See Figure 64)

The size of the energy director should be related to the size of the areas to be joined. Its height should be at least .008 inches, or 1/10 the joint width. The director's width at its base should be at least 20% the joint width. Two or more energy directors should be used on thick-walled joints, when the director height of 1/10 the joint width is greater than .020 inches and the sum of their two heights should equal 1/10 the joint width. The optimum energy director angle at the peak is 90°.

Energy directors are employed in other joint configurations such as, the "step joint" and the "tongue and groove joint". Joint strength, appearance, difficulties to incorporate in mold or part and materials determine the types of joints also. Tongue and groove joints have good strength characteristics, but the need to maintain clearance on the joint sides make it more difficult to mold.

The step joint molds readily, and provides a strong, well aligned joint with a minimum of effort. The fill between joint surfaces should not be so tight that vibratory motion is restricted. Energy director joints are not recommended for use with crystalline materials such as acetal, nylon, and thermoplastic polyester. These materials tend to crystalize before sufficient heat is generated at the weld joint. With the exception of the joint area immediately adjacent to the energy director, only spotty welds will result.

The molten material that becomes exposed to the air will also degrade giving poor welds. With crystalline material having sharply defined melting points, the shear or scarf joints are recommended. These types of joints produce a telescoping action and the heat is contained within the weld area.

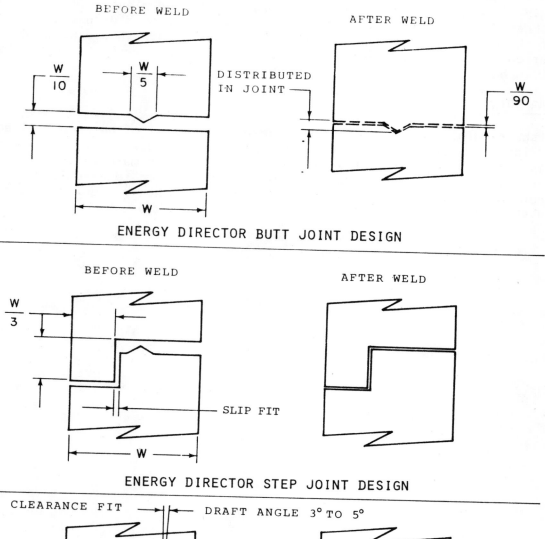

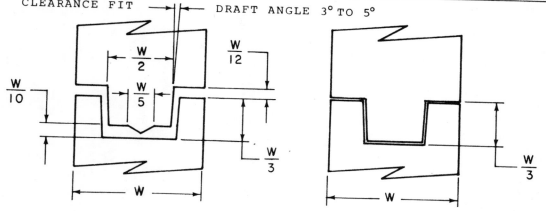

FIGURE 64

SCARF JOINT:

When designing a scarf joint, the dimension of the part should be approximately 6% larger than its counter part on the second piece. If the wall thickness is .025 inches or less, an angle of 60° should be used. If the wall thickness is .060 inches or more, an angle of 30° should be used. Intermediate angles are recommended for wall thickness as between .025 and .060 inches.

SHEAR JOINT:

The shear joint construction uses an interference fit of between 0.01 and 0.02 inches on straight walls. This type of joint usually provides hermetic seals and strengths up to 100% of the strength of the material. Because a shear joint moves larger amounts of material, the time to weld is usually higher than conventional joints. When strength is not a requirement, a bead shear joint may be utilized.

(See Figure 65)

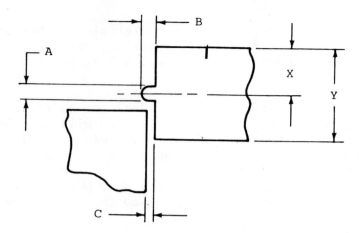

WHERE:
A = 0.01 TO 0.05 IN.
B = 0.01 TO 0.05 IN.
C = 0.0 TO 0.01 IN.
TYPICALLY X = ½Y

BEAD SHEAR DESIGN

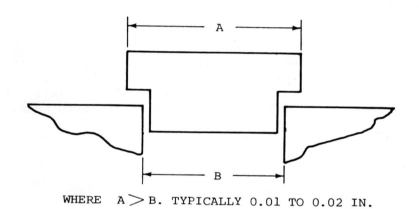

WHERE A > B. TYPICALLY 0.01 TO 0.02 IN.

SHEAR JOINT DESIGN

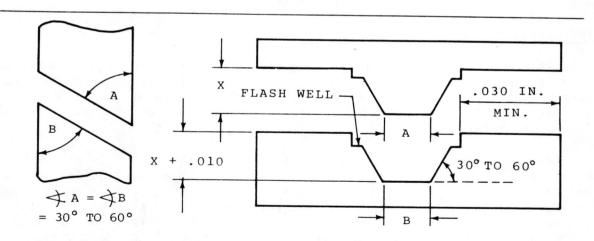

SCARF JOINT DESIGN

MODIFIED SCARF DESIGN

FIGURE 65

MATERIALS:

Ultrasonic bonding requires that at least one of the materials be thermoplastic. Materials such as acrylics, acetyles, polyolefins, polyamids, polycarbonates, polyvinyls, polysulfones, polystyrenes, S.A.N. and many others, may be bonded.

The more rigid materials are usually easier to bond. Many not so rigid materials bond better when mixed with fillers. Some fillers will aid the weldability of the material, while others will hamper it. Glass usually improves the weldability. However, glass will not bond, so therefore, the higher the concentration of glass, the less material is available to be bonded. Glass fillers up to 30% can generally be used, but, because of its abrasive quality, tool wear will increase proportioned to the percentage of glass used.

Mold release on the other hand, will restrict the welding. It works as a lubricant and it reduces the friction that is necessary to be present to produce a good bond. Colorants and lubricants can also alter the weldability of the parts. High rubber content in the material will interfere with its weldability.

Different materials can be welded together if their chemical composition is similar and if their melting points are within $25°F$. Materials that have low melt temperatures, and greater modules of elasticity, co-efficient of friction and thermal conductivity, the easier it is to weld.

Those materials that are hydroscopic such as nylon, polycarbonate, and polysulfone, should be dried prior to welding.

OTHER TECHNIQUES:

There are other methods of bonding, using ultrasonics than the ones previously mentioned. These methods allow for the joining of thermoplastics with non-thermoplastics or even non-plastics.

INSERTIONS:

This is the process if encapsulating metal components into plastic material. A cavity slightly smaller than the metal component is either molded in the part or machined into a non-molded part. The dimension of this cavity is determined by the end use requirement, and in the case of standard inserts, the manufacturers recommendations.

For inserting standard threaded inserts or studs, the insert is placed in a cavity. The cavity is usually tapered part of the way, to guide the part and to allow the molten plastic an area to settle in. There is an interference fit between the insert and the walls of the cavity, again, this depends on the manufacturers recommendations and the desired end use requirements. The insert is then ultrasonically driven into the cavity.

The sonics causes the plastic in contact with the insert to melt and flow around and under the knurles, undercuts and slots on the insert body. These knurles, undercuts and slots determine the pullout and torque strengths.

For instance, deep vertical undercuts, or knurles will yield strong pull out strength.

Certain small cross sectional metal components may be driven without the aid of a hole or cavity, but flash is usually produced by the displaced plastic. Strength can be gained with flat components by designing the part to be inserted with barbs. Assembling in this method usually takes less than a second, and in many cases more than one component can be inserted simultaneously.

Insertion eliminates costly insert molding, or expensive secondary operations to fasten one or more components to a plastic component. Screws, may also be inserted ultrasonically. This is accomplished by driving a screw into an undersized cavity, using an ultrasonic tool (horn). When the screw is removed, there are perfect threads in the plastic that allows the screw to be fastened again. Using ultrasonics, the screw may be inserted and the components attached at the same time, saving secondary operations. Due to the fact that ultrasonics do not create stress in the material, rejects also diminish.

Insertion has become a popular method of joining metal and plastic components, and is used for the following reasons:

1. When two or more components (one of which is plastic), are to be assembled, and sometime in the future may require disassembling, without destroying the plastic.

2. When a threaded hole is required in a plastic part, no drilling or tapping is required.

3. When permanent installation of metal part in plastic is required.

STAKING OR HEADING:

Staking or heading is the process of forming the plastic ultrasonically over, and/or around another object(s) for the purpose of securing the two or more objects together. The other parts that are being secured may or may/not be plastic material. Assembly with this method usually takes less than one second.

One of the most widely used staking techniques, is the use of a plastic boss(es) molded on the part. This boss(es) is used to locate the various parts to be assembled. Ultrasonically, a head is formed on the boss to produce a plastic rivet. This plastic rivet will secure the parts permanently. The parts that are being assembled may be any material. The one carrying the boss, must be __THERMOPLASTIC__. This technique is also a good method for joining dissimilar plastics.

It is obvious that by joining components in this manner, costly fasteners are eliminated. Screws, washers, snap rings and similar hardware are not necessary. Also, the assembly time is much faster than conventional assembly techniques. Staking is used quite frequently in the auto industry for joining instrument panels, grills, and other instruments.

There are two basic rules of thumb for stud size. If a low profile is required, the head resulting from the stake is 1.5 times the diameter of the stud. The stud should extend beyond the top layer by .6 times the diameter prior to staking. *(See Figure 66)*

The second rule is for a standard profile, one in which the finished stud is two times the diameter and its height is .5 times its diameter. This is accomplished when the starting stud extends 1.6 times the diameter above the top layer. By varying this formula, all other values between can be accomplished. The diameter of the stud is determined by the ultimate strength required.

There need not be a stud to accomplish a staking operation. Material may be moved ultrasonically from one part to trap a second component. A tip or horn can be designed to move the material to capture the second component. This allows economic assembly techniques to be used where it is not feasible to use or incorporate bosses for staking. *(See Figure 67)*.

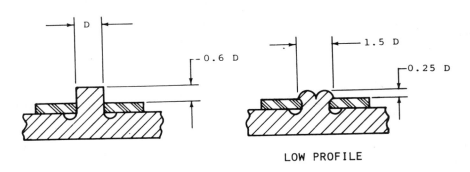

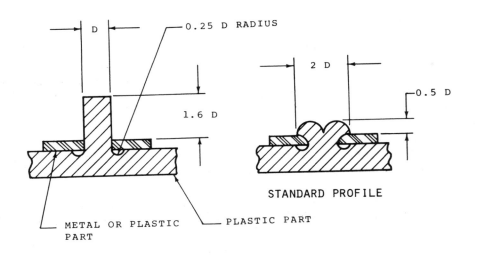

ULTRASONIC STAKING

Figure 66.

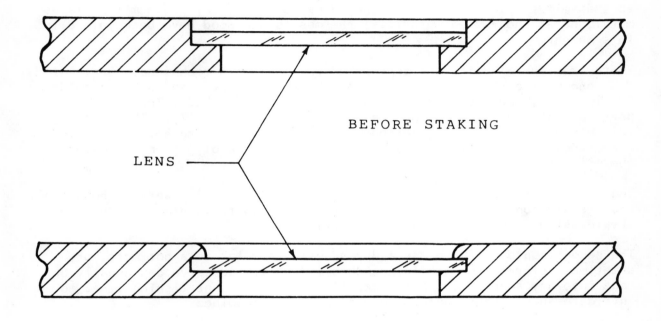

CROSS-SECTION OF A LENS CAPTURED IN
A PLASTIC HOUSING USING STAKING

FIGURE 67

ACTIVATING ADHESIVES:

This is the process of using ultrasonics to re-activate, activate, or cure adhesives. The application of adhesives can be messy, time consuming, and sometimes curing must be done either by heat, or long periods of air drying. Through the use of ultrasonics, this condition can be greatly improved, and a significant reduction in the manufacturing costs can be realized.

The parts being joined do not have to be thermo-plastic. Almost any material can be joined in this manner. The materials being joined must be able to transmit the sonic energy to the adhesive interface. Such materials as metal, paper, non-woven materials, thermosets and wood have been joined in this manner.

There are several different approaches to using this technique. One, is that one part or both are pre-coated with the adhesives and allowed to dry. For instance, the ribbon on the lamp shade is pre-coated. Then, the two parts are brought together and re-activated ultrasonically. The advantage this has over conventional methods, is that the parts are not wet, therefore, they can be controlled and there is no chance of getting adhesive where it doesn't belong.

The method insures that the heat, pressure, and time necessary to bond the two, remain constant from part to part. These controls are difficult to obtain using conventional manual methods.

The second approach is to use a dry adhesive similar to a hot melt, or sometimes referred to as "linear Adhesive". This adhesive can usually be made into any shape from what looks like a monofiliment fishing line, to intricate shapes that conform to the parts being joined.

The advantage of this method, is that only the adhesive needed is used, there is no waste. The shelf life of this adhesive is usually longer and it's easy to automate. The heat, pressure, and time controls also apply for linear adhesive.

The third technique is to cure adhesives ultrasonically. Certain adhesives that have long curing times set up much faster using ultrasonics. In many cases, curing times of hours, have been reduced to seconds.

CUT AND SEAL:

This is the process of using ultrasonics to cut and/or seal synthetic material. The material subjected to this process may be cut only, or it may also be bonded simultaneously.

In cutting operations, the cutting of synthetics by knives, slitters, etc., generate heat and causes ragged edges. There are some operations that require a secondary bonding of the cut material at the edge, one example is woven rug backing. Using ultrasonics, the material can be cut with a smoother edge than with knives at the rate up to 200 to 300 feet per minute.

SPOT WELDING:

This is the process of ultrasonically bonding two pieces of plastic. The process is performed by placing one piece of plastic upon a second sheet. The ultrasonic horn used for this process usually has a pilot. This pilot will penetrate the top sheet, and into the bottom sheet, to a depth of 1/2 the thickness of the top sheet. This is considered a standard tip, of course, variation of this is possible.

The standard tip produces a head, having a diameter of three times the thickness of the top layer. The length of the protruding tip is one and one half times the thickness of the top layer. A release with a radius of 1/4 the thickness of the top layer has sufficient volume to form the displaced material into a neat clean ring. The outside surface of the bottom sheet remains unmarked.

A weld formed by staking migrates depending upon the material. By placing a series of these stakes close to each other, a hermetic seal can be obtained. In Europe, plastic beer barrels are welded together hermetically in this manner very successfully.

This process allows sheet material and thermo-formed components to be assembled economically. Glues, double back tapes, rivets, screws, etc., can be eliminated by assembling ultrasonically. Many molded parts are also assembled using staking when it is not feasible to install weld beads, posts or other means of joining ultrasonically.

(See Figure 68)

TOOLS (HORNS):

Tools or horns are the part of the ultrasonic system that comes in contact with the pieces to be bonded. The horn is designed to vibrate at its half wave resonant frequency, 20 KHZ is most common. When the tool vibrates, it moves out and in at both ends, this is called amplitude. The center of the tool is called the nodel point. At this point, there is no motion of amplitude of the tool but, the stress concentration at this point is the greatest.

Amplitude is measured as peak to peak displacement of the tool. This motion can be altered by changing the mass distribution of the tool, or, by altering the input amplitude to the tool.

The amplitude at the convertor or transducer is usually fixed. By changing the ratio of the tools output, or gain, as it is called, the amplitude can be increased or decreased within the limits of the tool material being used. The desired amplitude is determined by the shape of the part, the material used, and the type of joint used.

The material that the tools are made from, should have two main characteristics - they should have good acoustical and mechanical properties. Titanium, aluminum, monel and some alloy steel, are common materials used to produce tools. These materials have relatively low acoustical impedance and high fatigue strength.

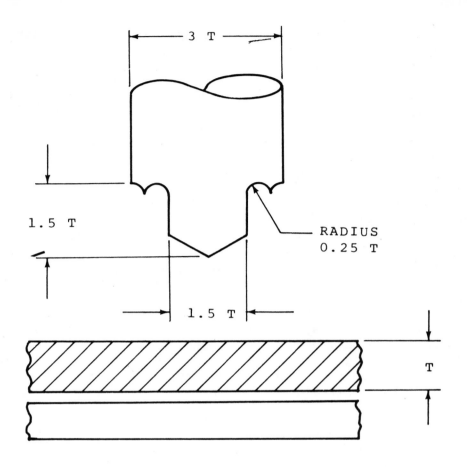

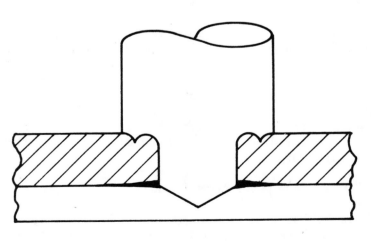

FIGURE 68

The acoustical impedance of a particular tool is derived by the product of.. the materials density, the speed of sound, and the cross sectional area of the tool involved. On the other hand, fatigue strength of the material is determined by its maximum velocity or amplitude that it can utilize without cracking. This velocity is determined by dividing the metal's fatigue strength by the acoustical impedance of the material. Tools with amplitudes up to 2.5 mils generally are made of aluminum, while those with gain above 2.5 mils are usually made from titanium.

Aluminum - is a popular material because the basic price of aluminum is less expensive than either titanium or monel and it is easier to machine. However, because aluminum is much softer, it tends to wear faster in highly abrasive operations. It also has a tendency to impart oxides on the materials it comes in contact with. It is recommended that tools made from aluminum be either chrome plated, or hard coated for longer life and to eliminate oxides.

Titanium - is the best all around material for tools. It costs a bit more, but that's usually absorbed by its longer life. Where high gain is required, titanium must be used.

Monel - is usually used when carbon or another material must be brazed on the tool. Monel will accept brazing more readily than other tool materials.

Steel - is used for low gain applications where abrasion is a problem, such as cut and seal, or insertion type applications. In many cases, tool life can be extended by the use of a tip. A tip is a threaded portion either steel or titanium, that screws into a horn. The tip may wear, but the tool will remain intact. It is more economical to replace the tip rather than the entire tool. Tips are normally used for staking, spot welding, and insertion type applications.

TOOL SHAPES:

Tools have various shapes depending upon the material used, the gain required, and the parts being bonded. There are 6 basic shapes as follows:

1. Stepped Tool or Double Cylinder Tool:

    This design consists of a tool with two different sections, each having uniform cross sectional areas. The transmition from one section to another is made near the nodel point by a small radius. Stepped tools provide the greatest stroke magnification ratio with the smallest changes in cross-section areas. It also has the highest peak stress during vibration. Monel or aluminum are conservatively used for strokes up to 3½ mils, titanium up to 6 mils.

    This tool can be used for any high gain application including welding, staking and swaging.

    Large diameters require longitudinal slots to inhibit other modes of vibration. These slots introduces stress peaks which limit the vibration stroke to lower volumes.

2. **Exponential Tool:**

   The cross section area of this tool is tapered exponentially between input and output ends. Stress peaks are lower because of more gradual taper, as compared to the stepped cylinder type. Magnification is equal to the ratio of input and output diameters. Monel and aluminum cones are conservatively used for strokes up to 3-½ mils, titanium up to 5 mils. This type of tool is generally used for applications requiring high force and low amplitude, such as insertion.

3. **Catenoidal Tool:**

   This tool's cross-section is tapered with a catenoidal variation between the input and output ends. For a given diameter ratio, it has a magnification greater than for the exponential tool, but less than the stepped cylinder. These tools are most suitable for welding and staking.

4. **Conical or Linear Tool:**

   This tool's cross-sectional area tapers linearly from input to output ends. This type of tool is generally used for low magnification, up to about 2 mils. These tools have low stress and low gain. They are generally used for applications requiring low amplitude and very high force.

5. **Cylindrical or Annular Tools:**

   Cylindrical or Annular tools can have a tapered cross-section which is hollow at one end. Hollow tools are generally used for parts requiring circumference contact only. These tools may also be solid, if their diameter is over 4 inches, they usually have slots to break up radial stresses.

6. **Rectangular, Blade or Bar Tools:**

   Rectangular tools have cross sections which may be tapered similar to stepped tools. The blades are stepped at the nodial point. These tools may be made from either bar or round stock. When a blade is over 3.5 inches in width, longitudinal slots are required. This type of tool has magnification equal to the ratios of input to output areas. These tools are used for bonding, multi-staking, multi-insertion, cut and seal, and a multitude of applications.

Tools must be installed on the transducer with the proper torque as recommended by the manufacturer of the equipment. The tool should fit good against the transducer body and there should be a minimum of slop between the threaded stud and its mating cavity. Tips must be torqued up per specifications. Good coupling is very important, it protects the system and insures that the application be done properly and consistently.

(See Figure 69)

SUMMATION:

This has been a very brief description of the use of ultrasonics for bonding. It has not meant to be all encompassing. Each day, we find new and unique applications for ultrasonics. Besides all the new and exciting applications in the Manufacturing environment, we have found numerous applications in the Dental and Medical fields.

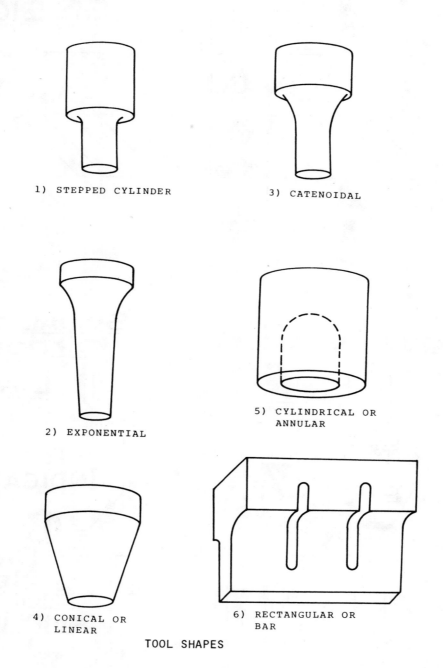

FIGURE 69

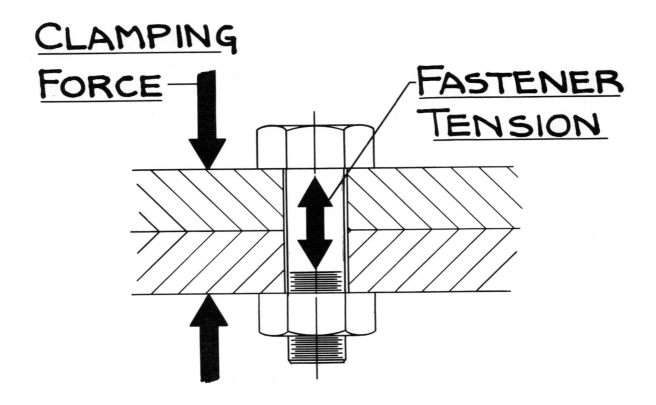

FIGURE 70

FIGURE 71

Tensioning Threaded Fasteners

Tightening operations performed on a threaded fastener are essentially for the purpose of tensioning the fastener to provide a clamping force on the members of the assembly being joined. A number of tightening techniques are available, but they vary greatly in their inherent ability to tension the fastener properly or to provide for adequate monitoring to indicate that the desired tension has been achieved. These techniques range from simple static torque control, to dynamic torque monitoring and control, through bolt-stretch control, and, ultimately, to systems that dynamically sense the actual tension of the fastener as it is tightened and automatically cease rotation when the desired tension is reached. Advantages and limitations of each technique are discussed as to achieving tensioning reliability and monitorability while obtaining maximum performance from a given threaded fastener.

INTRODUCTION

Nearly all tightening operations performed on threaded fasteners have a single primary objective. That is tensioning the fastener to some specific value that assures the proper clamping force on the parts being joined in the assembly. *(fig. 70)*. It is important -- even critical -- for this force to be carefully controlled to assure proper operation, integrity and long life of the assembly. If a joint is too loose, the assembly will fall apart or operate erratically. If it is too tight, the mating parts may be damaged or the fastener can fail, either during assembly or -- of even greater concern -- after a few stress cycles in operation.

The traditional paramter used to control and monitor threaded fastener clamping force has been the torque required to tighten the fastener. Although convenient and useful in many cases, torque-related controls do have limitations -- because torque is only an indirect indicator of the fastener tension.

Torque indications are often confounded by misleading variables -- like screw-thread and bolt-head friction, burrs, dirt, condition of the members being joined, presence or absence of lubricants, etc. It is not our intention here to dismiss torque indications as entirely impractical, since they are often adequate in many applications where the required clamping force is not critical. However, many applications require more positive methods where the actual tension on the fastener is the only production parameter that can assure reliable clamping. Two ways of achieving this are: (1) By controlling and determining the finite amount of bolt stretch or strain achieved in the tightening operation. Strain is, of course, directly proportional to tension among fasteners of the same size and material, within the elastic limit. (2) By controlling and monitoring the fastener stress or tension directly.

Although relatively new, both of these methods are available today for production operations. And there's nothing fancy or exotic about them. They both rely on engineering principles and hardware that have served us for years.

In the following sections we will take a close look at the advantages, limitations and relative reliability of the various fastener tensioning techniques commercially available to industry. The discussion will include:

1. Impact wrenches

2. Stall-type torquing motors

3. Stall motors with integral torque monitoring

4. Stall motors with dynamic torque feedback to shut off the tool at a predetermined torque level

5. "Turn of the nut" techniques where the bolt is stretched a specific distance by controlling degrees of tightening rotation after a threshold torque is reached

6. Direct fastener tensioning methods where the yield point of the fastener is automatically sensed and achieved

IMPACT WRENCHES

Impact wrenches that rely on a present maximum repetitive torque to tighten the fastener are the least reliable of all tensioning devices. The dynamics of the torquing action are subject to extreme variation, and the accepted practice of verifying tightened fastener torque has been with a manual torque wrench, which is not a true indication because it includes stick-slip friction at the threads and under the bolt head. This "breakaway" torque required to restart the fastener usually varies considerably from the true or dynamic torque that was required to tighten the fastener in the first place. The hand wrench technique also is entirely dependent on the skill and judgement of the person using the tool -- a highly variable situation from mechanic to mechanic and even from time to time in the same individual. In conclusion, impact wrenches should only be used in the most uncritical situations where a torque scatter up to $\pm 25\%$ is not objectionable and where reliable verification is not important.

STALL-TYPE TORQUING DEVICES

Stall-type torquing devices are far superior to the impact wrenches we have just discussed. A wide variety of high-quality air-powered tools are available that will deliver a torque repeatability within $\pm 10\%$ -- providing the tool is properly lubricated and maintained and that the air supply doesn't fluctuate appreciably. The output torque of these tools is normally infinitely adjustable within limits.

A limiting factor in using these tools is the reliance on torque as a measure of fastener tensioning. However, they are satisfactory for many applications and have a reliability of $\pm 10\%$ torque scatter. Unfortunately, the added unreliability of static checking with a manual torque wrench makes it impractical to be confident that this degree of control can actually be achieved in production. Design tolerance of ± 15 to 20% would be more

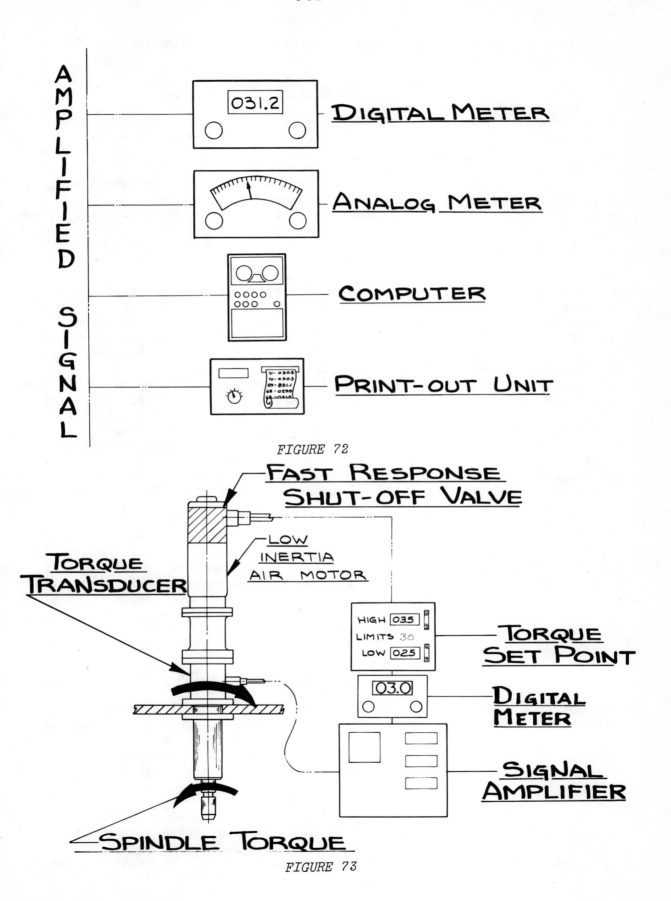

FIGURE 72

FIGURE 73

realistic using stall-type torquing motors with static torque checking.

TORQUE-MONITORED STALL MOTORS

If we take the same stall-type torquing device just described and add a transducer and necessary electronics, we have a system that can measure torque dynamically. The torquing tool retains its $\pm 10\%$ scatter, but the verification has now become extremely accurate. Dynamic monitoring eliminates the errors inherent in a hand torque wrench; no inspector as such is required; the system makes the same "judgement" every time; and torque is monitored during actual rundown of the fastener -- the most accurate torque indication possible. Furthermore, the torque output signals can be channeled to part marking, automatic reject or other control warning systems.

Figure 71 shows a typical electronic torque inspection system. "A" is the reaction type strain gage torque transducer. It is called "reaction type" because it senses the reaction torque on the air motor housing. The reaction torque is equal and opposite to the torque applied to the spindle and fastener. The signal is amplified by circuit card "B", then fed into comparator circuitry "C" to verify that peak dynamic torque falls within the acceptable torque range. Combining the comparator output signals with logic, "low torque", "high torque" and "all spindles to torque" signals can be provided. High-level analog torque output signals are available for use in several ways. Some examples are to drive a peak-reading analog, or digital-meter, print-out or computer-data handling system *(fig. 72)*. The output signal can also initiate part marking systems, automation reject, flag setting mechanisms, etc. The point here is that once you have a system capable of providing a torque level signal for every fastener in an assembly, appropriate interfacing with the assembly operation provides a comprehensive torque assurance system.

The limitations on this fastener tensioning technique are that the governing parameter is still torque and, of course, the stall motor still has an inherent torque scatter of $\pm 10\%$. But at least we now know exactly what the actual torque being developed is and can make appropriate adjustments without interfering with production. We can also reduce overdesign in many cases because of the high degree of assurance dynamic torque monitoring provides.

TORQUE FEEDBACK TO CONTROL STALL MOTORS

Why not go one step further and feed the torque monitoring signal back to the motor, and when a predetermined torque has been achieved, automatically stop fastener tightening? The torque control system seen in *figure 73* does exactly that, and many production assembly operations have employed this basic equipment in single- and multiple-spindle set-ups over the past few years.

The objective was to develop a system that would reduce torque scatter to $\pm 5\%$ or better using the proven stall motor and monitoring transducer, coupled with a shut-down device. The order of dynamic response required for such a system obviously has to be extremely fast. By examining torque

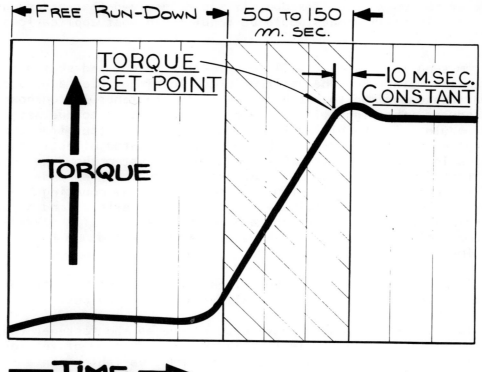

FIGURE 74

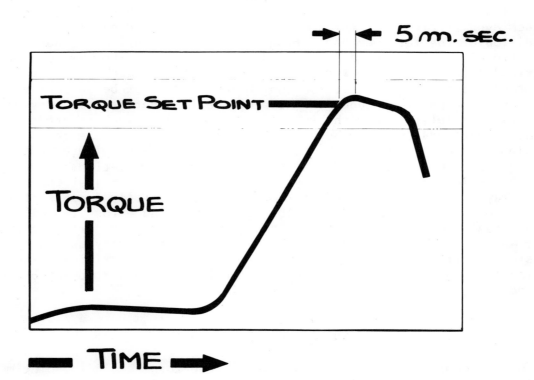

FIGURE 75

vs. time traces on an oscilliscope, it was determined that a response time of less than 10 milliseconds would be required in most applications. *(See fig. 74)*

The system seen in *figure 73* will both monitor torque and shut off the air supply to the motor when the desired torque level has been achieved. The reaction transducer monitors the torque as the fastener is run down. From the transducer, the signal is amplified and fed into a torque set-point comparator module. At the dialed-in torque level, the output from the comparator energizes an ultra-fast air shut-off valve mounted directly to the back of the low-inertia air motor. The valve in the system shown has a response of less than 5 milliseconds. By contrast, the typical response time of commercially available valves is four times as long. By mounting the valve directly to the motor to eliminate any accumulator effect, and by using a low-inertia motor, the torque increase to the fastener will cease 5 to 6 milliseconds after the shut-off valve has been signaled. A typical torque-versus-time trace is shown in *figure 75*.

In addition to shutting off the motor, the transducer torque signal is also used for monitoring and actuating quality assurance devices.

All previous systems discussed have used torque as the fastener tensioning parameter, yet torque in itself is only an approximation of tension. To review briefly, the reason for this is that many variables can affect the torque/tension relationship. These include variations in fastener hardness, friction between the head of the fastener and the parts being clamped, and the friction in the threads. Low hardness levels coupled with low friction can bring a bolt to failure -- or near failure -- before the required torque is achieved. Frictional influences can vary widely, depending on the surface condition of the fastener and mating surfaces and lubricants that are either intentionally or accidentally applied to those areas. For example, residual cutting fluids, presence of rust or other corrosion, coatings on the fastener to prevent corrosion or friction (which may or may not be uniformly applied), and surface conditions under the head of the fastener are all very critical. This is especially true with high-strength fasteners such as grade 5 and grade 8 fasteners used in the automotive industry. In these cases, the unit loading under the head of the fastener often approaches 100,000 psi, and many unusual things take place at these pressures. Sometimes a burnishing effect develops that will greatly reduce friction, or -- just the opposite -- it may cause a seizing effect that will substantially increase friction. When we consider that only about 10% of the torquing effort typically goes into actual tensioning, and the rest is spent on friction, it is easy to see that we are trying to control fastener tension with a very poor grasp on tension itself. So, where clamping force becomes more critical, where high reliability in accurate tensioning is required, there are advanced production methods that are less dependent on friction.

"TURN OF THE NUT" TECHNIQUE

In the "turn of the nut" technique, or "torque turn" as it is sometimes called, the fastener is tightened to a low threshold level, using torque as only a starting parameter *(fig. 76)*. After a predetermined threshold torque is reached -- a value just sufficient to seat the fastener and the

TURN OF THE NUT

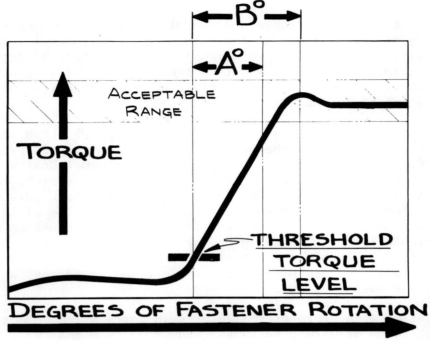

FIGURE 76

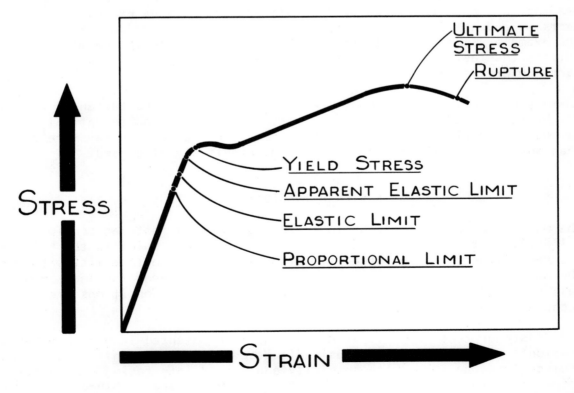

FIGURE 77

mating parts -- the spindle (and fastener) is rotated an additional predetermined number of degrees. The specific amount of angular rotation required after threshold is usually determined experimentally, either by measuring bolt elongation or by strain gaging the bolt.

In rotating the bolt through a predetermined number of degrees, we are using the threads of the fastener as the threads of a micrometer. The bolt thread has a specific pitch, and by rotating it a fixed number of degrees, the thread is going to advance a finite distance. This will stretch the fastener and compress the mating parts a specific amount, producing uniform tension from one joint to the next.

Looking at a classical stress/strain diagram *(fig. 77)*, the characteristic curve remains essentially the same when strain is replaced by rotation *(as in fig. 76)*. Rotation and stress are directly proportional within the elastic range of the fastener. In the "torque turn" system, rotation can be programmed to cease at any point along the stress/strain curve -- at A° of *figure 75*, for example. If the fastener is taken to the yield point, stress (tension) tends to level off and additional rotation can take place without an appreciable increase in clamping force. So by going to the yield point, or past it, there is a plateau region where we have a range of strain (rotation) that provides highly repeatable clamping forces, as long as the fasteners and mating parts are uniform.

The limitation with "turn of the nut" is in parts which do not mate properly because of rough or warped surfaces. Any condition that will cause <u>apparent</u> seating at threshold torque when full seating has not actually been accomplished will affect the reliability of this system. A false threshold reading can cause the subsequent angular rotation to be expended seating the parts -- and not actually stretching the fastener to produce proper tension. For example, in a shallowly tapped hole, the threshold torque might be sensed only as the fastener bottomed in the hole, never actually engaging the mating parts to provide a clamping load.

In many critical applications, where qualified parts and fasteners are used, the "torque turn" technique for bolt tensioning provides an entirely reliable method of assuring proper clamping force. It is certainly far more reliable than any of the torque-dependent systems previously discussed.

DIRECT FASTENER TENSIONING TECHNIQUES

The most advanced development in achieving threaded fastener tensioning reliability utilizes the actual elastic-limit stress point in the bolt as its control of clamping force. It also has the inherent abilities to obtain the maximum possible performance from any given threaded fastener -- and to tell us if the fastener, mating parts or other joint conditions are within acceptable quality limits.

One "fastener tensioning" system that has been patented incorporates a conventional air motor and previously discussed electronic monitoring and control devices, together with specially engineered interfacing components *(fig. 78)*. The unique way in which these elements are combined and the roles they are assigned give the system its advanced capabilities. The system is

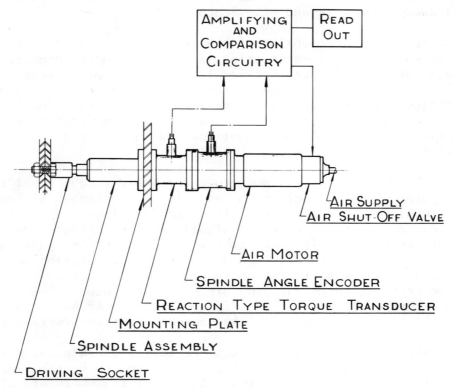

FIGURE 78

FASTENER TENSIONING

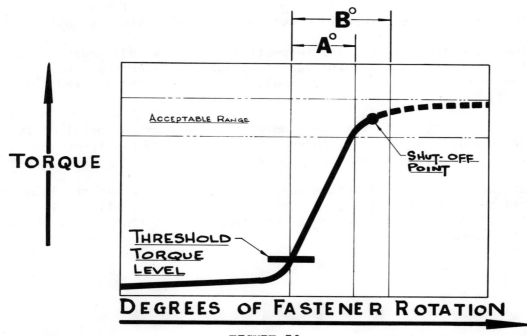

FIGURE 79

therefore evolutional rather than developmental, since it is based on existing hardware, software and experience.

The "fastener tensioning" technique is basically quite simple. As the fastener is rotated by the air motor, signals from a standard angular position encoder and a torque transducer are fed into an integrating electronic circuit which senses initial yielding of the threaded fastener by detecting a change in torque rate proportional to degrees of rotation. The instant yield occurs, and torque no longer increases at a rate proportional to rotation *(fig. 79)*, the high-speed shut-down valve is energized to stop the air motor from rotating.

In practice, the fastener is run down to a threshold torque level that assures the joint is solidly clamped. As rotation continues without interruption, a signal from the angular position encoder (mounted on the air motor spindle) is integrated with the signal from the spindle transducer. Associated electronic circuitry constantly compares the increase in rotation to the increase in torque. When an increment of rotation occurs with a corresponding incremental torque change that is a predetermined amount less than the previous increment, the air motor shut-down valve is actuated and stops any further rotation. In effect, the fastener has been tightened to a tension level corresponding to the knee of the stress/strain curve.

In order to establish that the bolt stressed to its yield point has not been overtightened and is an acceptable bolt, the system checks the final torque and angular displacement. This action establishes that both are within permitted maximum and minimum values, thereby providing assurance that the bolt, stressed to its elastic limit, meets dimensional and material specifications established by the product designer. The system will also signal a bad fastener -- perhaps an entire lot that has been improperly heat treated, or was incorrectly allocated to the wrong assembly station. Bad mating parts, shallowly tapped holes, etc. are also detected. Burrs, lubricants and other surface variables are effectively negated as tensioning influences.

It is important to note that the inspection system is independent from the control system, and that all the quality assurance interfaces previously discussed in connection with other monitoring systems can be tied into this one as well.

By taking the fastener to its yield point, the maximum potential of the fastener is utilized. In many cases, this could mean specifying a less expensive, lower grade or smaller size fastener.

We should mention that the system need not necessarily clamp the joint at the yield point of the fastener. There are means to establish a lower point on the stress/strain curve as the clamping criterion -- say 90% of yield, for example.

CHAPTER 9

Inspection Functions

Inspection has always been an important factor in the productivity of any assembly process. The physical characteristics of the components determine the acceptability of the completed assembly. Incorrect components, shortage of components, damaged components, components which will not fit together, plague every assembly operation.

In manual systems, the inspection is accomplished by the (until now) ultimate inspection device....Man. In automatic assembly systems, that key manual inspector is not present, and this function must be filled by automatic inspection.

Assembly system inspection functions can be divided into four general groups. In ascending order of sophistication they would be:

1. Qualifying inspections checking for rough parameters of shape, size, orientation, color, function, etc. Usually these inspections are made prior to putting the part into the assembly system where they could cause equipment stoppage, jamups, etc. Situated immediately before and independent of the assembly system, or in feed tracks having provisions for removing the undesired parts, they can eliminate the potential malfunction before it costs productive time and other losses.

2. Probing for parts presence, position, orientation, function or security of fastening, usually to a dimensional accuracy on the order of .015 is used to verify previous assembly functions and set up memory for rejects and future functions.

3. Gaging for dimensional integrity includes accuracies down to light wave lengths. Other gaging functions include hardness, force, pressure, velocity, light, displacement, torque, etc.

4. Functional Testing of 2 or more parameters which are interrelated in the assembly acceptability has become highly sophisticated. Selective assembly may require several characteristics, each individually within tolerance. Functional tests involving flow, pressure and stroke are another example. Finally, leak testing by many techniques, fits this category.

Virtually every kind of inspection technique has been automated from the simplest profile gage on a conveyor line to computerized, complete engine performance testing. The massive flow of data possible with today's inspection components challenges the peripheral equipment capability to process, interpret and print out the results. Some of the basics and a few sophisticated examples follow:

All automatic gaging, measuring or testing can usually be classified as digital or analog types. The simplest are devices such as the limit switch, for position; the electric eyes, for part presence; and pressure switches. They all are the type which provide a contact closure or simple "yes-no" type answer. Either the switch closes or it does not. They are simple, repeatable, reliable, quick responding and generally economical.

Little difficulty is encountered in setting up, repair, or replacement of this type of digital gage.

The analog types are used to convert some physical parameter into usable electronic output. The transducer sense such parameters as displacement, velocity, light, force, or pressure; and converts them to electronic signals. The transducer signal is then usually converted by an amplifier into a usable voltage or current output to drive a classifier, output meter, or input to the assembly machine's logic. Many times the analog signals are converted into digital "yes-no" signals.

Within the last twenty years much of the instrumentation has evolved from the laboratory to the production floor. Many of these devices formerly required skilled technicians, frequent calibration and a protected environment. The space age has developed many of these systems into reliable, linear, rugged measuring devices. The aerospace work has also contributed much to the telemetry of the data and interfacing these systems to computers.

There are some devices used in assembly systems which are not exactly digital or analog. Among these are the lasers. Generally, they use the interference fringes and count them as a function of wave lengths of light. This count is then summed up and displayed as a digital readout of length or computer input.

The laser has the unique property of needing little or no calibration. The velocity of light (and wave length) vary only slightly in air due to changes in atmospheric pressure, humidity and temperature. For more industrial applications, these result in errors so small (millionths of an inch) they can be neglected.

There are many considerations in applying these systems to assembly machines. Will the system measure what is required? Is it linear over the tolerance of the part parameter? Is it sensitive enough? These are the common measurement system parameters of resolution, linearity. There are other process requirements to consider in selecting the system.

What about response time and the production rate? Overload protection? Maintenance requirements -- What wears out how often? - How easy to repair? - Cost of repair? Is the system rugged enough to survive in its operating environment? What are the calibration requirements? Are there toxic or radiation hazards? Sometimes these questions have been answered most economically by setting up a prototype station in the existing assembly line. This may also provide a good training device for maintenance personnel and operators.

An automatic machine receiving incorrect parts will produce reject parts cycle after cycle, hour after hour, all day until stopped. This is not zero productivity; this is negative productivity. Bad parts whether oversized, undersized, or damaged, usually cause misloads and jams in automatic equipment. A jammed feeder track can starve a machine and without adequate inspection following the loading operation, missing key components will not be detected. Therefore, a major factor in inspecting any component is to check the part presence in the assembly as close to the loading operation as possible. This is necessary because if other components are to be loaded over the missing one, they very often will cover up the fact that the first one is missing.

This type of inspection of incoming parts, parts coming to the machine, to determine their shape, orientation, color, size, function, is the first step of adding inspection to an automatic assembly process. However, the key to improving productivity is involved with the decision as to what action is to be taken when an inspection is not passed. Once again, at its simplest, the process can be stopped and the man alerted to the problem. He can then make a decision as to what to do with the faulty component. However, this slows up the entire assembly process and the productivity suffers. The technique that is widely employed is to isolate the faulty partial assembly and to add no further components to it. This partial assembly is then isolated in a reject chute or location somewhere on the machine. This allows for the part to be disassembled and its components recycled through the machine. In this way, the machine cycle remains constant with only one cycle of time lost to a reject assembly.

Let's examine in detail some options available when inspection identified an unacceptable component or assembly. As already discussed, the machine can remember the reject condition and isolate the bad assembly. The bad assembly can then be reworked or discarded as justified by the cost of the components versus the cost to repair.

From a machine control standpoint, this requires machine memory to be added. The machine must know that a part at a given station is good or reject before it works on the part. From a mechanical standpoint, this requires that the stations can be "locked out" and prevented from operating on reject assemblies. It also requires a dual unload system for segregating good assemblies from rejects. Thus the cost of the inspection may be minimal within itself, but the cost to maintain a high level of productivity and still accommodate rejects can be substantial.

To aid in diagnosing a major problem, inspection stations are often given a "sequential reject count and stop system". In this way, if tooling is broken or a part feeding track is jammed and good assemblies cannot be produced without manual intervention, the machine operator can be alerted. If three or any selected number of consecutive rejects are created, the machine is stopped and the operator alerted through the use of a warning light.

The discussion has been limited to inspection to insure that a part that should have been loaded is in fact in the assembly. Inspection of this type is common in most assembly systems.

Another problem is that associated with parts of the wrong size, shape or orientation finding their way into the feeding or loading system. When there is serious danger of this occuring, steps must be taken to prevent the part from being loaded. This is especially true if the faulty part affects the final function of the assembly. Finding this condition at final function testing can be very expensive. Not finding it until the part fails in the field can be extremely serious.

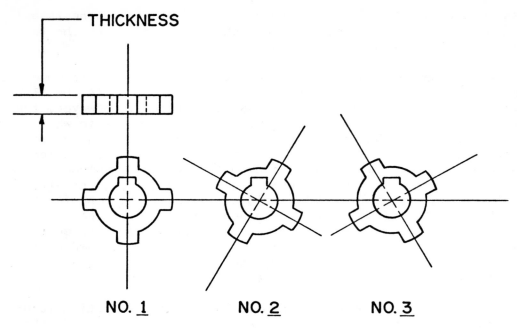

GEAR TYPES

FIG. 1

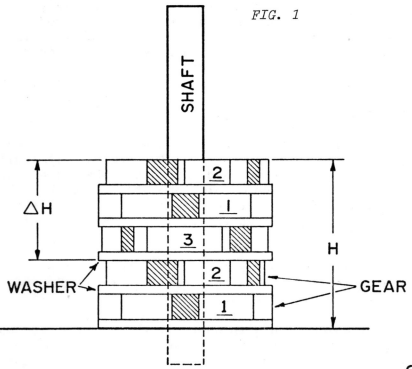

STACK OF ALTERNATE GEARS AND WASHERS

FIG. 2

There are many ways of determining the condition of the parts as they are being fed to the machine. A check which is used includes forcing the part through a profile check and determining the orientation, size and shape. In other systems, gages are installed on feeding tracks to insure that the parts have exactly the correct size. In this way, most reject parts can be prevented from entering the machine feeding system at all.

If on occasion bad parts are put into the assembly and are detected later, a number of ways can be utilized to handle this condition. On non-synchronous machines, a repair loop can be introduced so that the part is diverted from the normal assembly process, manually repaired at an operator's pace on a repair loop, and re-introduced into this system.

The most sophisticated systems require a kind of selective assembly in order to produce satisfactory finished assemblies. Selective assembly is required when the parts making up an assembly must be matched, dimensionally or functionally, to produce a quality final assembly. For instance, a ten thousandths inch washer may fit into a given housing, whereas another housing may require a twelve thousandths inch washer. In this case, both the housing and the washers must be gaged to a high level of accuracy in order to make the desired selection. The following example demonstrates the flexibility of such a selective assembly system.

EXAMPLE OF SELECTIVE ASSEMBLY

The mechanism produced on this selective assembly system is a tightly dimensioned drive gear. The assembly, which can be produced in three different lengths, is made up of alternate gears and spacers to produce a kind of spiral gear drive. *(See Figures 1, 2, 3 & 4)*

In manual assembly, a Gear #1, referring to *Figure 1*, is loaded to a shaft and gaged for thickness. A decision is made as to whether a five thousandths or a seven thousandths washer is added in order to maintain the tolerance on the length of the assembly. The correct washer is added and Gear #2 is placed into the assembly. The assembly is again gaged and a decision is made as to whether a five or a seven thousandths washer is necessary and introduced into the system. Finally, Gear #3 is put onto the assembly and gaged and the same decision made as to the thickness of the washer required. This process is repeated time and time again up to 150 times to produce a necessary gear train. This particular part must be held to very close tolerances both for the overall length of the stack and the incremental length of the stack. *(Figure 2)*

As a manual assembly process, this is a slow, tedious assembly to work on. The necessity of constant gaging and selecting of components produces a high level of tedium to the operator and mistakes are very common. Even with an experienced operator, a complete assembly may take as much as a half an hour to produce. More serious than the time required to produce the assembly is the reject rate which approached 50 percent.

The supplier proposed and provided a two machine system for producing this assembly. The first machine inspected the gears and loaded them into magazines according to the orientation of the key to the gear teeth *(Figure 1)*. The thickness was determined by a LVDT gage and parts that

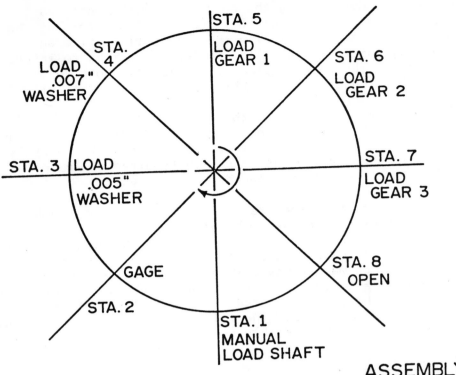

FIG. 3

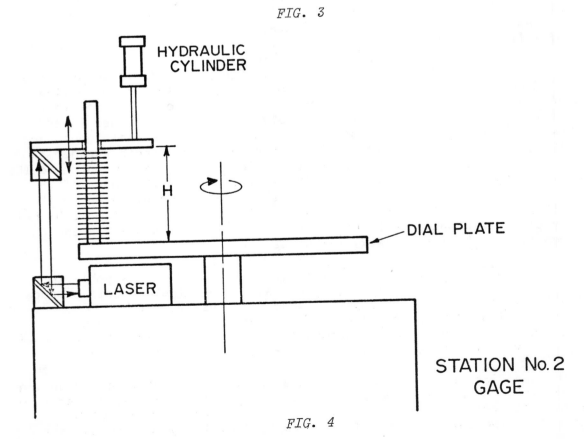

FIG. 4

exceeded the tolerance were automatically removed from the system. The acceptable gears were subsequently indexed to other stations on the machine where they were loaded into the correct magazine for the key gear orientation (Figure 2). These magazines of pre-inspected and selected gears were then available for loading into the final assembly machine.

The sequence of the final assembly machine is shown in *Figure 3*. An empty shaft was loaded at Station one and subsequently indexed to Station two, where it was gaged with a laser interferometer. A decision was made as to what subsequent component should be loaded, either a five thousandths or a seven thousandths washer, or Gear one, two, or three. The correct washer was loaded to the assembly, then Gear #1 was loaded to the assembly. The part was subsequently indexed around the machine back around to the gage. The decision was made as to whether a five or a seven thousandths washer was required on this pass. The correct washer was then loaded, indexed again to the gage where the decision was again made as to what washer to load. The part was then indexed to Station seven and Gear #3 was loaded. This process was continued until the complete stack up of parts had been loaded to the shaft. The laser interferometer and the press, which pressed the components to a final height, is shown in *Figure 4*.

The mini-computer controlling this system, interfaced to the laser interferometer and to all of the load stations, provided a number of features which resulted in a very productive machine. The first thing that the computer did when an empty shaft was loaded at Station one and subsequently indexed in station two, was to determine if the zero reference for that particular fixture and dial plate location was correct. Each fixture location on the dial plate was given a reference zero dimension and was checked every time a new shaft was put into the system. In this way, the computer could ascertain that the shaft was in fact empty.

Because the computer continually knew what the actual height of the stack should be for a given number of gears and washers, a partial assembly could be introduced into the system at any time. In this way, a partial assembly could be introduced into Station one, indexed to the gage station, and the gage station could be used to determine what was the next correct component to put into the assembly.

The computer monitored both the overall stack height and the incremental height of the last three components going into the system. If a faulty component was introduced into the system and the overall tolerance or the incremental tolerance was not maintained, the computer printed out a message on the Teletype terminal and alerted the operator through the use of a red light. The machine did no further work on that assembly. At his leisure, the operator could return to the machine, ascertain that the part in fixture "X" was not being worked on, check the Teletype message which included the dimensions of the last three gaging operations, and determine how many parts to remove from the assembly. The operator then stopped the machine and removed the offending components, initiated the cycle by telling the computer that this was now a partial assembly, and the assembly was re-gaged and reinitialized into the system. The operator could remove only a few parts or could remove a large number of parts and the computer would always reinitialize at the gage station and pick up the right sequence to make a satisfactory assembly.

In the same way, when the assembly reached its final complete length, the computer printed out (<u>Fixture "X" assembly complete</u>) and alerted the operator with a horn. If the operator did not respond immediately, the part continued to cycle through the machine with no further work being performed on that assembly.

The operator could allow the machine to cycle until all fixtures had completed assemblies or he could immediately remove the one completed assembly and load an empty shaft and tell the computer it was now looking at an empty assembly.

Other features on the machine through the computer control, include a three consecutive reject count and stop. If a station made three consecutive rejects, then the machine was stopped and the operator alerted. This was a rather complicated feature to add because the computer had to remember that it was supposed to load Gear #1 and that it failed to load Gear #1 in order to produce a consecutive reject count feature.

In the same way, the computer had a "try-again" feature on rejects. If a given fixture required a Gear #2 and did not receive it, the computer would direct the machine to "try again" to load Gear #2. It would try to do this three times and if it failed to get Gear #2 after three tries, it would assume that there was something wrong with the shaft or with the assembly and stop work on the assembly.

When the computer received a signal that the part to be gaged was a new assembly or a partial assembly, the computer advanced the press slowly because it had no way of knowing how long a stack it was looking for. It pressed into the stack, stalled out and then referenced that dimension. The stalling of the cylinder was determined through the laser interferometer. The computer read the stack height and when three consecutive readings were determined over a period of 50 milliseconds, the computer determined that the press had been stalled.

One of the features available with computer control is production data. The computer on this machine printed out "on demand" and at the end of the shift, the production data for that period of time. Included in the report is the time of day and the number of cycles run, the number completed assemblies, the number of stoppages due to problems within the machine, etc.

This type of selective assembly system would be impossible to implement without computer control and when the computer is introduced in this system, many side benefits are realized.

<u>ADVANTAGES OF SELECTIVE ASSEMBLY</u>

The advantages of selective assembly are as follows:

1. A machine can utilize parts with loose machining tolerance.

2. The machine can utilize a process with looser tolerance.

3. The machine can utilize a high percentage of parts with wide range of tolerance.

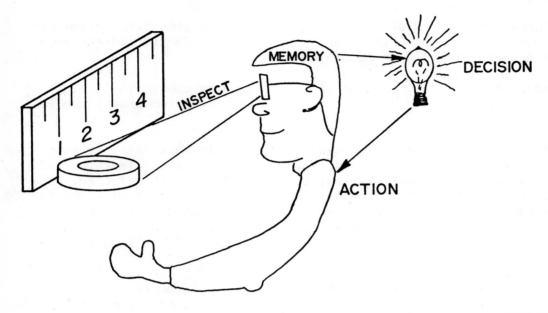

INSPECTION SEQUENCE

FIG. 5

4. Rejects are reduced to a minimum by fitting components together to make adequate assemblies.

The looser machine tolerances of the parts is an important factor because the cost of components is directly proportional to the tolerance required in the manufacturing or machining process.

The selective assembly machine will accept parts coming to it with a wider range of tolerances and therefore the cost of the components can be reduced to a minimum.

Final functional inspection is very important in many assembly processes. If for instance, a shaft is bound up or an assembly leaks, or if it vibrates during functional testing, the assembly must be rejected. The range of functional tests that can be run on assemblies is as wide and diverse as the number of assemblies which can be assembled. A functional test is attractive because it usually tells the condition of a large number of components in the assembly without inspecting each of the single components.

Inspection is rapidly becoming a requirement in automatic assembly. In order to be cost effective; however, inspection must have some characteristics shown in *Figure 5*.

The inspection or gaging operation itself is only as effective as the "skill" of the machine. The machine must be fitted with the best sensors and tooling to make the desired measurements. Just as a manual inspector is only as good as his gages, micrometers and tools, a machine is only as good as the transducers and sensors which it employs. Also, a skilled inspector makes optimum use of his tools. Similarly a machine designed and built by an experienced builder has the "skill" to perform precise inspection operations. Secondly, the inspector must be intelligent enough to make the correct decisions. The requirement for additional machine intelligence has made programmable control very widely used in automatic assembly systems. Programmable controllers, mini-computers, and micro-processors will continue to expand in usage. The buyer of automatic assembly systems should insure that his system is intelligent enough to make thecorrect decision. Thirdly, all of the inspection skill and intelligence is wasted if the required action cannot be accomplished. If the man lacks the physical dexterity, faculty, or support facility to take the required action, then his ability accomplishes nothing. Similarily, if the assembly machine lacks the tooling accuracy, flexibility, or reliability to satisfactorily perform, then precise inspection skill and superior intelligence is useless.

Inspection in automatic assembly can show dramatic increases in productivity by keeping the machine cycling efficiently, by reducing rejects, by providing a means of repairing rejects if they should occur and through selective assembly can reduce the cost by relaxing otherwise stringent components specifications.

The assembly process usually can be reduced to three steps or phases. First, individual piece parts are loaded. Second, these parts may be oriented or fixed in place by subsequent parts or fasteners. Finally, a sub-assembly or final assembly is completed and ready for "functional testing".

During the loading process, piece part inspection or gaging may take place. These processes are dimensional gaging, profile checking, hardness testing, etc. The in-line gaging will either divert the parts not meeting the inspection criteria or use the measurement information for selecting subsequent mating parts (selective assembly).

An example of this type system is shown in *Figure 6*. The wobble plate is part of the shaft sub-assembly. The thrust bearing washers are selected for a fit of the nearest .0005 inch washer. The wobble plate is rotated and the two surfaces are measured. This measurement is used in a computer calculation to match the proper piston, ball and shoe dimensions. The selection of one of 30 shoes of .0001 inch increments assures the best possible assembly. A spin test further downstream assures a good assembly with properly selected parts.

A programmable controller runs the machine. Gross checks are by limit switches while the measurement transducers are LVDT's. A computer is used for solving the equations for the wobble plate and storage of the dimensions required for selective assembly. Machine cycle time is 8.7 seconds, or 413 parts per hour gross.

As a sub-assembly is being completed, the more advanced assembly machines check for proper placement of individual components. Subsequent stations are locked out, or not allowed to load good parts into reject assemblies. This feature almost makes the assembly machine a gage in itself.

Many experienced operators can sense by the frequency and type of reject which vendor or die or mold produced the problem component. Another example of gaging or testing during sub-assembly is leak testing. This type of test varies from flow measurements of pressure difference across a nozzle or orifice, to pressure change as a function of time, to the most sophisticated mass spectrometer.

Figure 7 is an example of an assembly machine employing a mass spectrometer. This particular application required several significant steps prior to building the machine. First, the reliability and maintainability of this equipment had to be demonstrated. This was accomplished by a prototype station set-up and run in the customer's plant - on line and at anticipated cycle time. This approach highlighted some equipment shortcomings such as electrical noise immunity and cross-correlation problems. Most important was the learning curve and confidence level acquired. This machine also incorporates balancing and machining operations. It operates at 10 seconds per cycle or 360 assemblies per hour gross rate.

1. Rollover
2. Gage and load cylinder
3. Separate cylinder
5. Open station
6. Lubricate assembly
7. Open station
9. Load No. 1 piston
10. Orient No. 1 and load No. 2 piston
11. Orient No. 2 and load No. 3 piston
12. Check 3 piston subassemblies
13. Load tube and close cylinder
14. Open station
16. Unload rejects
17. Rollover and wash fixture

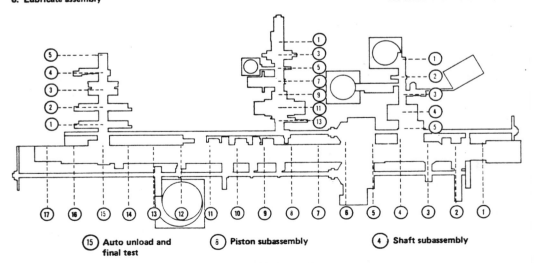

(15) Auto unload and final test
(8) Piston subassembly
(4) Shaft subassembly

FIGURE 6

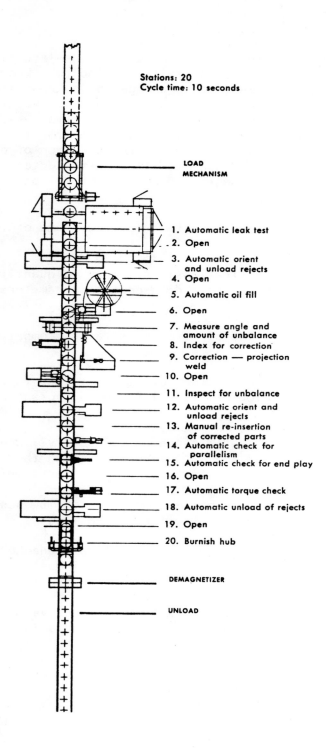

FIGURE 7

Another sophisticated system is shown in *Figure 8*. The part is an automotive master brake cylinder. The testing, in addition to part presence and torque checks, is primarily leak testing. The change in pressure as a function of time is the method used.

A unique feature of the testing stations is self-calibration. Periodically the transducers are automatically coupled to a known volume at atmospheric pressure. The volume is then changed by a known amount causing a calculated pressure change. This pressure change provides a transducer (strain gage) output voltage. The voltage is digitally stored in computer memory for use in calculating equivalent voltages for the high and low test limits. Leak rate specifications are a pressure drop of one inch of water per second. A total of five leak tests and a stroke test is made on each part. This machine cycle time is 7.8 seconds or 461 parts per hour gross rate.

In determining the type of system required, some of the questions which should be asked are:

1. What are the test specifications? Are they in quantitative, scientific units? Do they include part and inspection parameters such as cleanliness, temperature?

2. Do the specifications include time constraints compatible with the assembly cycle times?

3. What output from the inspection device or station is required -- simple accept/reject, or analog or digital for data storage or statistical analysis.

4. What is the measuring system range and linearity? Does it have the resolution required for the part tolerance? Can the system survive overloads and recover without reset or re-calibration?

5. Is the measuring system response time, the tooling positioning, and inspection process time compatible with the machine cycle time?

6. Does the measuring system require special maintenance techniques? Are maintenance personnel qualified for this equipment or is some training required?

7. What about safety? Are toxic gases or radioactive materials involved? Are there high cost consumable gases or elements involved?

8. What about calibration? Are masters required? Do the masters require rechecking? How often is re-calibration required? How long does it take? Who is qualified to do it?

9. Where in the process should the testing take place? What do we do with the rejects?

10. How repeatable is the product's inspection parameters?

Inspection systems, gaging and functional testing are now highly reliable processes which are becoming more and more a part of automatic assembly systems. The degree of sophistication is, as usual, a cost/performance trade-off. The trend is toward the completely automated systems and automatic factory.

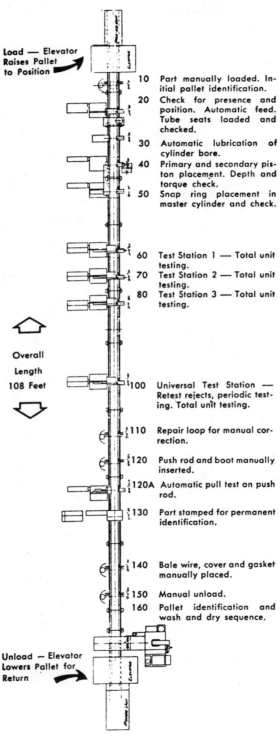

FIGURE 8

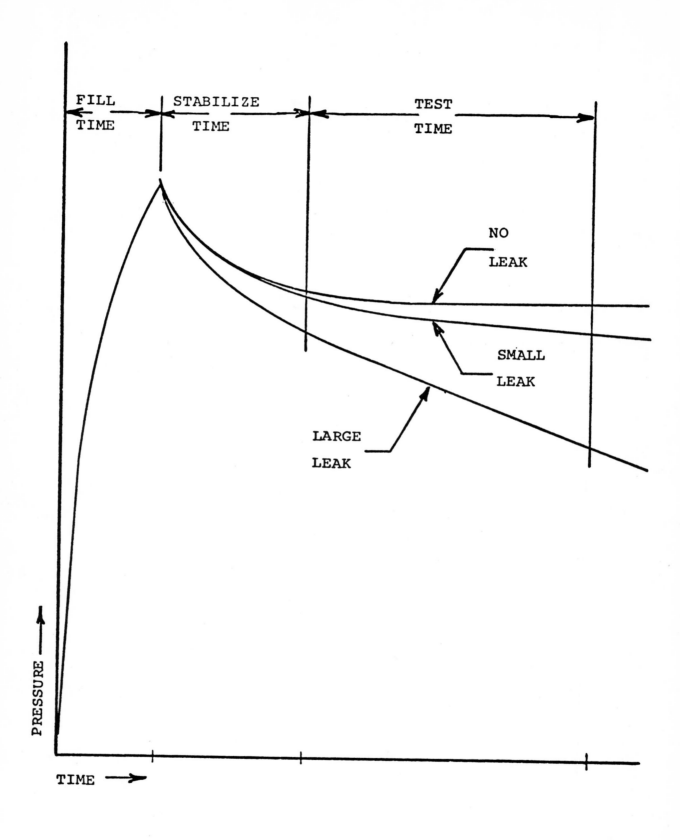

FIGURE 9

LEAK TEST METHODS

Three methods used for leak tests on automatic assembly machines are:

1. Pressure loss or pressure decay,

2. Mass flow meter, and

3. Mass spectrometer.

THE PRESSURE LOSS OR PRESSURE DECAY method uses a sensitive pressure transducer to measure pressure changes in the part being tested. The transducer frequently consists of a strain gage mounted on a diaphram which deflects under pressure.

Pressure loss or pressure decay instrumentation serves for a wide range of tests for leaks with flow rates of .01 cc/min. to 1000 cc/min. A small part with a volume less than 1 cubic inch (an internal passage in a carburetor, for example) can be tested in 1.5 to 2 seconds, while a larger part with a volume of 10 or 20 cubic inches requires from 3 to 10 seconds to test.

Pressure loss or pressure decay methods can be used with pressure above atmospheric, or with vacuum. Test pressures of 10 to 30 psig are commonly used.

When air is used to fill a chamber, some pressure change is experienced due to the expansion of the air and resulting temperature change. This pressure change, as well as changes due to a part arriving at the test station hot from previous operations (welding, etc.) can be nulled out with electronic compensators. The same compensators can also be used to nullify the effects of porous castings soaking up supply air, or of rubber and plastic parts ballooning out as the pressure increases. Typical test cycle conditions are shown in *Figure 9*. Comparative leak rates and detection methods are shown in *Figure 10*.

THE MASS FLOWMETER METHOD utilizes the measurement of pressure drop across an orfice or a hot-cold anemometer, and is generally used for measuring flow rates greater than 1000 cc/min. Either pressure or partial vacuum (down to about 5 psia) may be used. The speed of response of the transducer is generally much faster than the speed at which the test air flow reaches equilibrium, and so test-time becomes dependent upon establishing the required air flow as rapidly as possible. Test cycle times from 2 to 10 seconds are possible on many parts.

THE MASS SPECTROMETER method is used for detecting very small leaks and has a range from .01 cc/min to .000001 cc/min. or smaller. The equipment used for this method is very expensive compared to the other two methods.

The product to be tested is enclosed in an air-tight chamber and a tracer gas such as helium is introduced into the product. The air in the enclosure is pumped out by a vacuum pump and when a high vacuum is reached, a portion of the residual atmosphere is valved into a mass spectrometer

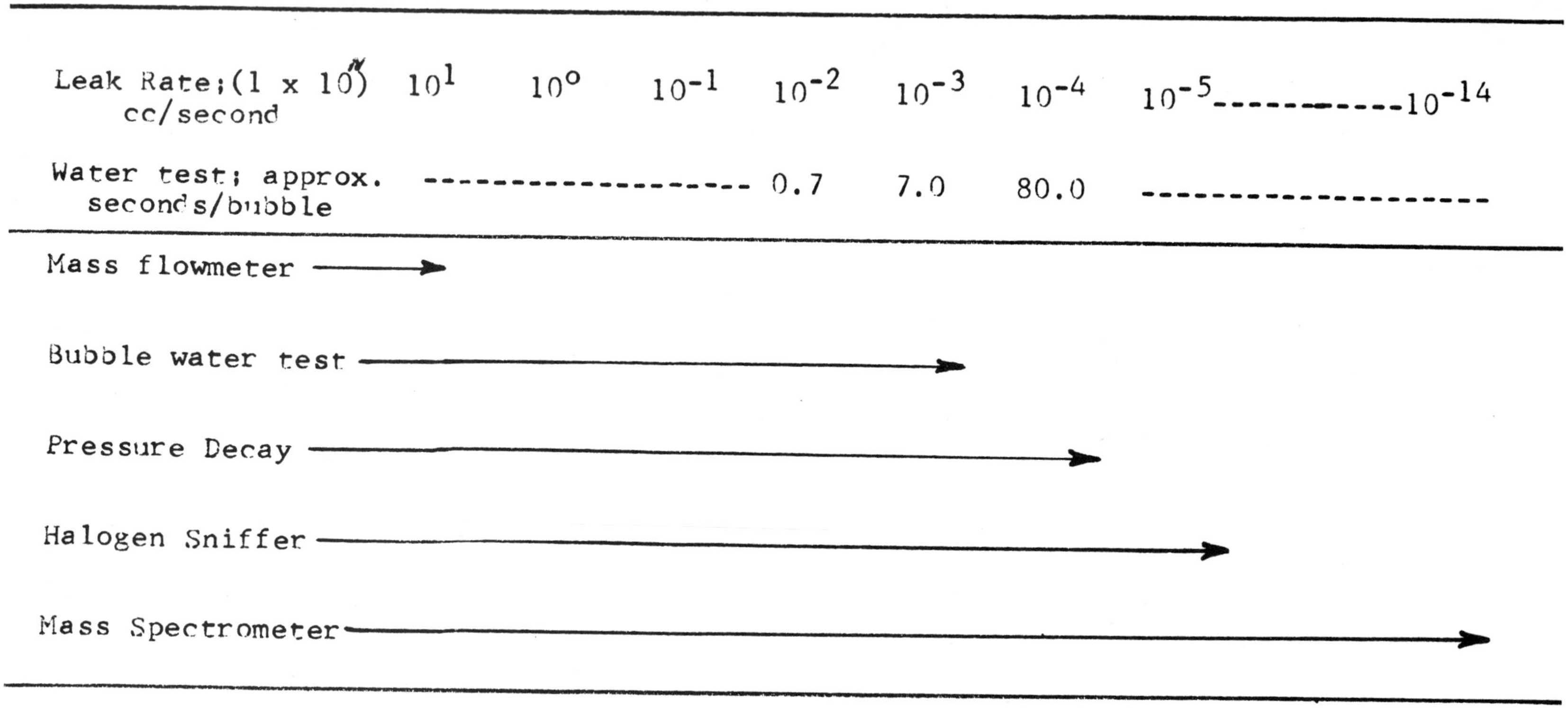

| Leak Rate:(1×10^n) | 10^1 | 10^0 | 10^{-1} | 10^{-2} | 10^{-3} | 10^{-4} | 10^{-5} -------- 10^{-14} |
cc/second

Water test; approx. -------- 0.7   7.0   80.0 -----------
seconds/bubble

Mass flowmeter ———→

Bubble water test ————————————→

Pressure Decay ———————————————————→

Halogen Sniffer ———————————————————————→

Mass Spectrometer ——————————————————————————————→

FIGURE 10. APPROXIMATE COMPARATIVE LEAK RATES AND DETECTION METHODS.

tube. This tube ionizes the gas molecules which are then accelerated toward a target area. A magnetic field influences the trajectory of the ionized particles, and causes only the desired (helium) particles to strike the target. The resultant electrical charge on the target then becomes a measure of the amount of helium leaking out of the product.

Test cycle time is largely dependent upon the time required to pump down a vacuum on the product to be tested. In a recent application, a transfer machine operating on a ten-second cycle tests an automobile torque converter assembly. The actual test cycle requires seven seconds after the part is in position.

Outputs and Displays

A variety of operator displays and electrical control outputs are available.

1. Analog Display

A voltmeter connected to the output of the transducer amplifier will display the stimulus seen by the transducer. Usual scale units are "percent of transducer full scale", or direct engineering units (psi, for example). The meter face may be green or red banded to help an operator quickly determine accept or reject status. Analog displays are inexpensive, reliable, visible under most lighting conditions, and require little maintenance.

2. Digital Displays

A digital readout voltmeter connected to the output of the amplifier will provide a digital display of either percent of transducer range or direct engineering units. They are three or four times as expensive as an analog readout and some types require occasional lamp or display tube replacement. Some types are hard to read in variable lighting situations. The least significant digit displayed may flicker back and forth between two or more numbers, and this condition can cause operator annoyance and indecision.

3. Go, No-Go Lamp Display

Indicating lamps may be connected to limit control detectors to display Go, No-Go or Low, Go, High information to an operator. Limit control detectors are inexpensive and can provide control signals for machine control as well as the visual indication.

4. Track and Hold Display

Any of the three displays described above can be improved by a "track-and-hold" circuit which samples the amplifier output at some point in time, and then holds this display rather than continuing to follow the output of the transducer. For Go, No-Go lamp displays, the track-and-hold circuit can be built as part of the relay control panel; for analog or digital display, an additional electronic module is needed.

5. Computer Interfaces

The output of the transducer amplifier may be fed to a computer via an analog-to-digital converter. The outputs of a Go, No-Go detector or track-and-hold circuit also may be used as computer inputs; however, the computer can easily perform the Go, No-Go and track-and-hold functions itself.

The outputs of the computer can be machine control; analog, digital lamp, or graphic display; audio alarm; printout; data signals for other computers; and data storage on tapes or cards for processing at a later time.

The computer outputs can be used in several ways to increase productivity and to insure product quality. Some of these ways are:

a. The record-keeping function will show trends and changes to the system that will pin point problem areas such as faulty piece parts entering the assembly process.

b. If a series of tests are performed, the computer program can be set up to provide instructions for repair area personnel.

c. The accurately recorded test data from every part manufactured can be reduced and analyzed to provide information for product design changes and re-designs.

Automatic Calibration and Automatic Verification

Fully automatic calibration consists of readjusting the output of the test instrument to a known value when a given physical stimulus is applied to the transducer. This can be done with a computer controlled test station. Built-in standard leaks or variable volumes are connected to the transducer to supply the reference input, and the computer recalculates the coefficients of the conversion curve. If the new coefficients are within limits, the production cycle continues using the new coefficients. If they are out of limits, production is halted and alarms are sent out. Machine running time, parts count, and the amount of change in coefficients between calibration attempts can be used to adjust the schedule for the next calibration.

Fully automatic calibration requires several seconds of time to pressurize and stabilize the reference applied to the transducer. Sometimes it is possible to utilize the non-productive part of the machine cycle (such as parts transfer time) to perform the calibration.

Self verification is used on non-computer controlled test instruments since these machines cannot automatically readjust the equipment output. Self verification consists of connecting a known stimulus to the transducer and comparing the output to known limits. If the test instrument is still within limits, production continues. If the instrument is out of limits, the machine sends out an operator alarm.

TYPES OF BALANCING OPERATIONS

Balancing operations are of two types, static or dynamic. Static balancing requires only the application of a single correction weight to the rotor. In what the trade refers to as dynamic balancing, two corrections weights are needed, each in proper positions in separate planes spaced some distance apart and perpendicular to the rotational axis.

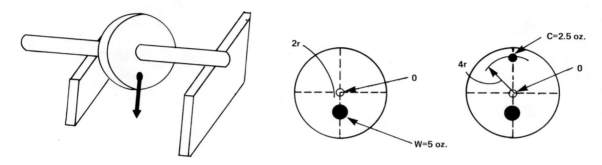

FIGURE 11   FIGURE 12

STATIC BALANCING

Static unbalance, sometimes called force unbalance, when of appreciable magnitude, is observed when the unbalanced part is mounted on horizontal knife edges, as in *Fig. 11*. The part will roll until its heavy or unbalanced point reached the lowermost position.

If the disc of *Figs. 11 and 12* is unbalanced only by the 5 ounce weight "W" which is 2 inches from the rotational axis "O", the amount of unbalance is 5 times 2 or 10 ounce inches. This unbalance of 10 ounce inches may be corrected by a weight of known value placed opposite from the heavy point and at such a distance from the axis that the product of weight and distance will also equal 10. Thus the weight "C" of 2.5 ounces placed 4" from "O" will correct the unbalance caused by "W" since 2.5 ounces x 4 inches also equals 10. The 10 ounce inches of centrifugal force exerted by "W" is opposed by the 10 ounce inches of correction weight "C", and the disc is in perfect *static* balance.

DYNAMIC BALANCING

The presence of dynamic unbalance, also called moment unbalance, is shown only when the part is rotating. This type of unbalance is caused by two equal weights at equal distance from the rotational axis and located on opposite sides and opposite ends of the rotating part as in *Fig. 13*. If the rotor is not restrained in bearings, the equal and opposite centrifugal forces produced by these weights cause the axis to generate two cones the points of which meet. Suppose the thickness of the disc of *Fig. 12* is increased until it becomes a cylinder as illustrated in *Fig. 14*. If the correction weight "C" in *Fig. 14* is applied diametrically opposite "W" but on the opposite end of the cylinder, the part will be in *static* balance. However, when the cylinder is rotated the equal centrifugal forces of "W" and "C" will cause the ends of the cylinder to move in opposite directions. The piece will be *dynamically* unbalanced. In this circumstance the additions of weight "W1" and "C1" will introduce a moment equal and opposite to that of the original weights "W" and "C". The piece will then be in dynamic balance which can be obtained, as in this instance, only by the addition of weight in each of two planes perpendicular to the rotational axis and spaced a distance apart.

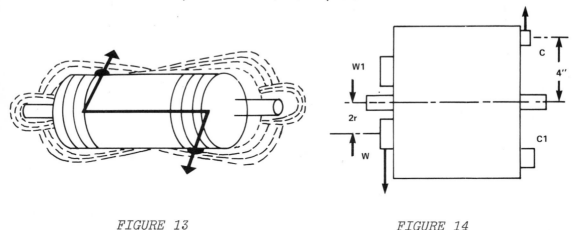

FIGURE 13   FIGURE 14

COMBINED STATIC AND DYNAMIC (2 PLANE) BALANCING

Rotating parts ordinarily have both static and dynamic unbalance. This is illustrated by *Fig. 15* where the two equal weights near the ends cause dynamic unbalance and the third weight introduces static unbalance. Such combined static and dynamic unbalance can be corrected by weights placed in two different planes perpendicular to the axis of rotation.

Each transverse section of a part will be unbalanced. Let "W" in *Fig. 16* represent the unbalance in one such section. And let planes "L" and "R" be those in which correction can be most conveniently added or removed without effecting desired qualities in the part. Now, by adding weights in planes "L" and "R" it is possible to neutralize the centrifugal force of "W".

In *Figure 16* the heavy spot "W" happens to be nearer correction plane "L" than it is to "R". Now if the unbalance "W" of 6 ounce inches is in the relative position shown, it may be compensated for by the two correction weights "C_L" of 4 ounce inches and "C_R" of 2 ounce inches. Note "C_L" and "C_R" are of dissimilar sizes and that, while their sum equals "W", the larger correction weight "C_L" is placed in the correction plane nearest to "W".

In similar manner, corrections could be made in planes "L" and "R" for the unbalances in each of the other transverse sections of the part. And then, an effect identical with the resultant effects of all these individual "L" or "R" corrections can be produced by a single correction weight. And so, with two corrections of proper size and position, one in each of two planes, the body is corrected for both static and dynamic unbalance.

It is this two-plane balancing which corrects both static and dynamic unbalance, which is generally called dynamic balancing by the trade.

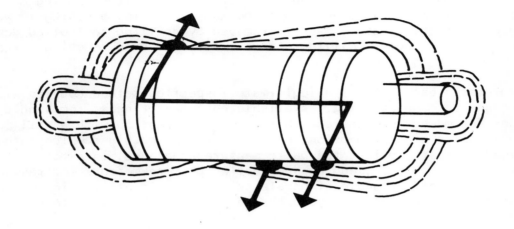

FIGURE 15

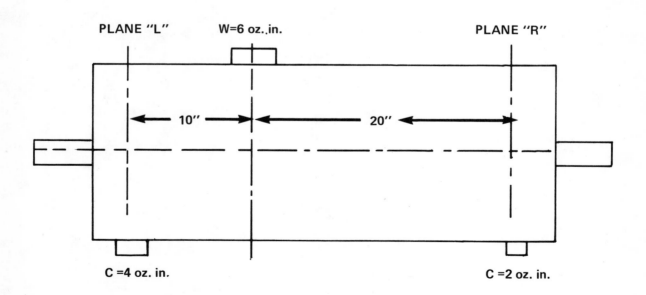

FIGURE 16

Balancing is another inspection method frequently automated. Any unbalanced rotating part can cause excessive bearing wear, noisy operation, structural failure of the device and irritation, (indeed, high fatigue), to the users of the equipment. Parts from jet engine rotors to windshield wiper motors must be balanced. A description of the types of balancing follow.

Correction of unbalance conditions are made by either adding mass or removing mass. Mass is added by welding known weights, pressing in weights, adding weld material, etc. Mass is removed by drilling controlled depths and quantities of holes, milling off material, and even removing weight by laser.

Automated balancing is a legitimate inspection method coupled with part change in manufacturing. Available equipment ranges from single stations for measuring unbalance with manual correction with a drill or spot weld up to high production, multiple station machines which automatically load the part, orient the part, measure unbalance, index the part to the correction position, add or subtract weight, re-measure for residual unbalance, orient for angle of unbalance, mark or identify point of maximum unbalance, segregate parts beyond the capability of the unit to correct, and unload.

CHAPTER 10

ANCILLARY FUNCTIONS

Virtually any manufacturing technique can be incorporated into an assembly machine or system - and vice versa. Whether the combination is labelled an "assembly system", "machining system", "grinding system" or whatever, probably depends upon which technology is predominant. Perhaps "manufacturing system" is the most descriptive tag, especially when the system takes a part from raw material and processes it into a finished product.

Ancillary, defined as supplementary, is used in this book on assembly as denoting anything other than strict assembly. A cross-section of these techniques is listed in *Figure 1* and examples of a few such applications are illustrated in *Figures 2* through

One "manufacturing system" example which incorporates six different techniques, machining, grinding, assembly, balancing, washing and gaging, is shown in *Figure 2*. This equipment produces 240 disc brake rotors per hour complete from casting to finished assembly with one or two direct labor plus material handlers - essentially replacing a production department. The washing and gaging, (not shown), are the final operations after which the product is shipped to the automobile assembly line.

Other examples of specific non-assembly operations are illustrated on the following pages. A few comments on problems which typically arise in the process of applying such techniques follow the examples.

Many non-assembly functions are incompatible with the assembly process unless special provisions are made. For example:

1. Metal cutting chips and grinding abrasives can have a serious effect on the wear qualities of the assembly mechanisms. They may tend to "load-up" fixtures and assembly heads causing misalignment, jamming and other malfunctions. Work pieces such as automotive transmissions and engines may have to be thoroughly washed as a part of the system. Fixtures may also require frequent cleaning. Some systems remove the work pieces to off-the-line metal removal modules by means of "pick-and-place" devices to minimize and contain contaminants.

2. Welding operations such as arc, spot, projection, etc., may damage parts and/or equipment with flash and spatter. Shields may be required.

3. Strong electrical fields such as found in resistance welding tend to leave a residual magnetism in some materials. Automatic de-magnetizing may be necessary for precision assemblies, pallets and fixtures.

4. Adhesives, greases and other viscous fluids tend to "gum-up" precision equipment and require special care in control, application and removal.

5. Electron beam applications require special shielding to protect operators from stray radiation.

6. Laser equipment must have special operator protection from the often-invisible beams.

Conversely, the equipment itself may require special protection from a hostile factory environment such as:

1. Sensitive electronic gear such as mass spectrometers, gages, controllers, computors, etc., may need shielding and/or isolation from plant electrical supplies and nearby equipment.

2. Induction heating equipment often demands cooling water controlled for temperature, contaminents, etc.

3. Factory compressed air supplies frequently have excessive moisture or pressure variations which must be controlled by means of filters, accumulators and boosters.

4. Air-conditions, humidity-controlled environments are often specified for gaging equipment, powder dispensing devices, etc.

5. Isolation from vibration is necessary to the proper function of some lasers, gaging and alignments devices.

The list of "do's and don'ts" is as endless as the number of potential functions and their variables. A sophisticated system design should only be approached by a team of manufacturing, mechanical, electrical, electronic, pneumatic and hydraulic engineers if all the potential trouble points are to be anticipated.

Metal-removing, cutting
Turning
Milling
Breaching
Drilling
Reaming
Boring
Laser removal
Electron beam

Metal removing, abrasive
Grinding
Honing
Lapping
Superfinishing

Fabricating
Spot welding
Seam welding
Projection welding
Arc welding
Brazing
Soldering

Identifying
Stamping
Roll marking
Transfer marking
Paintmarking
Ink marking
Electro-etching
Pressure sensitive labelling
Coding
Magnetic marketing

Metal forming
Swaging
Hot-upsetting
Stamping
Spinning
Roll-forming
Burnishing
Wire forming
Magnetic forming

Molding
Plastic injection
Metal injection
Ceramics molding
Rubber molding

Heat Treating
Induction hardening
Induction normalizing
Flame hardening
Laser hardening
Electron beam hardening

Packaging
Boxing
Plastic bagging
Paper wrapping
Encapsulating

Miscellaneous
Magnetizing or de-magnetizing
Gasket making or forming
Balancing

Figure 1
TYPICAL ANCILLARY FUNCTIONS

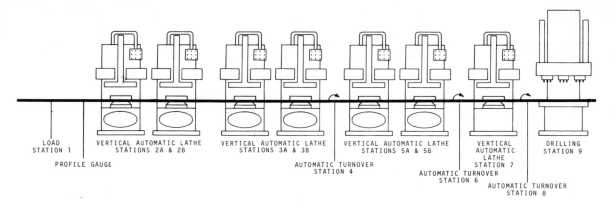

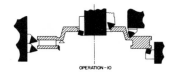 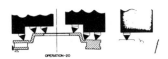 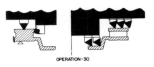

Figure 2A.
An 8 station dial index table is automatically loaded with the partially machined rotor. With 2-piece rotors the hub is manually loaded. Automatic stations hopper load the wheel mounting bolts and press them home with a 60 ton hydraulic press. Assembled rotors are automatically unloaded back to the conveyor.

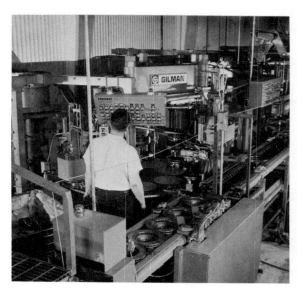

Figure 2B.
Bearings are automatically oriented and fed from two hoppers to the upper and lower heads of the assembly press. Pressure sensors detect too little or too much pressing force and indicate rejects. This on-the-line station requires that the part be supported off the conveyor carrier to absorb pressing forces.

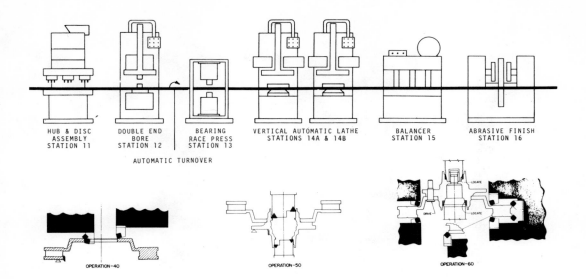

Figure 2C.
A simple turnover device is used at four points along the machine sequence. Note the "ratchet" type escapement at the left which "meters" parts to the turnover unit.

Figure 2D.
Double-end boring of the rotor race cavities is required to insure concentricity of the bearings. Parts are loaded and unloaded by the pick-and-place arm to the driving chuck. Boring heads above and below are mounted on common, tubular "ways", also to insure accurate boring.

Figure 2E.
Drilling, reaming and co'boring the five wheel mounting bolt holes is accomplished on this multiple-station, multiple-spindle drilling module. Pick-and-place mechanism is not on the machine in this photo.

Figure 3.
First operation machining on a cast iron flywheel. Note the "pick-and-place" load/unload device which removes the part from the fixture-less "power-and-free" conveyor, loads the turning module and unloads the part back to the conveyor. Note, also, the chips which are later removed by blow-off and wash. Lathe may be equipped with curtains or automatic doors for chip control.

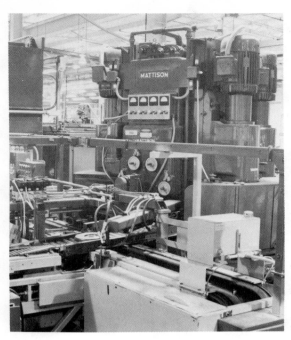

Figure 4.

Flywheel grinding on off-the-line automated machine. This is the last operation on the system and parts are unloaded to a gravity roll conveyor.

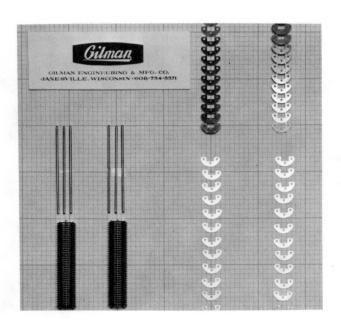

Figure 5.

Stamping of razor head blades may be performed adjacent to the assembly machine, parts left in the web and separated as they are assembled. This process will minimize handling and feeding malfunctions as well as in-process inventory. Disadvantage is that the press must be slowed down to assembly machine speed.

Figure 6.
Tab rolling four rows of tabs simultaneously fastens this transmission converter turbine assembly together at 360 parts per hour automatically on an indexing line.

Figure 7.
Roller burnishing the converter hub bearing diameter automatically on a synchronous conveyor at 360/hour removes nicks or scratches acquired during the manufacturing process.

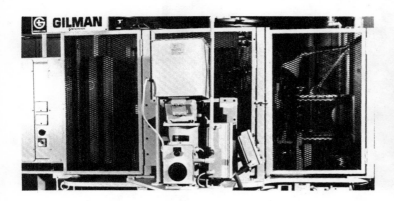
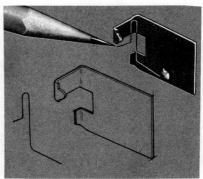

Figure 8.

A .006 wire spring is automatically formed from coil stock, cut off and welded to a "swing plate" on a dial index machine adjacent to a switch final assembly non-synchronous system. Other wire forms, coil springs, etc., may frequently be incorporated into the assembly machine through the use of many standard wire-forming modules.

Figure 9.

Magnetic forming is used to "shrink" the copper ring around the rubber boot and ball joint assembly to achieve a permanent, tight fit. Performed as part of a 1200/hr assembly machine, the operation is virtually instantaneous with the application of the powerful magnetic field.

Figure 10.
Spot welding a weight correction strip to the O.D. of this transmission coverter turbine assembly completes the balancing operation. Coil stock is fed to a metered length under control of a balance measuring station at the previous position, cut to length formed to a compound curvature, magnetically picked up and applied to the turbine shell and automatically spot welded at 360 parts/hour.

Figure 11.
Induction annealing is automatically performed on a ring gear at parts per hour.

Figure 12.

Laser hardening selected areas of this cast iron steering gear housing to 58-62 Rc produces the wear qualities desired without the distortion and subsequent machining required in conventional methods. Part goes directly on through assembly.

Figure 13 A.   and Figure 13 B.
Marking by hot transfer at 3600 parts/hour is part of the assembly process on these automotive odometer parts. Roll leaf - paint pigment tape unreels from the spool on the right and ten characters are hot transferred to the odometer dials in a rolling action as the assembly table indexes.

 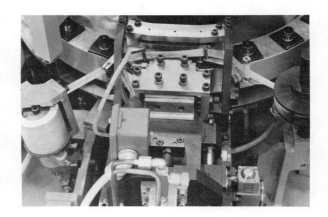

Figure 13A.   Figure 13 B.

CHAPTER 11

CONTROLS

DEFINITION: Systems or devices which exert a restraining, governing or directive influence.

A machine control system is one which directs the performance of the machine in some pre-determined manner. This direction is usually accomplished through some media in which energy has been stored such as electricity or pressurized fluids, including air and hydraulic fluid. As part of the control process, the energy associated with the control media is converted to mechanical motion, thus creating the necessary control function.

Familiar devices which convert control media energy to mechanical motion are the rotary devices, such as electric and fluid motors, and linear devices such as solenoids and fluidpowered cylinders.*(Fig. 1)* Often electric and fluid power devices are combined in the same system. For instance, a modest amount of electric power can be converted to a large amount of mechanical force through a solenoid-operated valve, moving the spool valve to release or restrict the flow of a pressurized fluid which in turn moves the piston in a cylinder great distances with large force.

Another, very fundamental type of control often used in automatic assembly, and pre-dating electric and fluid controls by at least 50 years, is straight mechanical control. We are all familiar with the old "foot-operated" sewing machine. The control system on a treadle-operated sewing machine was purely mechanical. All of the many motions were powered, directed and controlled by cams. The advantages of mechanical control are associated with the high-speed, smooth operation, controlled acceleration and deceleration of motions, and, if properly maintained, extraordinarily long life. The advantages are apparent in the sewing machine example where even after electrical control became available, the only change to the system was an added electric motor to replace the foot-operated treadle.

Automatic assembly systems use mechanical sequencing when high speed operation and low-mass parts are involved. Therefore, you will see examples of cam-controlled machines of the single-station, dial, indexing in-line and power-and-free types utilizing cam control for much of the station operation. High-speed, continuous assembly machines require mechanical control because of the rate of operation and over-lapping of motions.

The major disadvantage of mechanical control lies in its inflexibility. If the control sequence must be changed, a time consuming and often very awkward change must be made in the cams and cam positions to re-sequence the machine. This factor has caused electrical and fluid power controls to be incorporated into cam-controlled machines.

Another factor governing the type of control to be used is the decision between "sequenced" controls and "simultaneous" controls. The most common fluid-power and electrical controls which govern a series of motions are arranged to actuate the first motion, check to insure its completion, actuate the second motion, check its completion, etc. In complex, high-

FIGURE 1

<u>Electrical Control</u> - *Those devices which sense and/or alter electrical current to cause a specific motion or sequence of motions to occur.*

Relay: A device which senses current change in one circuit to affect the current flow in a second circuit. The energy is provided by a voltage source, either AC or DC. Because mechanical motion is associated with the operation of the relay (see sketches below), relays are called electro-mechanical or electro-magnetic devices.

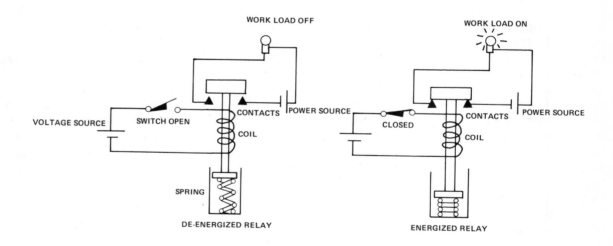

The "work load" illustrated may be an indicator light, an electric motor, linear or rotary air or hydraulic device. Usually only one motion is controlled from each circuit.

FIGURE 2

Cam-controlled device provides for one input signal to operate a "down-up-back-down-up-forward" sequence replacing a series of switches, relays, cylinders, etc. Furthermore, motions overlap for a time savings over the sequential requirements of separate controlling and actuating devices. In the "pick-and-place" device illustrated, a small air cylinder operates the jaws, but under certain conditions, this too, may be cam-operated.

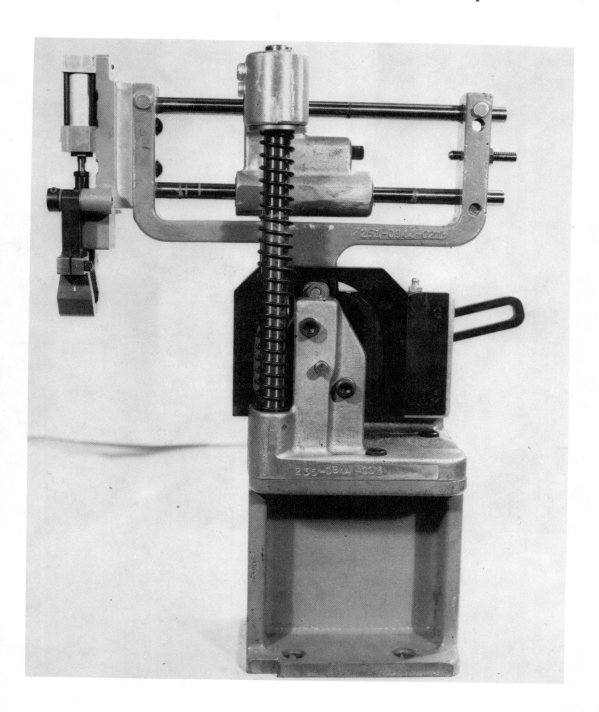

FIGURE 3

Limit switch in the lower right hand corner tells the control that a part is ready and seated in the transfer pocket. In series with this switch will be another to tell that the blade loading arm is in the back position and still another switch to determine that the unload area is vacant. Additional sensors may be added to insure that blades are ready in the magazine to the left, that the safety guards are closed, etc.

speed equipment, substantial time is lost in "reading" the machine status and reacting to initiate the next function. Cam operation, on the other hand, can often initiate the next function before the first is completed without danger of equipment smash-up due to its mechanical control interlock. *(Fig. 2)*

CONTROL FUNCTIONS

The purposes of the various components of a control system may be arbitrarily broken down into ten categories to assist the manufacturing engineer in understanding the system.

1. Actuate
2. Inspect
3. Count and Stop
4. Remember
5. Lock-out
6. Interlock
7. Indicate
8. Record
9. Compute
10. Interface

1. Actuation of a motion is the basic function of controls. First, something must tell our control that it is time to go to work, usually a sensor such as a limit switch which says a pallet or part is in position, needs a part and operates a valve or solenoid to actuate the functions' motive power. *(Fig. 3)* For example, a simple assembly function would be a straight line motion removing a part from spring-loaded jaws in a supply track and placing it into the assembly on the machine transfer. Such a basic action can be accomplished mechanically, electrically or hydraulically. But what happens if there are no parts in the track to assemble? What happens if the part does not get assembled?

2. Inspecting for parts presence in the feed track *(Fig. 3)* and for the actual placement of the part in the assembly is probably the second basic control function. It is vitally important to know these two items of information. Such "presence-checks" are usually made by limit switches on a go-no go basis although the trend is toward more reliable, noncontact devices such as proximity switches, photo-electric devices, magnetic detectors, etc. More sophisticated inspections may check additionally for dimensional integrity, hardness, leak rates, or any of a multitude of quality parameters. However, information accumulated is worthless unless some use is made of it. What are we going to do with the data? Are we going to let the machine go on making bad assemblies at its high rate of speed? Are we going to stop the machine and wait for its operator to do something about it? The latter would defeat the purpose of the automatic assembly concept if the malfunction is temporary and the former would only make a lot of scrap very quickly. So another control function is required.

3. The third function to be added is a "count-and-stop" device which will permit the machine to keep running if the malfunction is temporary. Usually they can be set to count any pre-determined number up to fifty.

FIGURE 4

The panel in front of and above the operator contains a memory system consisting, in this case, of latching relays arranged to be actuated depending upon whether the assembly at that station is "good" or "reject" according to the parts presence sensors. This information is transferred from station relay to station relay, "travelling" along the machine with the parts, telling each successive station to operate or lock-out. At the unload station the information is used to actuate a reject chute as required. Indicator lights inform the operator of the status of all stations and reset switches permit him to correct a bad assembly and reset the memory. Other lights may indicate whether the station completed its down and up strokes, etc. The smaller panel at the right is the operator's pushbutton station for start, stop, manual, jog, and other functions. Cycle start, stop and emergency stop pushbutton station for start, stop, manual, job, and other functions. Cycle start, stop and emergency stop pushbutton boxes are positioned at convenient intervals along the front of the machine.

Generally, the value of the part will indicate whether 3 or 10 or more bad assemblies can be passed along for rejection at the end of the machine. If the malfunction repeats beyond the pre-set number, the device will stop the machine, turn on a red light, signal the operator, or whatever function the user prefers. But what about the bad parts made and passed along on the first 3 or 5 or 10 cycles? What are we going to do about them?

4. A "memory" system *(Fig. 4)* may be added to our controls sequence as a fourth function to store the fact that a bad assembly is in the mill. This may be mechanical "flags" on the pallets, latching relays transferring their data as the parts progress through the machine, magnetic coding of pallets or memory transferred through a programmable controller or computer.

   But now another control function requirement becomes evident. What are we going to do with the stored data? We can automatically segregate bad parts from good assemblies at the unload station readily enough and we have not penalized the machine production because of temporary malfunction. But what about all the stations between the malfunction and unload? Are we going to add good parts to a bad assembly?

5. A "Lockout" feature on all assembly stations can, through communication with the memory, automatically prevent any station subsequent to the malfunction from operating. No additional work will be performed on that part only as desired and programmed into the memory system. Usually, the lockout is electrical by disabling the actuation function of such stations, although cam operated stations may have a solenoid operated lockout pin which prevents the motion from occurring. The cam still goes through its cycle, but an air cylinder, or "air spring" in the linkage is collapsed to absorb the motion without moving the tooling.

6. The "Inter-lock" system of control design probably first was applied to cam-operated equipment due to the desirability of always stopping the machine on the low point of the cams. This was accomplished by attaching a cam-switch to the machine cam shaft and setting the adjustable cams on the switch as required. Another, even more important use for the cam switch soon became obvious. It is highly desirable that the machine controls incorporate a means of determining that each station in fact _does_ move forward the proper distance and _does_ move back to the proper position at the proper _time_ on every cycle. Failure to do so might be caused by parts jamming, foreign matter in the system or machine failure, any of which should be detected and remedial action taken. With the addition of an interlock cam-switch mechanism it became a simple matter of adding banks of cam switches, one for each station function, timed with the cam shaft to tell the control system that all was well or that malfunctions had occurred and where. But what do we do with this information? The obvious use is to stop the machine if the malfunction is apt to damage it, and this of course is done. But the majority of the information obtained has other uses as follows.

7. To indicate what is happening to the assembly machine system, particularly complex ones, is a necessary - even mandatory - function. An indicator panel is included with most modern assembly machines with light

FIGURE 5

One form of equipment performance recording is this counterbank. Total cycles and good parts produced indicate the machine cycle performance. A running time counter may be added for overall efficiency. Additional counters may be added for each critical station to permit determination as to where efficiency is being lost. Use of a programmable controller or computer offers more complete data, printout, computations and anticipation of problem areas.

indications of the status of all motions at all times. In the event of a machine stoppage, the point of the malfunction is obvious as well as the action required, something which can save the operator untold hours in the course of a day. Signal lights may be turned on, horns may be blown, and todays electronics even offer a synthetic voice announcing over a loudspeaker the station and the function needing attention! Other devices in almost unlimited variety offer inspection readouts, status indicators, etc. But all these are status indicators as of a given moment and something more is required in todays record-happy environment.

8. Recording of the assembly system performance is becoming increasingly important *(Fig. 5)*. Production control has uses for production records. Departmental performance evaluation requires up-time and rate records. Quality control needs permanent records for reasons of warranty and government mandated records. Maintenance can benefit from machine performance records. The method of recording may be as simple as a bank of electric counters indicating the machine total cycles, good parts produced and malfunctions at each station. On the other end of the scale, tapes may be required for later analysis, printouts etc. by computer.

9. It may be necessary to compute the performance of the assembly and/or the machine on the job. An assembly may consist of measurement of several parameters and a computation made to determine which of many components is required to make an acceptable part in selective assembly. Performance of an assembly may only be measured by computer analysis. It may be necessary to integrate temperature, pressure, flow, etc., to determine whether given a part meets the specifications. Computers are increasingly applied to assembly machines to meet quality requirements and performance requirements not possible in any other way.

10. Interfacing is another control function which is getting greater attention. Interfacing with central computors for management information is one feature which should be considered when approaching the project. Interfacing with other machines for production and quality control reasons requires compatible equipment.

TYPES OF CONTROLS

Automatic assembly machine controls may be divided into six main categories as follows:

CONVENTIONAL CONTROLS	ELECTRONIC CONTROLS
1. Mechanical	4. Solid state
2. Fluid power	5. Programmable
3. Electro-mechanical	6. Computer

Most or all the functions of assembly machine controls can be performed by any of the above types of controls but the chances are that any sophisticated machine will have some of each. The combinations which offer the optimum performance will probably differ with every controls engineer. We will look at the principle characteristics of each.

CONVENTIONAL CONTROLS

1. <u>Mechanical controls</u> are those devices such as levers, gears, cams, pulleys and linkages which cause a specified motion or sequence of motions to occur. They are most commonly recognized in the form of cams which may appear a barrel cams, face cams, plate cams, etc. The major advantages lie in the simplicity of design, ease of maintenance, high speed operation, long life, ease of visual recognition of malfunctions of many sets of limit switches, relays, solenoic valves and cylinders.

    The major disadvantages lie in their inflexibility and the problems associated with any change of sequence or motion in the assembly machine. For single-purpose, high speed machines, however, the cam remains the most reliable answer.

    Mechanical controls are used not only for actuation, but are often used for memory in the form of memory pins on pallets set in various ways to identify models, rejects, categories, etc. Mechanical probes may be used to identify and check presence of parts. Lockout functions are performed mechanically as well as interlocks. Mechanical counters are a common indicating and recording method. Even some computations may be mechanically accomplished. The oldest form of machine control, dating back before mechanical speed governors, mechanical control should not be overlooked in this day of exotic devices.

2. <u>Fluid power</u> controls are those which sense and/or alter the flow or pressure of fluids, gases or liquids, to cause a specific motion or sequence of motions to occur.

    Pneumatic controls, as the name implies, use compressed gas as the source of potential energy. Air is the most common.

    Hydraulic controls use liquids under pressure, commonly petroleum based fluids developed for high pressure uses although water is sometimes used.

    Logic circuits and some computing circuits have been used in both pneumatic and hydraulic controls.

    Flexibility and low cost are generally associated with pneumatic controls while hydraulic controls are used where higher forces are involved.

    The popularity of fluid controls stems from the availability of standard, "shelf" components in a wide range of functions at relatively low cost.

FIGURE 6

A portion of a simple ladder diagram for an indexing table control. Symbols and circuitry format are being used for relay controls, solid state, programmable control and industrial mini-computers.

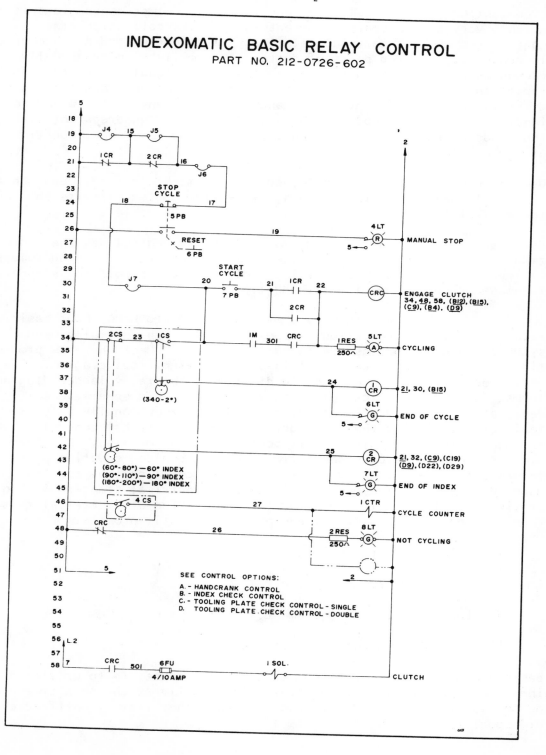

3. The term "electro-mechanical" controls is used here to apply to relays and associated electrical components as compared with the solid state systems. They are defined as those devices which sense and/or alter electrical current to cause a specific motion or sequence of motions to occur.

    A conventional electrical relay consists of an electromagnetic switch which a very small, control current energizes a coil which closes contacts capable of carrying relatively large currents. Any number of independent, isolated pairs of contacts, normally open or normally closed as desired, are activated by one input signal. Any number of inputs in sequence and/or in parallel may be combined to offer logic circuits capable of determining "and", "or", or any of the other logic functions. Large, complex machine tools may have hundreds, or even thousands of relays in their panels controlling many, many functions of the machine.

    Used in conjunction with many other electro-mechanical devices such as limit switches and micro-switches, extremely complex circuits challenge the reliability of the components to maintain operable equipment.

    For years, the standard for automatic assembly control was the conventional electromagnetic relay. The basis for control utilizing relays is found in the Joint Industrial Console (J.I.C.) electrical standards for mass production equipment, EMP-1-67.

    This standard has everything you will ever want to know about basic relay control for production equipment. All of the normally used symbols and descriptions are found in Appendix "A" of the mass production specs. and in Appendix "B" is a sample electrical diagram. The form used in this sample electric diagram is the relay ladder diagram format for the electrical elementary drawing. This schematic as developed over the years has proven to be an exceptionally valuable tool for identifying the construction of the electrical control as well as identifying the logic that is controlling the machine. No other piece of documentation such as flow chart or Boolean equation provides for as concise and clear definition of the actual control for the machine *(Fig. 6)*. The purpose of the JIC electrical standard is "to provide detail specifications for the application of electrical systems to mass production industrial equipment which will promote:

    1. Safety to personnel.

    2. Uninterrupted production.

    3. Long life of the equipment.

    4. Easy and low cost maintenance.

The basic purpose of the electrical control, of course, is to provide a machine which has a maximum productivity level while at the same time providing for safety, ease in maintenance and the other items identified by the JIC specification.

The electrical control that has developed through the years involves using relays at 120 volt AC. Many reasons are given for this voltage level.

First, the 120 volt AC signal is a voltage level commonly found in most industrial plants.

Secondly, the 120 volt signal is familiar to the electricians and does provide for a certain amount of sparking of contacts which in turn tends to keep them clean and free from contamination.

Problems exist with the relays, however. For instance, they are slow. Since they are energized with an AC signal, the shortest response time that can be generated to a signal from the machine is 16.7 milliseconds. In addition, if care is not taken, sneak paths can be generated through the hardwiring of relay contacts and cause erroneous actuations to occur.

ELECTRONIC CONTROLS

It might be well to pause at this point and get some definitions used in this relatively new field of controls. Since the WWII development of the transistor, new applications have developed and new terminology has arisen.

<u>Solid state</u> controls, those with no moving mechanical parts, are used in application from simple relay replacement to highly sophisticated computer control centers. (This covers everything that follows except NC tape control.)

<u>NC</u>, numerical control, was developed using tape punched with holes in desired patterns for motion control on multi-axis machining systems. It was convenient, understandable, and worked well under very hostile environments. It is generally not considered for automatic assembly applications.

<u>CNC</u>, computer numerical control, developed as a replacement for tape control as the requirements for higher speeds demanded more sophisticated controls. Originally involving a dedicated computor, this approach was aimed at motion control.

<u>DNC</u>, direct numerical control, was originally envisioned as a time-sharing, broad-scale, plant coordination and optimization of machining systems. Again, it was developed with motion control in mind.

<u>PC</u>, programmable control, was originally developed for assembly machine sequence control as opposed to motion control. (The term is used for computer-technology devices, but one could go back hundreds of years to the first music box with its replaceable metal drums or discs which could be "programmed" with various combinations of pins or projections to produce the desired sequence of musical notes. It was a programmable, mechanical, digital, binary device!)

<u>Computer Control</u> and programmable control are increasingly alike, differing mainly in capability and complexity. At this point, motion control and sequence control are both possible and practical.

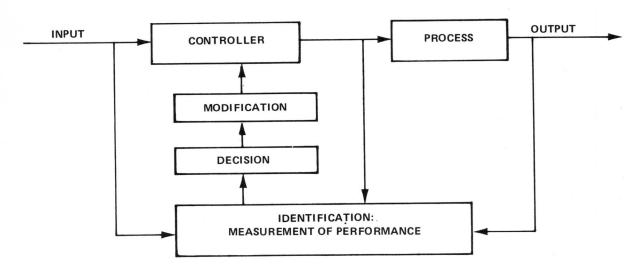

Figure 7.
THREE FUNCTIONS required for optimization type adaptive control are: identification (measurement of performance), decision, and modification.

Figure 8.
PROGRAMMING PANEL, like this one from Modicon, is an important piece of support equipment for system setup, interrogation of inputs and outputs, and general troubleshooting.

<u>Adaptive Control</u> *(Fig. 7)* is a term applied to systems which sense environmental conditions and adjust the machine accordingly *(Fig. 7)*. A military aircraft gun control system senses air speed, altitude, temperature, barometric pressure, etc., and "adapts" the aiming of the weapon to suit. A butting machine senses such data as part dimension, temperature, cutting rate, etc., and "adapts" the machine settings as required. The assembly machine which is gaging several parameters in selective assembly "adapts" the machine choice of parts categories to achieve the desired assembly specification.

<u>Diagnostics</u> *(Fig. 8)* are peripheral equipment items or programs designed to indicate system status. Systems can be engineered to automatically detect malfunctioning areas, alert the operator and print out or display visually or audibly the point of difficulty.

<u>Peripherals</u> include devices external to the controller or computer such as programming panels, typewriters, video tubes, some diagnostics, input and output devices, etc. *(Fig. 9)*

4. <u>Solid State</u>: Devices which modify a large current in one circuit with a small current in another circuit with no moving parts. Also devices which sense a presence, motion, pressure or other condition and modify a current with no moving parts. Devices based on transistor technology are used as electronic switches and sensors.

   The failure rate of electro-mechanical devices probably was the prime reason for the development and acceptance of solid state control devices. Once installed and cycled for a few hundred hours, the failure rate is extremely low and the ultimate life of the unit is virtually unlimited.

   Though more costly, solid state devices have provided much higher speeds in switching, space savings and energy savings over conventional relays. Plug-in designs make for rapid replacement if required.

   Unfortunately, when used as a mere replacement for the conventional relay, the solid state switch still retains all the problems of a "hardwired" relay circuit when changes and alterations are required. For the increasing demands in flexibility, speed and complexity, the next step forward in controls required a substantial breakthrough.

5. <u>Programmable controllers</u> are defined in NEMA Standards as "digitally operating electronic apparatus which use a programmable memory for the internal storage of instructions for implementing specific functions such as logic, sequencing, timing, counting, and arithmetic to control, through digital or analog input/output modules, various types of machines or processes". *(Fig. 10 & 11)*

   PCs, as they are known, have significantly influenced the direction in which automatic assembly has progressed through the 1970's. Conversely, automatic assembly has influenced the evolution of the programmable controller. Without PCs, automatic assembly would not be nearly as sophisticated as it is today. For instance, the level of testing and gaging so common in today's assembly systems would not be practical without Pcs. Similarly, some features presently provided

Figure 9.
MONITORING SYSTEM by Cross Fraser provides diagnostics information and data on tool change, cycle times, and production counts.

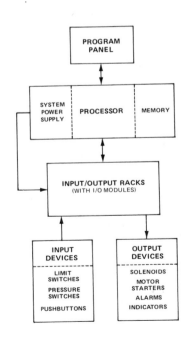

Figure 10.
BASIC COMPONENTS is a PC system. The central processing unit makes all decisions based on memory instructions and feedback from inputs and outputs.

Figure 11.
PROGRAMS FOR THE MAXIMISER PC can be loaded into memory in less than 1 1/2 minutes. Simple ladder diagram language is used.

Figure 12.
A MICROPROCESSOR CHIP which performs all logic functions is the heart of this 1806 PCS microcomputer.

with programmable control such as shift-register memory, high speed counters and analog inputs, which are not as much in demand in other manufacturing areas as they are in automatic assembly, might not be part of PCs today.

In the late 1960's, computer control was being widely touted as the next generation of control for automatic assembly systems. Those were the days when the PDP-8, IBM's 1800 and Honeywell's 316 were the major minicomputers available for control in assembly systems. A number of these systems were introduced and laid much of the groundwork for the systems that were to follow.

In 1969, a specification was written by the Hydra-matic Div. of General Motors Corporation for a new and different kind of control called "programmable control". Hydra-matic had experience with computer control on test stands and the advantages of the controls were obvious. However, the requirement for sophisticated software preparation and the problems associated with hardware and computer mainframes not suited for the industrial environment made it clear to Hydra-matic that a new generation of controls was required. From this specification the programmable controller was born. Two responses in particular, the DEC PDP-14 and the Modicon 084 were created to satisfy the specification.

All this was made possible by the development of the "micro-processor" chip at about this time. With the process called "large scale integration", dramatic reductions in size and cost have been made. LSI is the industrial process involving optically reducing a large pattern to a tiny chip having the equivalent of thousands of transistors in a space half the size of a stick of gum *(Fig. 12)*. The computer of the 1940's which occupied the space of an 800 cubic foot room could now be reduced to the size of a package of cigarettes.

"Dedicated" micro-processors are permanently "burned-in", or "wired" by the manufacturer to perform only a pre-selected series of functions, such as a hand calculator. "Programmable" micro-processors may be re-programmed many times according to the needs of the user.

WHY WERE PROGRAMMABLE CONTROLLERS ACCEPTABLE WHEN MINICOMPUTER CONTROL WAS NOT?

There were several reasons. The first reason has to do with the hardware. The minicomputer CPU, (central processing unit), was not physically packaged to survive in an industrial environment. Similarly, the CPU was configured to be interfaced via either serial or parallel interfaces to devices such as teleprinters, typewriters, printers, CRT's, etc. The problems associated with introducing high voltage, noisy and very transient electrical signals into this interface were impossible. The programmable controller took steps to avoid this packaging problem. The CPU's were designed to exist in hot, contaminated environments satisfactorily.

The first interfaces that were introduced were 120 volt AC signal converters. Both input and output converters as described by the Hydramatic specification were provided and the noise isolation required was realized.

Perhaps the major reason that programmable controllers were acceptable; however, was in the area of software. Minicomputer control required that the control program be written either in assembly level language or in real-time Fortran. This type of software requires an experienced programming professional in order to obtain good software. More to the point was the problem of adapting or changing this program to specific control problem or to re-sequence the machine in the industrial environment. Few if any of the end users of this type of equipment had programming staffs sufficiently trained to handle the problem of debugging or reprogramming the minicomputer control.

The programmable controller on the other hand used one of two techniques. The first technique required Boolean Equations. In some instances, a control orientated language such as "Test Input" or "Set Output" was required. Finally, the main software technique that really made the programmable controller succeed was the introduction of a relay ladder format that provided the programmable controller with a familiar means of programming. The relay format provided the machine builder with a method of generating logic that was familiar with him and to his servicement. It also provided the end user with a tool for changing, updating, and editing the program once the machine was installed in his plant.

Almost immediately, the programmable controller found a major usage in the area of automatic assembly. Early in it's history, it was used primarily as a straight relay replacement. Benefits associated with the programmable control in the early stages were associated with economies in wiring of the machine and probably most significantly in the area of machine debug. It was rapidly discovered that a complete new program could be entered in a matter of minutes where as to rewire a relay panel would require days or even weeks. In addition, the programmable controller proved to be a reliable piece of equipment operating in a factory environment, if treated with a certain amount of care.

In the early 1970's, the requirement for more features in programmable controllers provided a slow but steady evolution which has continued to date. No longer was mere relay replacement adequate. Hardware interfaces were added which provided for analog to digital and digital to analog signals. This allowed the programmable controllers to be interfaced directly to analog electronic test equipment. In addition, pulse output cards for driving stepping motors, binary coded decimal (BCD) inputs for interfacing to digital volt meters and high-speed counter cards were introduced.

Probably as important as any of these interface devices was the development of arithmetic capability. Arithmetic capability in the programmable controller meant that simple calculations could be done within the controller. Testing and gaging processes could be controlled while utilizing mathematics for the very first time.

It was about this time that the requirement on automatic assembly equipment for testing ang gaging reached an important level. The increased emphasis on product safety required that assemblies be tested before they left the assembly machine to insure proper and safe performance. Dimensional criticality and a requirement to provide more reliable products both forced more testing and gaging onto the assembly machines. Special interfaces and arithmetic capability made the programmable controller a key part of the growth of testing and gaging in automatic assembly. When the requirement came, the programmable controller was available.

In addition to these kinds of features, extra additions were added, such as printers, CRT's, etc. A programmable controller in this way provides for a permanent copy of test results and on-line diagnostics of a machine or controller faults.

Thus the programmable controller, born of computer and spage-age technology, came after the computer to fill a void in industrial requirements. It offered solid state reliability, computer speeds of operation, comparatively low cost, easier to maintain than conventional controls plus a flexibility through its' programmability that proved extremely attractive to users of complex assembly equipment.

However, as the PCs evolved it became obvious that the increased equipment sophistication they encouraged would outstrip even their capability. Automatic assembly systems, too, evolved to the point where more testing and more mathematics were required than was practical with the PC. The next step required the "brains" of the computer.

6. <u>Computers</u> were born in the 40's, went through puberty in the 50's, became educated in the 60's, and are settling down to compact, low cost sophistication in the 70's.

   The general concept of a computer is that of a giant memory capable of processing data at fantastic speeds and operating on tiny voltages and currents. As such they are extremely vulnerable to the factory environment and virtually useless in handling the units of power required for machine controls. (An oversimplification, of course.)

   Scaled-down versions of the computer, called "mini-computers", were developed in the 60's to handle smaller functions than their sires.

   By 1975, a product phenomon known as the "microcomput r" became available from a development known as a "microprocessor". This is a single, tiny chip of silicon containing an integrated circuit equivalent to thousands of transistors. When configured to include memory and input/output circuitry, it becomes a microcomputer. This is the development which made possible so many of our spage-age toys such as hand-held computers.

   Both minicomputers and microcomputers are used in automatic assembly system controls. The microcomputer consists of a central processing unit (CPU); the microprocessor chip; random access memory (RAM) which serves as the scratch-pad or dynamic portion of the memory; programmable ready-only memory (PROM) where the applications program is stored; an input-output decoder; clock generator; address latches; control and interrupt logic; and buffers for data storage.

   The minicomputer is structured somewhat differently. It also has a CPU, but this unit is usually made up of many integrated circuits designed in a way that makes it much faster than the CPU in a micro-

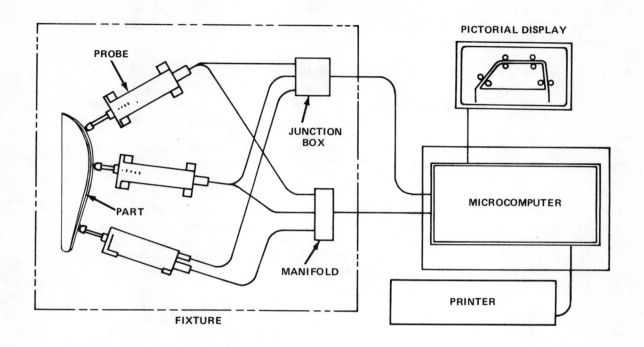

Figure 13.

TYPICAL GAGING SYSTEM *for sheet metal parts incorporates a microcomputer for processing binary signals generated by the probes.*

computer. However, industrial applications do not often require this high speed. Minicomputers also have much more random access memory than microcomputers, and RAM is used for program storage as well as computation.

A minicomputer is capable of approximately 10^5 additions per second, compared to about 10^4 for the microcomputer. And where the minicomputer can handle tasks simultaneously, a microcomputer can, at present, handle them only sequentially. On the other hand, a microcomputer can be used to upgrade a programmable controller, which is inherently rather simple and slow. It adds speed to the controllers capability and enables the controller to "think" while a power relay performs the more mundane tasks of output control.

As of this writing, a typical microcomputer costs about $400 compared to about $1500 for a minicomputer. And unlike hardwired logic, the micro required no logic diagrams or bench instruments for testing and troubleshooting. Instead, software and development hardware are used to debug the systems. Changes can be made by simply altering the programmable read-only memory.

For relatively high volume applications, where higher development costs can be amortized over many units, the microcomputer is carving out a well-deserved niche for itself. Applications are numerous, ranging from automatic wire-wrapping machines in the assembly of electronic units, to conveyor systems in foundries and on-line gaging systems for check automotive hoods, doors, quarter panels and underbodies. *(Fig. 13)*

The traditional method of inspecting automotive sheet metal parts is to check the first piece coming off the line for process and produce integrity, including correct dimensions, and then to repeat the process on a sampling basis at regular intervals. Because the checking is done off-line, and takes a relatively long time, the frequency of this checking is low compared to the production rate - once an hour contrasted with 500-600 pieces per hour. Therefore the sample may not be representative of what is being produced and will not catch and reject an occasional bad part.

In a typical system, the sheet metal parts are gaged by linear displacement probes. The microcomputer system scans the probes on a sequential, one-by-one basis and reads a binary digital signal. It determines whether the dimensions of each part are within tolerance and provides whatever statistical analysis is required. Accept-reject signal outputs are provided plus data printout and/or pictorial display as and when desired. One such system is installed in close proximity to welding lines in a NEMA 12 enclosure, exposed to the most severe environment with continuous heat, vibration, mechanical shock and electrical noise. The basic characteristics of a good microcomputer and associated electronic devices must provide immunity to radiated and conductive electrical fields.

Most microcomputers are used in dedicated applications where they perform an assigned task and no other. In larger installations, several microcomputers may be assigned different tasks and a minicomputer used for data handling. This frees the minicomputer for executive management jobs such as collecting and analyzing data, management reporting, and acting as a man-machine interface for a sophisticated device such as a CRT terminal. By freeing the minicomputer from detail work, expensive minicomputer time can be saved for larger number-crunching and data handling tasks.

Makers of microcomputers are developing hardware and software designed to simplify the job of engineering microcomputer-based systems. Much of this work involves specific families of microcomputers, but generalized equipment is beginning to emerge also. A complete system would include resident software, (including a monitor, assembler, debugger, and editor), a CRT terminal, high-speed printer, floppy disc drive, and PROM programmer. Such systems can range anywhere from $12,000 to $20,000 depending upon the peripherals.

The future of the microcomputer approach to control in automatic assembly is probably as impossible to predict as was the usage of the programmable controller at the early stages in it's history. In fact, it is becoming impossible to distinguish where the microcomputer-based programmable controller stops and a programmable controller of the earlier configuration starts.

One common element through all this, however, is the retention of the relay format programming language. It is not expected that this feature will change in the foreseeable future. When everything is considered, the relay format is an efficient way of providing logic documentation and sequence information about the control circuit. It is better, in most respects, than a flow chart. It is much more easily understood than a written sequence of operations. It is much more familiar than the Boolean Equation. All in all, it is a very fine method of identifying control logic.

SUMMARY

In the last three decades more innovations have taken place in the controls area of automated assembly than in any other. A whole new breed of controls engineers has evolved and their importance to the development of automatic assembly equipment cannot be over-emphasized. Merely staying abreast of the latest applications and technology is a time-consuming task for these front-line engineers.

Compare a few of the facets of controls status between the end of WWII and the present:

1. Automation controls usually accounted for 5-10% of the cost of a project vs. as much as 30-40% today.

2. Controls consisted of limit switches, relays, solenoid valves, etc. to merely turn-on and turn-off functions vs. today's sophisticated sensing, memories, computations, data processing, displays, inter-facing, etc.

3. Component sizes were on the order of 30-40 cubic inches for six contacts or so versus thousands of "contacts" in the size of a match head. Charge-coupled devices and magnetic bubble memories with as much as a million bits of data in a single chip are forecast.

4. Relay reaction times of tenths or hundreths of a second are being replaced with devices reacting in milli, micro and nano-seconds.

5. Assembly machines were hampered, delayed and failure-prine due to controls unable to service high-speed operations versus todays controls which are capable of much more than the machines can digest and utilize.

While we may have reached a plateau in the technological development of controls, and may not expect the same rate of progress to continue indefinitely, we can certainly look for much in the way of peripherals to use the known capability of control devices more efficiently.

A "Pandora's Box" of "toys" to play with, a superabundance of gadgets to apply, tends to lead engineers to extremes in which the potential usage of these new technologies is not justified by the results. The oldest tool of all - common sense - must be applied to avoid such extremes.

For example, the availability of control hardware to inspect unlimited parameters makes it a very tempting decision to check everything 100%, whether critical or not, when in fact, normal process capabilities may offer adequate control on many non-critical characteristics. Avoid "over-inspection".

Again, the availability of printouts tempt the manufacturing "bureaucrat" to insist on his own personal copy. This results in duplication of printouts, files, filing clerks, ad nauseum, when one master file might satisfy all needs.

Indeed, if permanent files of records are not mandatory, printouts should be limited to exceptions or cases where parts fail to meet specifications. Furthermore, some "statute of limitations" should govern how long permanent records should be kept.

Demands for statistical and diagnostic data must be tested with more common sense. It is possible to create an assembly system that does more controlling than assembly; that creates more paper than parts. It becomes very important that over-controlling, over-inspecting and over-recording, like any other factory waste, be kept under constant surveillance.

The manufacturing engineer has many marvelous new tools to use in the field of controls. His understanding of their uses, applications and limitations will determine the efficiency of his operation.

CHAPTER 12

DE-BUGGING

"De-bugging" is a coined, trade term which grew with the industry as a catch-all phrase to denote all the work which must be performed after an assembly system is built to correct design and manufacturing deficiencies.

Theoretically, de-bugging starts with the product design, is affected by concept selection, component design, tooling and all the areas heretofore described. If all the steps were planned to the optimum conditions, no de-bugging would be required. This is true in theory, but rarely, if ever, true in practice!

Assembly systems are usually "one-of-a-kind" projects. In some cases a quantity of identical machines will be built, but more than 3 or 4 is unusual. Perfection, or near-perfection in all of the preparations for a system is a worthy target, but not an economic practicality. Let's look at some of the problems.

<u>Parts</u> (product) design is seldom rigid. Especially in new product designs, changes can be expected as the project goes along. Frequently, little time is available for making system changes to accommodate these part revisions. Potential problem areas can readily be overlooked. On new parts, such items as stamping burrs, flatness, run-outs, surface finish, etc., can crop up unanticipated, which must be incorporated into the equipment capability planning.

<u>Design</u> of the equipment components cannot be expected to be perfect on the first go-around. Some deficiencies are to be expected. At the same time, there is a tendency among engineers who are conscientious to keep redesigning, without end, toward perfection. Some cut-off point at an optimum level must be imposed if the system is ever to be completed.

<u>Manufacturing</u> perfection has yet to be achieved. Again, a tendency to "over-manufacture" must be held within reasonable limits if an economically-feasible system is to be produced.

<u>Efficiency</u> of the completed system will never be 100%. Some acceptable level of efficiency must be targeted. In a production assembly system, de-bugging never really stops. It should be expected that a system in use for six months or a year will run more efficiently than when it was first built, or the personnel in charge of its operation are not really interested in getting the most out of their equipment. A general rule of thumb for acceptance of assembly machines would be 75% to 80% efficiency based on machine cycle time, i.e., machine efficiency only. Labor, material handling, maintenance availability, etc., should not be charged against the system at this point. This ratio will vary, of course, with the size and type of assembly.

The logistics of <u>material handling</u> alone limit the amount of pre-acceptance operation which is practical, particularly in the larger assemblies such as complete automobile bodies. Most assembly machines are built in some area remote from the production line and handling facilities such as conveyors, cargotainers, tubs, preceeding machines, etc., are not available. Continuous running of a system becomes a serious problem in areas not related to the performance of the machine itself.

"Final tuning" of a complex assembly system requires many thousands of cycles. This is usually practical only on the production line itself, a fact which is not sufficiently recognized in the industry.

An average assembly system may have 5,000-10,000 individual parts, many of them standard nuts-and-bolts types, but a great many specially-designed, one-of-a-kind components, too. The magnitude and complexity of modern equipment presents literally thousands of potential problems if not properly programmed and planned. To assemble a complete system without such foresight will surely be an exercise in frustration when the "button is pushed".

Due to the varieties of assembly machines, the fact that they are usually one-of-a-kind, and the enormous range of technical disciplines involved, a formal de-bugging procedure is possible only in generalities. Following is one approach in use, but first it is desirable to establish responsibilities and chain-of-command in order to avoid too many fingers in the pie, mistakes and inefficiencies.

Personnel

An <u>engineer-in-charge</u> of the project should be the sole point of decision and contact. Hopefully, the company or department building the equipment will have a one-on-one relationship with the buyer, or user, man-for-man, to minimize confusion and maximize intercourse. The engineer should have responsibility for

1. Completeness and quality of design.

2. Machine assembly procedures and/or sequence.

3. Supervision of final de-bug process.

4. Follow-up in the users' location to assure that all of the contractual agreements are met satisfactorily.

A <u>lead man</u>, assembler or setup man should be assigned the full responsibility for overseeing the actual assembly and de-bug under the supervision of the engineer. He should also accompany the machine to the users' site, oversee startup and final tuning to the agreed performance. Working as a partner with the engineer, he should perform the following functions.

1. Attend the initial project lineup meeting at which time he should be informed of:

    a. What the buyer or user expects to receive; production rates, functions, etc.

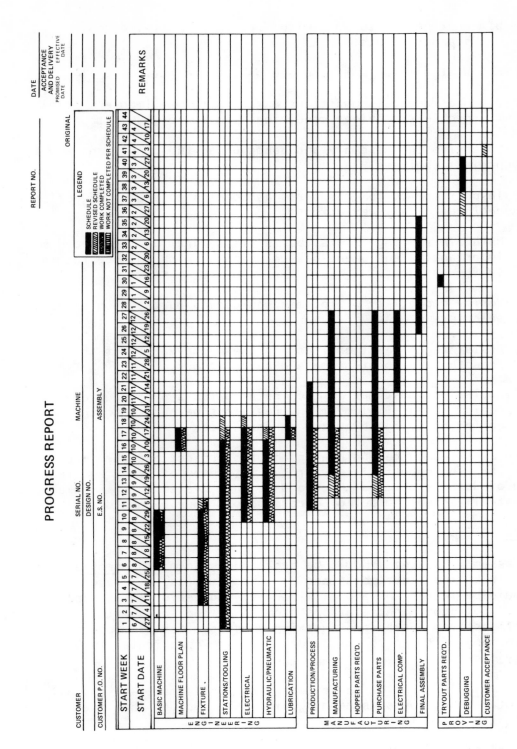

Figure 1. Typical bar chart schedule used for progress reports to the customer. In this case, a 41 week delivery promise is being reported at the end of the 17th week. Note that only 6 weeks is allowed for the last three phases of debug. This is practical in systems consisting of independent, self-contained modules which can be individually debugged to a high degree.

    b. What he has to work with; drawings, parts, gages, standards, etc.

    c. What timetable is imposed on the project.

    d. What unusual conditions exist which may impose special problems; ambient conditions, personnel, QC, etc.

2. Supervise the assembly and debug under the conditions set by the project engineer.

3. Inspect the customer's parts as soon as possible and notify the engineer of any deviations.

4. Maintain shortage lists, parts lists, drawing lists, and most importantly, equipment design changes which must be recorded by the engineer.

5. Verify capability of incoming purchased parts, such as parts selectors.

6. Assist the project engineer in the organization of the acceptance demonstration and carry it out.

7. Assist the customer in the installation and startup of the equipment and implement any revisions necessary to the equipment to reach quoted performance.

Operating as a team, the engineer plans the project and the lead man executes it. The opportunities for the interplay of ideas and redundant checking of details offer a check-and-balance system which should catch the potentially great number of errors before they become disasters.

Procedure *(Fig. 1)*

Complex assembly systems should be broken down into "manageable units" which could be individual stations or subassemblies. Wherever practical, they should be tested and a preliminary de-bug performed as separate entities. Self-contained stations from power and free systems, or independent stations from a standard indexing chassis offer the opportunity to eliminate many of the problems before they are complicated in assembly with many others.

Final assembly *(Fig. 2)* of these manageable units may take a small portion of the total debug hours, something like 15%. During this step the following items are involved.

1. Alignment of stations to fixtures.

2. Test feed rate of parts selectors.

3. Verify proper station timing and sequencing.

4. Adjustment of station motions and strokes.

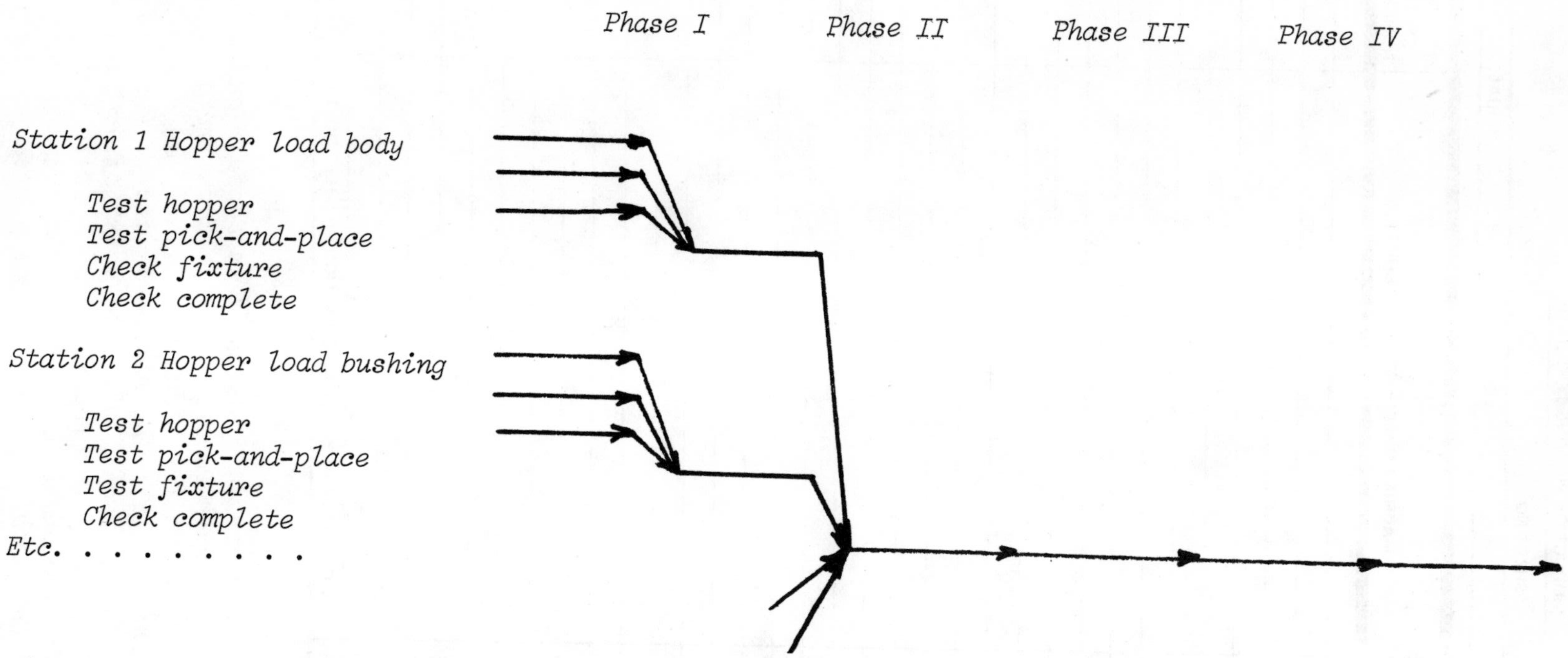

Phase I Phase II Phase III Phase IV

Station 1 Hopper load body

   Test hopper
   Test pick-and-place
   Check fixture
   Check complete

Station 2 Hopper load bushing

   Test hopper
   Test pick-and-place
   Test fixture
   Check complete

Etc. . . . .

Figure 2. Debugging time may be minimized by breaking the system down into "manageable units", or modules, which can be individually proven out to a high degree prior to assembly in the complete system.

This permits an overlap of functions in the time frame available. It also permits resolving isolated problems before they become hopelessly complicated in conjunction with many others.

PRODUCTION PROVING TEST DATA

TEST NO.　　　　　　　SHEET NO.　　　　　　　DATE

MACHINE TITLE　　　　　　　　W.O.　　　DES.　　　SERIAL

MACHINE STA. NO.	MACHINE DOWN TIME MIN.	SEC.	MACHINE MALFUNCTION DESCRIPTION	MACHINE CYCLE NUMBER
TOTAL			NOTE: SUMMARIZE ALL INFORMATION ON SHEET NO. 1	

Figure 3. *Typical data recording form used in Phase 2 and 3 of the debugging procedure. This approach records machine downtime and requires a careful description of the malfunctions involved if an accurate definition of the cause of the problem is to be obtained later. For example, it is not enough to record "feed track jam". The point of jam and reason for jam would have to be noted if the record keeping is to have any value. At the end of a given run, one hour perhaps, the running time losses are totaled by cause and the problems systematically attached beginning with the most difficult and proceeding to the smallest.*

PRODUCTION PROVING TEST DATA
SUMMARY SHEET

TEST NO.	SHEET NO.	W.O.—SERIAL NO.	DES. NO.

MACHINE TITLE	CYCLE TIME /MIN.
RECORDER'S SIGNATURE	NO. OF PEOPLE
MACHINE CYCLE COUNTER (FINISH)	FINISH TIME
MACHINE CYCLE COUNTER (START)	START TIME
MACHINE TOTAL CYCLES	TOTAL TIME
PRODUCTION COUNTER (FINISH)	IF PARTS MANUALLY LOADED
PRODUCTION COUNTER (START)	RECORD OPERATOR'S TOTAL:
TOTAL GOOD ASSEMBLIES	

STATION EFFICIENCY SUMMARY

STA. NO.	STATION NAME OR FUNCTION	COUNTER CYCLES	NO. OF STATION MALFUNCTIONS	% STATION EFFICIENCY

STATION STOPPAGES

STATION EFFICIENCY = $\dfrac{\text{TOTAL GOOD ASSEMBLIES}}{\text{TOTAL MACHINE CYCLES}}$ = _____ % EFFICIENCY

OVERALL EFFICIENCY = $\dfrac{\text{TOTAL GOOD ASSEMBLIES}}{\text{AVAILABLE TOTAL CYCLES}}$ = _____ % EFFICIENCY
(INCLUDING MANUAL OPERATIONS)

NET OVERALL MACHINE EFFICIENCY = $\dfrac{\text{TOTAL GOOD ASSEMBLIES}}{\text{AVAILABLE TOTAL CYCLES LESS MANUAL MISLOADS, ETC.}}$ = _____ % EFFICIENCY

Figure 4. Debugging test data is summarized on this form which offers an overall view of the system efficiency level. Areas of maximum losses become obvious. Note the "net overall machine efficiency" computation. It is necessary to delete the human factory, defective parts, and all other non-machine losses if a true picture of the debug level is to be obtained. Otherwise, the debug personnel will be spending time and effort in areas they cannot improve and which are properly a function of the "launch", or installation procedure when the ultimate operators, material handling, etc., are available.

5. Adjustment of limit switches and sensory devices.

6. Dry-cycling capability, smoothness and function.

One <u>debug</u> procedure in use is divided into four phases designed to progressively raise the machine efficiency to the required minimum while eliminating problems from gross to fine as they appear. Phase 1 is estimated to take an average of 30% of the total debug allowance and includes:

    a. Inspect the required number of customer parts to assure that they are within B/P and of sufficient accuracy for initial setup.

    b. Inspect machine index and transfer accuracy, critical fixture accuracy, and other machine alignment characteristics.

    c. Single-cycle individual stations, one after the other, and perform any alterations necessary to reach an 80% efficiency level on each.

Phase 2 is estimated to take approximately 25% of the total debug time on the average machine. In this step, the machine is cycled as a complete system, making the necessary adjustments to reach an 80% efficiency level over a minimum of 50 pieces consecutively. *(Fig. 3)*

Phase 3 uses another 25% of the debug time and requires cycling continuously and making the necessary adjustments and corrections to reach an 80% overall machine efficiency over a one hour continuous run or whatever the contract specified. *(Fig. 4)*

Phase 4 consumes the last 5% of debug time and involves demonstrating the system for the customer for the contracted time and performance. *(Fig. 5)*

The above debug procedure, simple as it is, may seem to be excessively detailed. However, it does offer a logical series of steps with the following advantages:

    1. Benchmarks are established which can be helpful in monitoring costs, timetable and management control.

    2. Problems must be defined and resolved as the project progresses instead of permitting bothersome details to be shelved until they come back in a landslide to haunt the engineer.

    3. Potential problems are isolated early in the game, the major items cropping out early, for a progressively finer performance improvement.

Debugging of automated assembly systems is often referred to as <u>equipment and product development</u>. The latter has been referred to many times in previous chapters, and rightly so, and again it is stressed that the optimum combination of manufacturing production costs and efficiencies can only be obtained if product design considerations are coupled with machine design considerations. An average project may involve tens of thousands of "variables" when all product dimensions, tolerances, specifications, etc., are coupled with the same "variables" in some 5,000-10,000 machine parts. Cooperation, coordination and common sense applied on a day-to-day basis can vastly improve the chances of an outstanding project completion.

The debugging process may be viewed as a 4-stage operation:

1. Observation

2. Definition

3. Correction

4. Verification

If all the previously described steps have been followed, the assembly system reaches the stage of final debug with all of the obvious, gross deficiencies weeded-out and corrected. At this point the "buttons are pushed", the starter is "mashed", and all the finer mis-matches come to light as the system comes to a screaming halt or the machine makes scrap at a fantastic rate. At this point, <u>observation</u> is the only tool. Usually this is visual, watching the equipment until the malfunction can be seen. Many times, however, the cause of the problem cannot be visually detected. Some of the tools for such observations are:

1. "Masters", or standard parts of known integrity are supplied by the customer which enable the lead man to observe misalignments too small for the unaided eye.

2. Alignment tools are frequently made by the machine builder for his own use in debugging. These can be gages for physical dimensions, standards for such functions as leak detection, etc.

3. Strip recorders having a multiplicity of pens for simultaneous display of motions or actions can be of tremendous help in spotting problems of interlocking actions or functions.

4. Strobe lights provide a "stop-motion" of rotation or repetitive motion functions.

5. Video cameras are used to record on tape. Often a malfunction occurs only once in a hundred cycles and the tape permits re-examination of that one malfunction many times once it occurs. Slow motion and stop motion are available.

6. High speed cameras offer 50,000 frames per second or more for detecting cam-bounces or similar ultra-high speed malfunctions.

7. Programming panels for programmable controllers offer automatic diagnostics to detect failure of limit switches or other control devices.

DATE_____

MACHINE SHIPMENT RELEASE

1. Work Order No._____ 2. Design No._____
3. Customer_____
4. Machine Description_____
5. On_____, A machine shipment release meeting was held at _____. In Attendance:

   A. <u>Customer</u>:   B. <u>Supplier</u>

   _____   _____
   _____   _____
   _____   _____
   _____   _____
   _____   _____

6. Requested ship date_____
7. Shipment conditions_____
_____
_____
_____
_____
_____
_____
_____
_____

8. Items to be completed before shipment_____
_____
_____
_____
_____
_____
_____
_____
_____

9. General Information_____
_____
10. Drawing/Manual Schedule:
   A. Sepias_____   B. Originals_____   C. Service Manuals_____
11. Copies to:

Figure 5. This machine release form records the conditions and status of the project at the time it is shipped from the build facility. This will usually run several pages and it should be very accurate and detailed for a permanent record.

Definition of the problem, using the data from the observations, is the key to efficient debug. Many times the true cause of a malfunction is not the obvious one and much time and money is wasted jumping to conclusions, not truly defining the problem. This is an analytical function ranging from a "horsesense" conclusion to one requiring computer models to solve.

Correction of the problem, or the "fix", may be as simple as stoning off the corner of a fixture or as complex as re-engineering a whole station. Most fixes are made on the lead man level based on common sense and mechanical aptitude, but an occasional "boo-boo" of a slightly catastrophic nature may have the project engineer in stitches for a long time.

Verification of the fix is as important as the fix itself. Many times the fix leads to a secondary problem which doesn't surface immediately and comes back to haunt the fixer in some other form at some later time.

SUMMARY

Acquisition of a complex assembly system, whether from your own facilities or from a supplier, which has a perfect, or near perfect efficiency, is economically impractical. The equipment would cost too much to justify and the engineers would never stop re-designing and improving.

Some predetermined level of operating efficiency should be established at which point the machine building stops and the production starts.

After production starts, machine performance should be monitored closely and those changes which can be made to substantially improve efficiency should be initiated for a running change.

If duplicate, or additional equipment is planned, the initial design and performance should be reviewed. Those areas performing satisfactorily should be left alone. Change and redesign should be confined to areas not coming up to expectation.

These are very obvious, very simple, logical, common-sense rules. Unfortunately, they are all too often ignored in the heat of excitment coincident with equipment projects.